T0335380

Series in Real Analysis – Vol. 13

THEORIES OF INTEGRATION
The Integrals of Riemann, Lebesgue,
Henstock–Kurzweil, and McShane

Second Edition

SERIES IN REAL ANALYSIS

Series in Real Analysis – Vol. 13

THEORIES OF INTEGRATION
The Integrals of Riemann, Lebesgue,
Henstock–Kurzweil, and McShane

Second Edition

Douglas S Kurtz
New Mexico State University, USA

Charles W Swartz
New Mexico State University, USA

World Scientific

NEW JERSEY · LONDON · SINGAPORE · BEIJING · SHANGHAI · HONG KONG · TAIPEI · CHENNAI

Published by

World Scientific Publishing Co. Pte. Ltd.

5 Toh Tuck Link, Singapore 596224

USA office: 27 Warren Street, Suite 401-402, Hackensack, NJ 07601

UK office: 57 Shelton Street, Covent Garden, London WC2H 9HE

British Library Cataloguing-in-Publication Data
A catalogue record for this book is available from the British Library.

Series in Real Analysis — Vol. 13
THEORIES OF INTEGRATION
The Integrals of Riemann, Lebesgue, Henstock–Kurzweil, and McShane
(2nd Edition)

ISBN-13 978-981-4368-99-5
ISBN-10 981-4368-99-7

Printed in Singapore by World Scientific Printers.

To Jessica and Nita, for supporting us during the long haul to bring this book to fruition.

Preface to the First Edition

This book introduces the reader to a broad collection of integration theories, focusing on the Riemann, Lebesgue, Henstock-Kurzweil and McShane integrals. By studying classical problems in integration theory (such as convergence theorems and integration of derivatives), we will follow a historical development to show how new theories of integration were developed to solve problems that earlier integration theories could not handle. Several of the integrals receive detailed developments; others are given a less complete discussion in the book, while problems and references directing the reader to future study are included.

The chapters of this book are written so that they may be read independently, except for the sections which compare the various integrals. This means that individual chapters of the book could be used to cover topics in integration theory in introductory real analysis courses. There should be sufficient exercises in each chapter to serve as a text.

We begin the book with the problem of defining and computing the area of a region in the plane including the computation of the area of the region interior to a circle. This leads to a discussion of the approximating sums that will be used throughout the book.

The real content of the book begins with a chapter on the Riemann integral. We give the definition of the Riemann integral and develop its basic properties, including linearity, positivity and the Cauchy criterion. After presenting Darboux's definition of the integral and proving necessary and sufficient conditions for Darboux integrability, we show the equivalence of the Riemann and Darboux definitions. We then discuss lattice properties and the Fundamental Theorem of Calculus. We present necessary and sufficient conditions for Riemann integrability in terms of sets with Lebesgue measure 0. We conclude the chapter with a discussion of improper integrals.

We motivate the development of the Lebesgue and Henstock-Kurzweil integrals in the next two chapters by pointing out deficiencies in the Riemann integral, which these integrals address. Convergence theorems are used to motivate the Lebesgue integral and the Fundamental Theorem of Calculus to motivate the Henstock-Kurzweil integral.

We begin the discussion of the Lebesgue integral by establishing the standard convergence theorem for the Riemann integral concerning uniformly convergent sequences. We then give an example that points out the failure of the Bounded Convergence Theorem for the Riemann integral, and use this to motivate Lebesgue's descriptive definition of the Lebesgue integral. We show how Lebesgue's descriptive definition leads in a natural way to the definitions of Lebesgue measure and the Lebesgue integral. Following a discussion of Lebesgue measurable functions and the Lebesgue integral, we develop the basic properties of the Lebesgue integral, including convergence theorems (Bounded, Monotone, and Dominated). Next, we compare the Riemann and Lebesgue integrals. We extend the Lebesgue integral to n-dimensional Euclidean space, give a characterization of the Lebesgue integral due to Mikusinski, and use the characterization to prove Fubini's Theorem on the equality of multiple and iterated integrals. A discussion of the space of integrable functions concludes with the Riesz-Fischer Theorem.

In the following chapter, we discuss versions of the Fundamental Theorem of Calculus for both the Riemann and Lebesgue integrals and give examples showing that the most general form of the Fundamental Theorem of Calculus does not hold for either integral. We then use the Fundamental Theorem to motivate the definition of the Henstock-Kurzweil integral, also know as the gauge integral and the generalized Riemann integral. We develop basic properties of the Henstock-Kurzweil integral, the Fundamental Theorem of Calculus in full generality, and the Monotone and Dominated Convergence Theorems. We show that there are no improper integrals in the Henstock-Kurzweil theory. After comparing the Henstock-Kurzweil integral with the Lebesgue integral, we conclude the chapter with a discussion of the space of Henstock-Kurzweil integrable functions and Henstock-Kurzweil integrals in $\mathbb{R}^n$.

Finally, we discuss the "gauge-type" integral of McShane, obtained by slightly varying the definition of the Henstock-Kurzweil integral. We establish the basic properties of the McShane integral and discuss absolute integrability. We then show that the McShane integral is equivalent to the Lebesgue integral and that a function is McShane integrable if and only if it is absolutely Henstock-Kurzweil integrable. Consequently, the McShane integral could be used to give a presentation of the Lebesgue integral which does not require the development of measure theory.

Preface to the Second Edition

The second edition of this text contains several additions and changes in Chapters 3, 4 and 5. In Chapter 3 on the Lebesgue integral, we have added material about the convolution product and spaces of Lebesgue integrable functions. As an application of the Fubini-Tonelli Theorems, the convolution product of two integrable functions is defined in Section 3.8.1. Approximate identities are defined and used with the convolution product to establish several approximation results, including the Weierstrass Approximation Theorem. In Section 3.9 on the space $L^1(E)$ of Lebesgue integrable functions, we have added examples of dense subsets of $L^1(E)$ and given applications including a proof of the Riemann-Lebesgue Lemma. We have included a change of variables theorem for the Lebesgue integral as a consequence of the Riesz-Fischer Theorem on the completeness of $L^1(E)$.

The notion of uniform integrability is introduced in Chapters 4 and 5 on the Henstock-Kurzweil and McShane integrals. New proofs of the major convergence theorems, the Monotone and Dominated Convergence Theorems, are given for both integrals based on the notion of uniform integrability. A new integration-by-parts result for the Henstock-Kurzweil integral is added and used to establish a version of the Riemann-Lebesgue Lemma for the Henstock-Kurzweil integral.

More exercised have been added.

Contents

Chapter 1

Introduction

1.1 Areas

Modern integration theory is the culmination of centuries of refinements and extensions of ideas dating back to the Greeks. It evolved from the ancient problem of calculating the area of a plane figure. We begin with three axioms for areas:

(1) the area of a rectangular region is the product of its length and width;
(2) area is an additive function of disjoint regions;
(3) congruent regions have equal areas.

Two regions are congruent if one can be converted into the other by a translation and a rotation. From the first and third axioms, it follows that the area of a right triangle is one half of the base times the height. Now, suppose that Δ is a triangle with vertices A, B, and C. Assume that AB is the longest of the three sides, and let P be the point on AB such that the line CP from C to P is perpendicular to AB. Then, ACP and BCP are two right triangles and, using the second axiom, the sum of their areas is the area of Δ. In this way, one can determine the area of irregularly shaped areas, by decomposing them into non-overlapping triangles.

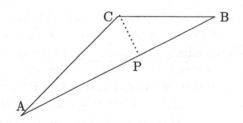

Figure 1.1

1

It is easy to see how this procedure would work for certain regularly shaped regions, such as a pentagon or a star-shaped region. For the pentagon, one merely joins each of the five vertices to the center (actually, any interior point will do), producing five triangles with disjoint interiors. This same idea works for a star-shaped region, though in this case, one connects both the points of the arms of the star and the points where two arms meet to the center of the region.

For more general regions in the plane, such as the interior of a circle, a more sophisticated method of computation is required. The basic idea is to approximate a general region with simpler geometric regions whose areas are easy to calculate and then use a limiting process to find the area of the original region. For example, the ancient Greeks calculated the area of a circle by approximating the circle by inscribed and circumscribed regular n-gons whose areas were easily computed and then found the area of the circle by using the method of exhaustion. Specifically, Archimedes claimed that the area of a circle of radius r is equal to the area of the right triangle with one leg equal to the radius of the circle and the other leg equal to the circumference of the circle. We will illustrate the method using modern notation.

Let C be a circle with radius r and area A. Let n be a positive integer, and let I_n and O_n be regular n-gons, with I_n inscribed inside of C and O_n circumscribed outside of C. Let a represent the area function and let $E_I = A - a(I_4)$ be the error in approximating A by the area of an inscribed 4-gon. The key estimate is

$$A - a\left(I_{2^{2+n}}\right) < \frac{1}{2^n} E_I, \qquad (1.1)$$

which follows, by induction, from the estimate

$$A - a\left(I_{2^{2+n+1}}\right) < \frac{1}{2}\left(A - a\left(I_{2^{2+n}}\right)\right).$$

To see this, fix $n \geq 0$ and let $I_{2^{2+n}}$ be inscribed in C. We let $I_{2^{2+n+1}}$ be the 2^{2+n+1}-gon with vertices comprised of the vertices of $I_{2^{2+n}}$ and the 2^{2+n} midpoints of arcs between adjacent vertices of $I_{2^{2+n}}$. See the figure below. Consider the area inside of C and outside of $I_{2^{2+n}}$. This area is comprised of 2^{2+n} congruent caps. Let cap_i^n be one such cap and let R_i^n be the smallest rectangle that contains cap_i^n. Note that R_i^n shares a base with cap_i^n (that is, the base inside the circle) and the opposite side touches the circle at one point, which is the midpoint of that side and a vertex of $I_{2^{2+n+1}}$. Let T_i^n be the triangle with the same base and opposite vertex at the midpoint. See the picture below.

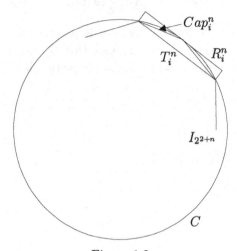

Figure 1.2

Suppose that cap_j^{n+1} and cap_{j+1}^{n+1} are the two caps inside of C and outside of $I_{2^{2+n+1}}$ that are contained in cap_i^n. Then, since $cap_j^{n+1} \cup cap_{j+1}^{n+1} \subset R_i^n \setminus T_i^n$,

$$a\left(T_i^n\right) = a\left(R_i^n \setminus T_i^n\right) > a\left(cap_j^{n+1} \cup cap_{j+1}^{n+1}\right),$$

which implies

$$\begin{aligned} a\left(cap_i^n\right) &= a\left(T_i^n\right) + a\left(cap_j^{n+1} \cup cap_{j+1}^{n+1}\right) \\ &> 2a\left(cap_j^{n+1} \cup cap_{j+1}^{n+1}\right) = 2\left[a\left(cap_j^{n+1}\right) + a\left(cap_{j+1}^{n+1}\right)\right]. \end{aligned}$$

Adding the areas in all the caps, we get

$$A - a\left(I_{2^{2+n+1}}\right) = \sum_{j=1}^{2^{2+n+1}} a\left(cap_j^{n+1}\right) < \frac{1}{2}\sum_{i=1}^{2^{2+n}} a\left(cap_i^n\right) = \frac{1}{2}\left(A - a\left(I_{2^{2+n}}\right)\right)$$

as we wished to show.

We can carry out a similar, but more complicated, analysis with the circumscribed rectangles to prove

$$a\left(O_{2^{2+n}}\right) - A < \frac{1}{2^n}E_O, \tag{1.2}$$

where $E_O = a(O_4) - A$ is the error from approximating A by the area of a circumscribed 4-gon. Again, this estimate follows from the inequality

$$a(O_{2^{2+n+1}}) - A < \frac{1}{2}(a(O_{2^{2+n}}) - A).$$

For simplicity, consider the case $n = 0$, so that $O_{2^2} = O_4$ is a square. By rotational invariance, we may assume that O_4 sits on one of its sides. Consider the lower right hand corner in the picture below.

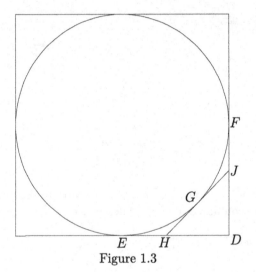

Figure 1.3

Let D be the lower right hand vertex of O_4 and let E and F be the points to the left of and above D, respectively, where O_4 and C meet. Let G be midpoint of the arc on C from E to F, and let H and J be the points where the tangent to C at G meets the segments DE and DF, respectively. Note that the segment HJ is one side of O_{2^2+1}. As in the argument above, it is enough to show that the area of the region bounded by the arc from E to F and the segments DE and DF is greater than twice the area of the two regions bounded by the arc from E to F and the segments EH, HJ and FJ. More simply, let S' be the region bounded by the arc from E to G and the segments EH and GH and S be the region bounded by the arc from E to G and the segments DG and DE. We wish to show that $a(S') < \frac{1}{2}a(S)$. To see this, note that the triangle DHG is a right triangle with hypotenuse DH, so that the length of DH, which we denote $|DH|$, is greater than the length of GH which is equal to the length of EH, since both are half the length of a side of $O_{2^{2}+1}$. Let h be the distance from G to DE. Then,

$$a\left(S'\right) < a\left(EGH\right) = \frac{1}{2}\left|EH\right|h < \frac{1}{2}\left|DH\right|h = a\left(DHG\right)$$

so that

$$a\left(S\right) = a\left(DHG\right) + a\left(S'\right) > 2a\left(S'\right),$$

and the proof of (1.2) follows as above.

With estimates (1.1) and (1.2), we can prove Archimedes claim that A is equal to the area of the right triangle with one leg equal to the radius of the circle and the other leg equal to the circumference of the circle. Call this area T. Suppose first that $A > T$. Then, $A - T > 0$, so that by (1.1) we can choose an n so large that $A - a\left(I_{2^{2}+n}\right) < A - T$, or $T < a\left(I_{2^{2}+n}\right)$. Let T_i be one of the 2^{2+n} congruent triangles comprising $I_{2^{2}+n}$ formed by joining the center of C to two adjacent vertices of $I_{2^{2}+n}$. Let s be the length of the side joining the vertices and let h be the distance from this side to the center. Then,

$$a\left(I_{2^{2}+n}\right) = 2^{2+n}a\left(T_i\right) = 2^{2+n}\frac{1}{2}sh = \frac{1}{2}\left(2^{2+n}s\right)h.$$

Since $h < r$ and $2^{2+n}s$ is less than the circumference of C, we see that $a\left(I_{2^{2}+n}\right) < T$, which is a contradiction. Thus, $A \leq T$.

Similarly, if $A < T$, then $T - A > 0$, so that by (1.2) we can choose an n so that $a\left(O_{2^{2}+n}\right) - A < T - A$, or $a\left(O_{2^{2}+n}\right) < T$. Let T_i' be one of the 2^{2+n} congruent triangles comprising $O_{2^{2}+n}$ formed by joining the center of C to two adjacent vertices of $O_{2^{2}+n}$. Let s' be the length of the side joining the vertices and let $h = r$ be the distance from this side to the center. Then,

$$a\left(O_{2^{2}+n}\right) = 2^{2+n}a\left(T_i'\right) = 2^{2+n}\frac{1}{2}s'r = \frac{1}{2}\left(2^{2+n}s'\right)r.$$

Since $2^{2+n}s'$ is greater than the circumference of C, we see that $a\left(O_{2^{2}+n}\right) > T$, which is a contradiction. Thus, $A \geq T$. Consequently, $A = T$.

In the computation above, we made the tacit assumption that the circle had a notion of area associated with it. We have made no attempt to define the area of a circle or, indeed, any other arbitrary region in the plane. We will discuss the problem of defining and computing the area of regions in the plane in Chapter 3.

The basic idea employed by the ancient Greeks leads in a very natural way to the modern theories of integration, using rectangles instead of triangles to compute the approximating areas. For example, let f be a positive function defined on an interval $[a, b]$. Consider the problem of computing the area of the region under the graph of the function f, that is, the area of the region $\mathcal{R} = \{(x, y) : a \leq x \leq b, 0 \leq y \leq f(x)\}$.

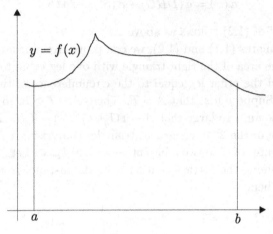

Figure 1.4

Analogous to the calculation of the area of the circle, we consider approximating the area of the region $\mathcal{R}$ by the sums of the areas of rectangles. We divide the interval $[a, b]$ into subintervals and use these subintervals for the bases of the rectangles. A *partition* of an interval $[a, b]$ is a finite, ordered set of points $\mathcal{P} = \{x_0, x_1, \ldots, x_n\}$, with $x_0 = a$ and $x_n = b$. The French mathematician Augustin-Louis Cauchy (1789-1857) studied the area of the region $\mathcal{R}$ for continuous functions. He approximated the area of the region $\mathcal{R}$ by the *Cauchy sum*

$$C(f, \mathcal{P}) = \sum_{i=1}^{n} f(x_{i-1})(x_i - x_{i-1})$$
$$= f(x_0)(x_1 - x_0) + \cdots + f(x_{n-1})(x_n - x_{n-1}).$$

Cauchy used the value of the function at the left hand endpoint of each subinterval $[x_{i-1}, x_i]$ to generate rectangles with area $f(x_{i-1})(x_i - x_{i-1})$. The sum of the areas of the rectangles approximate the area of the region $\mathcal{R}$.

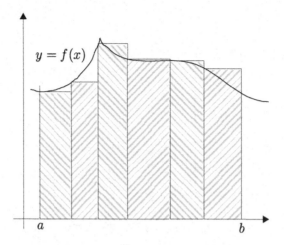

Figure 1.5

He then used the intermediate value property of continuous functions to argue that the Cauchy sums $C(f, \mathcal{P})$ satisfy a "Cauchy condition" as the *mesh* of the partition, $\mu(\mathcal{P}) = \max_{1 \le i \le n} (x_i - x_{i-1})$, approaches 0. He concluded that the sums $C(f, \mathcal{P})$ have a limit, which he defined to be the integral of f over $[a, b]$ and denoted by $\int_a^b f(x)\, dx$. Cauchy's assumptions, however, were too restrictive, since actually he assumed that the function was uniformly continuous on the interval $[a, b]$, a concept not understood at that time. (See Cauchy [C, (2) 4, pages 122-127], Pesin [Pe] and Grattan-Guinness [Gr] for descriptions of Cauchy's argument.)

The German mathematician Georg Friedrich Bernhard Riemann (1826-1866) was the first to consider the case of a general function f and region $\mathcal{R}$. Riemann generated approximating rectangles by choosing an arbitrary point t_i, called a *sampling point*, in each subinterval $[x_{i-1}, x_i]$ and forming the *Riemann sum*

$$S(f, \mathcal{P}, \{t_i\}_{i=1}^n) = \sum_{i=1}^n f(t_i)(x_i - x_{i-1})$$

to approximate the area of the region $\mathcal{R}$.

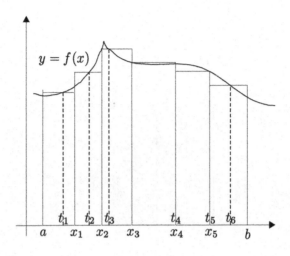

Figure 1.6

Riemann defined the function f to be integrable if the sums $S\left(f, \mathcal{P}, \{t_i\}_{i=1}^{n}\right)$ have a limit as $\mu\left(\mathcal{P}\right) = \max_{1 \le i \le n}\left(x_i - x_{i-1}\right)$ approaches 0. We will give a detailed exposition of the Riemann integral in Chapter 2.

The construction of the approximating sums in both the Cauchy and Riemann theories is exactly the same, but Cauchy associated a single set of sampling points to each partition while Riemann associated an uncountable collection of sets of sampling points. It is this seemingly small change that makes the Riemann integral so much more powerful than the Cauchy integral. It will be seen in subsequent chapters that using approximating sums, such as the Riemann sums, but imposing different conditions on the subintervals or sampling points, leads to other, more general integration theories.

In the Lebesgue theory of integration, the range of the function f is partitioned instead of the domain. A representative value, y, is chosen for each subinterval. The idea is then to multiply this value by the length of the set of points for which f is approximately equal to y. The problem is that this set of points need not be an interval, or even a union of intervals. This means that we must consider "partitioning" the domain $[a, b]$ into subsets other than intervals and we must develop a notion that generalizes the concept of length to these sets. These considerations led to the notion of Lebesgue measure and the Lebesgue integral, which we discuss in Chapter 3.

The Henstock-Kurzweil integral studied in Chapter 4 is obtained by using the Riemann sums as described above, but uses a different condition to control the size of the partition than that employed by Riemann. It will be seen that this leads to a very powerful theory more general than the Riemann (or Lebesgue) theory.

The McShane integral, discussed in Chapter 5, likewise uses Riemann-type sums. The construction of the McShane integral is exactly the same as the Henstock-Kurzweil integral, except that the sampling points t_i are not required to belong to the interval $[x_{i-1}, x_i]$. Since more general sums are used in approximating the integral, the McShane integral is not as general as the Henstock-Kurzweil integral; however, the McShane integral has some very interesting properties and it is actually equivalent to the Lebesgue integral.

1.2 Exercises

Exercise 1.1 Let T be an isosceles triangle with base of length b and two equal sides of length s. Find the area of T.

Exercise 1.2 Let C be a circle with center P and radius r and let I_n and O_n be n-gons inscribed and circumscribed about C. By joining the vertices to P, we can decompose either I_n or O_n into n congruent, non-overlapping isosceles triangles. Each of these $2n$ triangles will make an angle of $\dfrac{2\pi}{n}$ at P.

Use this information to find the area of I_n; this gives a lower bound on the area inside of C. Then, find the area of O_n to get an upper bound on the area of C. Take the limits of both these expressions to compute the area inside of C.

Exercise 1.3 Let $0 < a < b$. Define $f : [a, b] \to \mathbb{R}$ by $f(x) = x^2$ and let $\mathcal{P}$ be a partition of $[a, b]$. Explain why the Cauchy sum $C(f, \mathcal{P})$ is the smallest Riemann sum associated to $\mathcal{P}$ for this function f.

Chapter 2

Riemann integral

2.1 Riemann's definition

The Riemann integral, defined in 1854 (see [Ri1], [Ri2]), was the first of the modern theories of integration and enjoys many of the desirable properties of an integration theory. While the most popular integral discussed in introductory analysis texts, the Riemann integral does have serious shortcomings which motivated mathematicians to seek more general integration theories to overcome them, as we will see in subsequent chapters.

The groundwork for the Riemann integral of a function f over the interval $[a, b]$ begins with dividing the interval into smaller subintervals.

Definition 2.1 Let $[a, b] \subset \mathbb{R}$. A *partition* of $[a, b]$ is a finite set of numbers $\mathcal{P} = \{x_0, x_1, \ldots, x_n\}$ such that $x_0 = a$, $x_n = b$ and $x_{i-1} < x_i$ for $i = 1, \ldots, n$. For each subinterval $[x_{i-1}, x_i]$, define its *length* to be $\ell([x_{i-1}, x_1]) = x_i - x_{i-1}$. The *mesh* of the partition is then the length of the largest subinterval, $[x_{i-1}, x_i]$:

$$\mu(\mathcal{P}) = \max\{x_i - x_{i-1} : i = 1, \ldots, n\}.$$

Thus, the points $\{x_0, x_1, \ldots, x_n\}$ form an increasing sequence of numbers in $[a, b]$ that divides the interval $[a, b]$ into contiguous subintervals.

Let $f : [a, b] \to \mathbb{R}$, $\mathcal{P} = \{x_0, x_1, \ldots, x_n\}$ be a partition of $[a, b]$, and $t_i \in [x_{i-1}, x_i]$ for each i. As noted in Chapter 1, Riemann began by considering the approximating (Riemann) sums

$$S(f, \mathcal{P}, \{t_i\}_{i=1}^n) = \sum_{i=1}^{n} f(t_i)(x_i - x_{i-1}),$$

defined with respect to the partition $\mathcal{P}$ and the set of sampling points

$\{t_i\}_{i=1}^{n}$. Riemann considered the integral of f over $[a, b]$ to be a "limit" of the sums $S(f, \mathcal{P}, \{t_i\}_{i=1}^{n})$ in the following sense.

Definition 2.2   A function $f : [a, b] \to \mathbb{R}$ is *Riemann integrable* over $[a, b]$ if there is an $A \in \mathbb{R}$ such that for all $\epsilon > 0$ there is a $\delta > 0$ so that if $\mathcal{P}$ is any partition of $[a, b]$ with $\mu(\mathcal{P}) < \delta$ and $t_i \in [x_{i-1}, x_i]$ for all i, then

$$|S(f, \mathcal{P}, \{t_i\}_{i=1}^{n}) - A| < \epsilon.$$

We write $A = \int_a^b f = \int_a^b f(t)\, dt$ or, if we set $I = [a, b]$, $\int_I f$.

This definition defines the integral as a limit of sums as the mesh of the partition approaches 0.

The following proposition justifies our definition of and notation for the integral.

Proposition 2.3   *If f is Riemann integrable over $[a, b]$, then the value of the integral is unique.*

Proof.   Suppose that f is Riemann integrable over $[a, b]$ and both A and B satisfy Definition 2.2. Fix $\epsilon > 0$ and choose δ_A and δ_B corresponding to A and B, respectively, in the definition with $\epsilon' = \frac{\epsilon}{2}$. Let $\delta = \min(\delta_A, \delta_B)$ and suppose that $\mathcal{P}$ is a partition with $\mu(\mathcal{P}) < \delta$, and hence with mesh less than both δ_A and δ_B. Let $\{t_i\}_{i=1}^{n}$ be any set of sampling points for $\mathcal{P}$. Then,

$$|A - B| \le |A - S(f, \mathcal{P}, \{t_i\}_{i=1}^{n})| + |S(f, \mathcal{P}, \{t_i\}_{i=1}^{n}) - B| < \epsilon' + \epsilon' = \epsilon.$$

Since ϵ was arbitrary, it follows that $A = B$. Thus, the value of the integral is unique. $\square$

Remark 2.4   *The value of δ is a measure of how small the subintervals must be so that the Riemann sums closely approximate the integral. When we wish to satisfy two such conditions, we use (any positive number smaller than or equal to) the smaller of the two δ's. This works for a finite number of conditions by choosing the minimum of all the δ's, but may fail for infinitely many conditions since, in this case, the infimum may be 0.*

We consider now several examples.

Example 2.5   Let $a, b, c, d \in \mathbb{R}$ with $a \le c < d \le b$. Set $I = [c, d]$ and let χ_I be the *characteristic function* of I, defined by

$$\chi_I(x) = \begin{cases} 1 \text{ if } x \in I \\ 0 \text{ if } x \notin I \end{cases}.$$

Then, $\int_a^b \chi_I = d - c$.

Let $\mathcal{P} = \{x_0, x_1, \ldots, x_n\}$ be a partition of $[a, b]$. Let $[x_{i-1}, x_i]$ be a subinterval determined by the partition. The contribution to the Riemann sum from $[x_{i-1}, x_i]$ is either $x_i - x_{i-1}$ or 0 depending on whether or not the sampling point is in I.

Now, fix $\epsilon > 0$, let $\delta = \epsilon/2$ and let $\mathcal{P}$ be a partition of $[a, b]$ with mesh less than δ. Let j be the smallest index such that $c \in [x_{j-1}, x_j]$ and let k be the largest index such that $d \in [x_{k-1}, x_k]$. (If $c \in \mathcal{P} \setminus \{a, b\}$, then c is in two subintervals determined by $\mathcal{P}$.) Then, if $t_i \in [x_{i-1}, x_i]$ for each i,

$$S(f, \mathcal{P}, \{t_i\}_{i=1}^n) = f(t_j)(x_j - x_{j-1})$$
$$+ \sum_{i=j+1}^{k-1} (x_i - x_{i-1}) + f(t_k)(x_k - x_{k-1})$$
$$< \delta + (d - c) + \delta.$$

On the other hand,

$$S(f, \mathcal{P}, \{t_i\}_{i=1}^n) \geq \sum_{i=j+1}^{k-1} (x_i - x_{i-1})$$
$$= \sum_{i=j}^{k} (x_i - x_{i-1}) - \{(x_j - x_{j-1}) + (x_k - x_{k-1})\}$$
$$> (d - c) - 2\delta$$

so that

$$|S(f, \mathcal{P}, \{t_i\}_{i=1}^n) - (d - c)| < 2\delta = \epsilon.$$

Thus, χ_I is Riemann integrable and $\int_a^b \chi_I = d - c$.

Example 2.6 Define $f : [0, 1] \to \mathbb{R}$ by $f(x) = x$. Let $\mathcal{P} = \{x_0, x_1, \ldots, x_n\}$ be a partition of $[0, 1]$ and choose t_i so that $x_{i-1} \leq t_i \leq x_i$. Write $\frac{1}{2}$ as a telescoping sum

$$\frac{1}{2} = \frac{1}{2}(x_n^2 - x_0^2) = \frac{1}{2}\{(x_1^2 - x_0^2) + (x_2^2 - x_1^2) + \cdots + (x_n^2 - x_{n-1}^2)\}.$$

Then,

$$\left| S\left(f, \mathcal{P}, \{t_i\}_{i=1}^n\right) - \frac{1}{2} \right| = \left| \sum_{i=1}^n t_i \left(x_i - x_{i-1}\right) - \frac{1}{2} \sum_{i=1}^n \left(x_i^2 - x_{i-1}^2\right) \right|$$

$$= \left| \sum_{i=1}^n \left(t_i - \frac{x_i + x_{i-1}}{2}\right)\left(x_i - x_{i-1}\right) \right|.$$

Since $t_i, \dfrac{x_i + x_{i-1}}{2} \in [x_{i-1}, x_i]$, $\left| t_i - \dfrac{x_i + x_{i-1}}{2} \right| \leq |x_i - x_{i-1}| \leq \mu\left(\mathcal{P}\right)$. So, given $\epsilon > 0$, set $\delta = \epsilon$. Then, if $\mu\left(\mathcal{P}\right) < \delta$,

$$\left| S\left(f, \mathcal{P}, \{t_i\}_{i=1}^n\right) - \frac{1}{2} \right| \leq \sum_{i=1}^n \left| \left(t_i - \frac{x_i + x_{i-1}}{2}\right)\left(x_i - x_{i-1}\right) \right|$$

$$< \delta \sum_{i=1}^n \left(x_i - x_{i-1}\right) = \delta = \epsilon$$

since $\sum_{i=1}^n \left(x_i - x_{i-1}\right) = 1$. Thus, f is Riemann integrable on $[0, 1]$ and has integral $\frac{1}{2}$.

The Riemann integral is well suited for continuous functions, and can handle functions whose points of discontinuity form, in some sense, a small set. See Corollary 2.42. However, if the function has many discontinuities, this integral may fail to exist.

Example 2.7   Define the *Dirichlet function* $f : [0, 1] \to \mathbb{R}$ by

$$f\left(x\right) = \begin{cases} 1 \text{ if } x \in \mathbb{Q} \\ 0 \text{ if } x \notin \mathbb{Q} \end{cases}.$$

Let $\mathcal{P} = \{x_0, x_1, \ldots, x_n\}$ be a partition of $[0, 1]$. In every subinterval $[x_{i-1}, x_i]$ there is a rational number r_i and an irrational number q_i. Thus,

$$S\left(f, \mathcal{P}, \{r_i\}_{i=1}^n\right) = \sum_{i=1}^n f\left(r_i\right)\left(x_i - x_{i-1}\right) = \sum_{i=1}^n 0 = 0$$

while

$$S\left(f, \mathcal{P}, \{q_i\}_{i=1}^n\right) = \sum_{i=1}^n f\left(q_i\right)\left(x_i - x_{i-1}\right) = \sum_{i=1}^n \left(x_i - x_{i-1}\right) = 1.$$

So, no matter how fine the partition, we can always find a set of sampling points so that the corresponding Riemann sum equals 0 and another set so that the corresponding Riemann sum equals 1. Now, suppose f were

Riemann integrable with integral A. Fix $\epsilon < \frac{1}{2}$ and choose a corresponding δ. If $\mathcal{P}$ is any partition with mesh less than δ, then

$$
\begin{aligned}
1 &= |S\left(f, \mathcal{P}, \{q_i\}_{i=1}^n\right) - S\left(f, \mathcal{P}, \{r_i\}_{i=1}^n\right)| \\
&\leq |S\left(f, \mathcal{P}, \{q_i\}_{i=1}^n\right) - A| + |A - S\left(f, \mathcal{P}, \{r_i\}_{i=1}^n\right)| < \epsilon + \epsilon < 1.
\end{aligned}
$$

This contradiction shows that f is not Riemann integrable.

2.2 Basic properties

In the calculus, we study functions which associate one number (the input) to another number (the output). We can think of the Riemann integral in much the same way, except now the input is a function and the output is either a number (in the case of definite integration) or a function (for indefinite integration). We call a function whose inputs are themselves functions an *operator*, so that the Riemann integral is an operator acting on Riemann integrable functions. Two fundamental properties satisfied by the Riemann integral or any reasonable integral are known as *linearity* and *positivity*. Linearity means that scalars factor outside the operation and the operation distributes over sums; positivity means that a nonnegative input produces a nonnegative output.

Proposition 2.8 *(Linearity) Let $f, g : [a, b] \to \mathbb{R}$ and let $\alpha, \beta \in \mathbb{R}$. If f and g are Riemann integrable, then $\alpha f + \beta g$ is Riemann integrable and*

$$
\int_a^b (\alpha f + \beta g) = \alpha \int_a^b f + \beta \int_a^b g.
$$

Proof. Fix $\epsilon > 0$ and choose $\delta_f > 0$ so that if $\mathcal{P}$ is a partition of $[a, b]$ with $\mu(\mathcal{P}) < \delta_f$, then

$$
\left| S\left(f, \mathcal{P}, \{t_i\}_{i=1}^n\right) - \int_a^b f \right| < \frac{\epsilon}{2(1 + |\alpha|)}
$$

for any set of sampling points $\{t_i\}_{i=1}^n$. Similarly, choose $\delta_g > 0$ so that if $\mathcal{P}$ is a partition of $[a, b]$ with $\mu(\mathcal{P}) < \delta_g$, then

$$
\left| S\left(g, \mathcal{P}, \{t_i\}_{i=1}^n\right) - \int_a^b g \right| < \frac{\epsilon}{2(1 + |\beta|)}.
$$

Now, let $\delta = \min\{\delta_f, \delta_g\}$ and suppose that $\mathcal{P}$ is a partition of $[a, b]$ with $\mu(\mathcal{P}) < \delta$ and $t_i \in [x_{i-1}, x_i]$ for $i = 1, \ldots, n$. Then,

$$\left| S\left(\alpha f + \beta g, \mathcal{P}, \{t_i\}_{i=1}^n\right) - \left(\alpha \int_a^b f + \beta \int_a^b g\right) \right|$$

$$= \left| \left(\alpha S\left(f, \mathcal{P}, \{t_i\}_{i=1}^n\right) + \beta S\left(g, \mathcal{P}, \{t_i\}_{i=1}^n\right)\right) - \left(\alpha \int_a^b f + \beta \int_a^b g\right) \right|$$

$$= \left| \alpha \left(S\left(f, \mathcal{P}, \{t_i\}_{i=1}^n\right) - \int_a^b f \right) + \beta \left(S\left(g, \mathcal{P}, \{t_i\}_{i=1}^n\right) - \int_a^b g \right) \right|$$

$$\leq |\alpha| \left| S\left(f, \mathcal{P}, \{t_i\}_{i=1}^n\right) - \int_a^b f \right| + |\beta| \left| S\left(g, \mathcal{P}, \{t_i\}_{i=1}^n\right) - \int_a^b g \right|$$

$$< \frac{\epsilon |\alpha|}{2(1 + |\alpha|)} + \frac{\epsilon |\beta|}{2(1 + |\beta|)} < \epsilon.$$

Since ϵ was arbitrary, it follows that $\alpha f + \beta g$ is Riemann integrable and

$$\int_a^b (\alpha f + \beta g) = \alpha \int_a^b f + \beta \int_a^b g.$$

$\square$

Proposition 2.9   *(Positivity) Let $f : [a, b] \to \mathbb{R}$. Suppose that f is non-negative and Riemann integrable. Then, $\int_a^b f \geq 0$.*

Proof.   Let $\epsilon > 0$ and choose a $\delta > 0$ according to Definition 2.2. Then, if $\mathcal{P}$ is a partition of $[a, b]$ with $\mu(\mathcal{P}) < \delta$ and $t_i \in [x_{i-1}, x_i]$,

$$\left| S\left(f, \mathcal{P}, \{t_i\}_{i=1}^n\right) - \int_a^b f \right| < \epsilon.$$

Consequently, since $S\left(f, \mathcal{P}, \{t_i\}_{i=1}^n\right) \geq 0$,

$$\int_a^b f > S\left(f, \mathcal{P}, \{t_i\}_{i=1}^n\right) - \epsilon > -\epsilon$$

for any positive ϵ. It follows that $\int_a^b f \geq 0$. $\square$

Applying this result to the difference $g - f$ we have the following comparison result.

Corollary 2.10 *Suppose f and g are Riemann integrable on $[a,b]$ and $f(x) \le g(x)$ for all $x \in [a,b]$. Then,*

$$\int_a^b f \le \int_a^b g.$$

Suppose that $f : [a,b] \to \mathbb{R}$ and f is unbounded on $[a,b]$. Let $\mathcal{P}$ be a partition of $[a,b]$. Then, there is a subinterval $[x_{j-1}, x_j]$ on which f is unbounded. For, if f were bounded on each subinterval $[x_{i-1}, x_i]$, with a bound of M_i, then f would be bounded on $[a,b]$ with a bound of $\max\{M_1, M_2, \dots, M_n\}$. Thus, there is a sequence $\{y_k\}_{k=1}^\infty \subset [x_{j-1}, x_j]$ such that $|f(y_k)| \ge k$. Can such a function be Riemann integrable? Consider the following heuristic argument.

Fix a set of sampling points $t_i \in [x_{i-1}, x_i]$ for $i \ne j$, so that the sum

$$\sum_{\substack{1 \le i \le n \\ i \ne j}} f(t_i)(x_i - x_{i-1})$$

is a fixed constant. Set $t_j = y_k$. Then,

$$S(f, \mathcal{P}, \{t_i\}_{i=1}^n) = \sum_{\substack{1 \le i \le n \\ i \ne j}} f(t_i)(x_i - x_{i-1}) + f(y_k)(x_j - x_{j-1}).$$

Note that as we vary k, the Riemann sums diverge and f is not Riemann integrable. Thus, a Riemann integrable function must be bounded. We formalized this result with the following proposition.

Proposition 2.11 *Suppose that $f : [a,b] \to \mathbb{R}$ is a Riemann integrable function. Then, f is bounded.*

Proof. Choose $\delta > 0$ so that

$$\left| S(f, \mathcal{P}, \{t_i\}_{i=1}^n) - \int_a^b f \right| < \frac{1}{2}$$

if $\mu(\mathcal{P}) < \delta$. Fix such a partition $\mathcal{P}$ and sampling points $\{t_i\}_{i=1}^n$, and let $M = \max\{|f(t_1)|, |f(t_2)|, \dots, |f(t_n)|\}$ and $\Delta = \min\{x_1 - x_0, x_2 - x_1, \dots, x_n - x_{n-1}\} > 0$. Let $x \in [a,b]$ and let j be the smallest index such that $x \in [x_{j-1}, x_j]$. Let T be the set of sampling points $\{t_1, \dots, t_{j-1}, x, t_{j+1}, \dots, t_n\}$. Note that

$$|f(x)(x_j - x_{j-1}) - f(t_j)(x_j - x_{j-1})| = |S(f, \mathcal{P}, T) - S(f, \mathcal{P}, \{t_i\}_{i=1}^n)|$$

since the two Riemann sums contain the same addends except for the terms corresponding to the subinterval $[x_{j-1}, x_j]$. Further,

$$
\begin{aligned}
|S(f, \mathcal{P}, T) - S(f, \mathcal{P}, \{t_i\}_{i=1}^n)| &= \left| S(f, \mathcal{P}, T) - \int_a^b f \right. \\
&\quad \left. + \int_a^b f - S(f, \mathcal{P}, \{t_i\}_{i=1}^n) \right| \\
&\leq \left| S(f, \mathcal{P}, T) - \int_a^b f \right| \\
&\quad + \left| \int_a^b f - S(f, \mathcal{P}, \{t_i\}_{i=1}^n) \right| \\
&< 1.
\end{aligned}
$$

It follows that

$$
|f(x)| (x_j - x_{j-1}) < |f(t_j)| (x_j - x_{j-1}) + 1 \leq M(x_j - x_{j-1}) + 1
$$

or

$$
|f(x)| < M + \frac{1}{(x_j - x_{j-1})} \leq M + \frac{1}{\Delta}.
$$

Since x was arbitrary, we see that f is bounded. $\qquad \square$

2.3   Cauchy criterion

Let $\{x_n\}_{n=1}^\infty$ be a convergent sequence. Then, $\{x_n\}_{n=1}^\infty$ satisfies a Cauchy condition; that is, given $\epsilon > 0$ there is a natural number N such that $|x_n - x_m| < \epsilon$ whenever $n, m > N$. The proof of the boundedness of Riemann integrable functions demonstrates that the Riemann sums of an integrable function satisfy an analogous estimate. Suppose that f is Riemann integrable on $[a, b]$. Fix $\epsilon > 0$ and choose δ corresponding to $\epsilon/2$ in Definition 2.2. Let $\mathcal{P}_j = \left\{ x_0^{(j)}, x_1^{(j)}, \ldots, x_{n_j}^{(j)} \right\}$, $j = 1, 2$, be two partitions with mesh less than δ and let $t_i^{(j)} \in \left[x_{i-1}^{(j)}, x_i^{(j)}\right]$. Then

$$\left| S\left(f, \mathcal{P}_1, \left\{t_i^{(1)}\right\}_{i=1}^{n_1}\right) - S\left(f, \mathcal{P}_2, \left\{t_i^{(2)}\right\}_{i=1}^{n_2}\right) \right|$$

$$= \left| S\left(f, \mathcal{P}_1, \left\{t_i^{(1)}\right\}_{i=1}^{n_1}\right) - \int_a^b f + \int_a^b f - S\left(f, \mathcal{P}_2, \left\{t_i^{(2)}\right\}_{i=1}^{n_2}\right) \right|$$

$$\leq \left| S\left(f, \mathcal{P}_1, \left\{t_i^{(1)}\right\}_{i=1}^{n_1}\right) - \int_a^b f \right| + \left| \int_a^b f - S\left(f, \mathcal{P}_2, \left\{t_i^{(2)}\right\}_{i=1}^{n_2}\right) \right| < \epsilon.$$

Analogous to the situation for real-valued sequences, the condition that

$$\left| S\left(f, \mathcal{P}_1, \left\{t_i^{(1)}\right\}_{i=1}^{n_1}\right) - S\left(f, \mathcal{P}_2, \left\{t_i^{(2)}\right\}_{i=1}^{n_2}\right) \right| < \epsilon$$

for all partitions $\mathcal{P}_1$ and $\mathcal{P}_2$ with mesh less that δ, which is known as the *Cauchy criterion*, actually characterizes the integrability of f.

Theorem 2.12 *Let $f : [a, b] \to \mathbb{R}$. Then, f is Riemann integrable over $[a, b]$ if, and only if, for each $\epsilon > 0$ there is a $\delta > 0$ so that if $\mathcal{P}_j$, $j = 1, 2$, are partitions of $[a, b]$ with $\mu\left(\mathcal{P}_j\right) < \delta$ and $\left\{t_i^{(j)}\right\}_{i=1}^{n_j}$ are sets of sampling points relative to $\mathcal{P}_j$, then*

$$\left| S\left(f, \mathcal{P}_1, \left\{t_i^{(1)}\right\}_{i=1}^{n_1}\right) - S\left(f, \mathcal{P}_2, \left\{t_i^{(2)}\right\}_{i=1}^{n_2}\right) \right| < \epsilon.$$

Proof. We have already proved that the integrability of f implies the Cauchy criterion. So, assume the Cauchy criterion holds. We will prove that f is Riemann integrable.

For each $k \in \mathbb{N}$, choose a $\delta_k > 0$ so that for any two partitions $\mathcal{P}_1$ and $\mathcal{P}_2$, with mesh less than δ_k, and corresponding sampling points, we have

$$\left| S\left(f, \mathcal{P}_1, \left\{t_i^{(1)}\right\}_{i=1}^{n_1}\right) - S\left(f, \mathcal{P}_2, \left\{t_i^{(2)}\right\}_{i=1}^{n_2}\right) \right| < \frac{1}{k}.$$

Replacing δ_k by $\min\{\delta_1, \delta_2, \ldots, \delta_k\}$, we may assume that $\delta_k \geq \delta_{k+1}$.

Next, for each k, fix a partition $\mathcal{P}_k$ with $\mu\left(\mathcal{P}_k\right) < \delta_k$ and a set of sampling points $\left\{t_i^{(k)}\right\}_{i=1}^{n_k}$. Note that for $j > k$, $\mu\left(\mathcal{P}_j\right) < \delta_j \leq \delta_k$. Thus,

$$\left| S\left(f, \mathcal{P}_k, \left\{t_i^{(k)}\right\}_{i=1}^{n_k}\right) - S\left(f, \mathcal{P}_j, \left\{t_i^{(j)}\right\}_{i=1}^{n_j}\right) \right| < \frac{1}{\min\{j, k\}},$$

which implies that the sequence $\left\{ S\left(f, \mathcal{P}_k, \left\{t_i^{(k)}\right\}_{i=1}^{n_k}\right) \right\}_{k=1}^{\infty}$ is a Cauchy sequence in $\mathbb{R}$, and hence converges. Let A be the limit of this sequence. It follows from the previous inequality that

$$\left| S\left(f, \mathcal{P}_k, \left\{t_i^{(k)}\right\}_{i=1}^{n_k}\right) - A \right| \leq \frac{1}{k}.$$

It remains to show that A satisfies Definition 2.2.

Fix $\epsilon > 0$ and choose $K > 2/\epsilon$. Let $\mathcal{P}$ be a partition with $\mu(\mathcal{P}) < \delta_K$ and let $\{t_i\}_{i=1}^n$ be a set of sampling points for $\mathcal{P}$. Then,

$$|S(f, \mathcal{P}, \{t_i\}_{i=1}^n) - A|$$
$$= \left| S(f, \mathcal{P}, \{t_i\}_{i=1}^n) - S\left(f, \mathcal{P}_K, \left\{t_i^{(K)}\right\}_{i=1}^{n_K}\right) + S\left(f, \mathcal{P}_K, \left\{t_i^{(K)}\right\}_{i=1}^{n_K}\right) - A \right|$$
$$\leq \left| S(f, \mathcal{P}, \{t_i\}_{i=1}^n) - S\left(f, \mathcal{P}_K, \left\{t_i^{(K)}\right\}_{i=1}^{n_K}\right) \right| + \left| S\left(f, \mathcal{P}_K, \left\{t_i^{(K)}\right\}_{i=1}^{n_K}\right) - A \right|$$
$$< \frac{1}{K} + \frac{1}{K} < \epsilon.$$

It now follows that f is Riemann integrable on $[a, b]$. $\qquad\square$

In practice, the Cauchy criterion may be easier to verify than Definition 2.2 if the value of the integral is not known.

2.4 Darboux's definition

In 1875, twenty-one years after Riemann introduced his integral, Gaston Darboux (1842-1917) developed a generalization of Riemann sums and used them to characterize Riemann integrability. (See [D]; see also [Sm].) Let $f : [a, b] \rightarrow \mathbb{R}$ be a bounded function and let $m = \inf\{f(x) : a \leq x \leq b\}$ and $M = \sup\{f(x) : a \leq x \leq b\}$, so that $m \leq f(x) \leq M$ for all $x \in [a, b]$. Let $\mathcal{P} = \{x_0, x_1, \ldots, x_n\}$ be a partition of $[a, b]$, and for each subinterval $[x_{i-1}, x_i]$, $i = 1, \ldots, n$, define M_i and m_i by

$$M_i = \sup\{f(x) : x_{i-1} \leq x \leq x_i\}$$

and

$$m_i = \inf\{f(x) : x_{i-1} \leq x \leq x_i\}.$$

We define the *upper* and *lower Darboux sums* associated to f and $\mathcal{P}$ by

$$U(f, \mathcal{P}) = \sum_{i=1}^n M_i(x_i - x_{i-1})$$

and

$$L\left(f,\mathcal{P}\right) = \sum_{i=1}^{n} m_i \left(x_i - x_{i-1}\right).$$

Note that we always have $L\left(f,\mathcal{P}\right) \leq U\left(f,\mathcal{P}\right)$. In fact, since $m \leq f\left(x\right) \leq M$, we have

$$m\left(b-a\right) \leq L\left(f,\mathcal{P}\right) \leq U\left(f,\mathcal{P}\right) \leq M\left(b-a\right).$$

When $f \geq 0$, each upper Darboux sum provides an upper bound for the area under the graph of f and each lower Darboux sum gives a lower bound for this area.

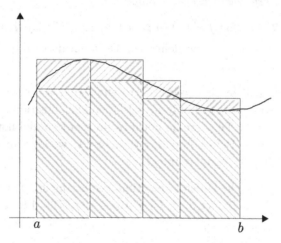

Figure 2.1

Example 2.13   Consider the function $f\left(x\right) = \sin \pi x$ on the interval $[0,3]$. Let $\mathcal{P} = \left\{0, \dfrac{3}{4}, \dfrac{4}{3}, 3\right\}$. Using calculus to find the extreme values of f on the three subintervals, we see that

$$U\left(f,\mathcal{P}\right) = 1 \cdot \left(\frac{3}{4} - 0\right) + \frac{\sqrt{2}}{2} \cdot \left(\frac{4}{3} - \frac{3}{4}\right) + 1 \cdot \left(3 - \frac{4}{3}\right) = \frac{29}{12} + \frac{7}{24}\sqrt{2}$$

and

$$L\left(f,\mathcal{P}\right) = 0 \cdot \left(\frac{3}{4} - 0\right) - \frac{\sqrt{3}}{2} \cdot \left(\frac{4}{3} - \frac{3}{4}\right) - 1 \cdot \left(3 - \frac{4}{3}\right) = -\frac{5}{3} - \frac{7}{24}\sqrt{3}.$$

Next, we define the *upper* and *lower integrals* of f by

$$\overline{\int}_a^b f = \inf \{U(f, \mathcal{P}) : \mathcal{P} \text{ is a partition of } [a, b]\}$$

and

$$\underline{\int}_a^b f = \sup \{L(f, \mathcal{P}) : \mathcal{P} \text{ is a partition of } [a, b]\},$$

both of which exist since the upper sums are bounded below and the lower sums are bounded above. It follows from the comment above that when $f \geq 0$, the upper integral gives an upper bound for the area under the graph of f, since it is an infimum of upper bounds for this area. Similarly, the lower integral yields a lower bound.

Definition 2.14   Let $f : [a, b] \to \mathbb{R}$ be bounded. We say that f is *Darboux integrable* if $\overline{\int}_a^b f = \underline{\int}_a^b f$ and define the Darboux integral of f to be equal to this common value.

Our main goal in this section is to show that a bounded function is Darboux integrable if, and only if, it is Riemann integrable, and that the integrals are equal. Thus, we do not introduce any special notation for the Darboux integral. Before pursuing that result, we give an example of a function that is not Darboux integrable.

Example 2.15   The Dirichlet function (see Example 2.7) is not Darboux integrable on $[0, 1]$. In fact, $L(f, \mathcal{P}) = 0$ and $U(f, \mathcal{P}) = 1$ for every partition $\mathcal{P}$, so that $\underline{\int}_0^1 f = 0$ and $\overline{\int}_0^1 f = 1$.

Let $\mathcal{P}$ be a partition. We say that a partition $\mathcal{P}'$ is a *refinement* of $\mathcal{P}$ if $x \in \mathcal{P}$ implies $x \in \mathcal{P}'$; that is, every partition point of $\mathcal{P}$ is also a partition point of $\mathcal{P}'$. The next result shows that passing to a refinement decreases the upper sum and increases the lower sum.

Proposition 2.16   *Let $f : [a, b] \to \mathbb{R}$ be bounded and let $\mathcal{P}$ and $\mathcal{P}'$ be partitions of $[a, b]$. If $\mathcal{P}'$ is a refinement of $\mathcal{P}$, then $L(f, \mathcal{P}) \leq L(f, \mathcal{P}')$ and $U(f, \mathcal{P}') \leq U(f, \mathcal{P})$.*

Proof.   Let $\mathcal{P} = \{x_0, x_1, \ldots, x_n\}$ be a partition of $[a, b]$ and suppose $\mathcal{P}'$ is the partition obtained by adding a single point, say c, to $\mathcal{P}$. Suppose $x_{j-1} < c < x_j$. Let M_i and m_i be defined as above. Set $M_j' = \sup \{f(x) : x_{j-1} \leq x \leq c\}$,

$M_j'' = \sup\{f(x) : c \leq x \leq x_j\}$, $m_j' = \inf\{f(x) : x_{j-1} \leq x \leq c\}$, and $m_j'' = \inf\{f(x) : c \leq x \leq x_j\}$. Since $m_j', m_j'' \geq m_j$, it follows that

$$m_j'(c - x_{j-1}) + m_j''(x_j - c) \geq m_j(c - x_{j-1}) + m_j(x_j - c) = m_j(x_j - x_{j-1}).$$

Since all the other terms in the lower sums are unchanged, we see that $L(f, \mathcal{P}') \geq L(f, \mathcal{P})$. Similarly, it follows from $M_j', M_j'' \leq M_j$ that

$$M_j'(c - x_{j-1}) + M_j''(x_j - c) \leq M_j(c - x_{j-1}) + M_j(x_j - c)$$
$$= M_j(x_j - x_{j-1}),$$

so that $U(f, \mathcal{P}') \leq U(f, \mathcal{P})$.

Finally, suppose that $\mathcal{P}'$ contains k more terms than $\mathcal{P}$. Repeating the above argument k times, adding one point to the refinement at each stage, completes the proof of the proposition. □

An easy consequence of this result is that every lower sum is less than or equal to every upper sum.

Corollary 2.17 *Let $f : [a, b] \to \mathbb{R}$ be bounded and let $\mathcal{P}_1$ and $\mathcal{P}_2$ be partitions of $[a, b]$. Then, $L(f, \mathcal{P}_1) \leq U(f, \mathcal{P}_2)$.*

Proof. Let $\mathcal{P}_1$ and $\mathcal{P}_2$ be two partitions of $[a, b]$. Then, $\mathcal{P} = \mathcal{P}_1 \cup \mathcal{P}_2$ is a partition of $[a, b]$ which is a refinement of both $\mathcal{P}_1$ and $\mathcal{P}_2$. By the previous proposition,

$$L(f, \mathcal{P}_1) \leq L(f, \mathcal{P}) \leq U(f, \mathcal{P}) \leq U(f, \mathcal{P}_2).$$

□

We can now prove that the lower integral is less than or equal to the upper integral.

Proposition 2.18 *Let $f : [a, b] \to \mathbb{R}$ be bounded. Then,*

$$\underline{\int_a^b} f \leq \overline{\int_a^b} f.$$

Proof. Let $\mathcal{P}$ and $\mathcal{P}'$ be two partitions of $[a, b]$. By the previous corollary, $L(f, \mathcal{P}) \leq U(f, \mathcal{P}')$, so that $U(f, \mathcal{P}')$ is an upper bound for the set $\{L(f, \mathcal{P}) : \mathcal{P} \text{ is a partition of } [a, b]\}$, which implies that

$$\underline{\int_a^b} f \leq U(f, \mathcal{P}').$$

Since this inequality holds for all partitions $\mathcal{P}'$, we see that $\underline{\int_a^b} f$ is a lower bound for the set $\{U(f, \mathcal{P}) : \mathcal{P} \text{ is a partition of } [a, b]\}$, and, consequently,

$$\underline{\int_a^b} f \leq \overline{\int_a^b} f$$

as we wished to show. $\square$

2.4.1 *Necessary and sufficient conditions for Darboux integrability*

Suppose that $f : [a, b] \to \mathbb{R}$ is bounded and Darboux integrable and let $\epsilon > 0$ be fixed. There is a partition $\mathcal{P}_L$ such that

$$\underline{\int_a^b} f - L(f, \mathcal{P}_L) < \frac{\epsilon}{2}$$

and a partition $\mathcal{P}_U$ such that

$$U(f, \mathcal{P}_U) - \overline{\int_a^b} f < \frac{\epsilon}{2}.$$

Let $\mathcal{P} = \mathcal{P}_L \cup \mathcal{P}_U$. Then,

$$\underline{\int_a^b} f - \frac{\epsilon}{2} < L(f, \mathcal{P}_L) \leq L(f, \mathcal{P}) \leq U(f, \mathcal{P}) \leq U(f, \mathcal{P}_U) \leq \overline{\int_a^b} f + \frac{\epsilon}{2}.$$

Since $\underline{\int_a^b} f = \overline{\int_a^b} f$, we see that $U(f, \mathcal{P}) - L(f, \mathcal{P}) < \epsilon$. As the next result shows, this condition actually characterized Darboux integrability.

Theorem 2.19 *Let $f : [a, b] \to \mathbb{R}$ be bounded. Then, f is Darboux integrable on $[a, b]$ if, and only if, for each $\epsilon > 0$ there is a partition $\mathcal{P}$ such that*

$$U(f, \mathcal{P}) - L(f, \mathcal{P}) < \epsilon.$$

Proof. We have already proved that Darboux integrability implies the existence of such partitions. So, assume that for any $\epsilon > 0$ there is a partition $\mathcal{P}$ such that $U(f, \mathcal{P}) - L(f, \mathcal{P}) < \epsilon$. We claim that f is Darboux integrable.

Let $\epsilon > 0$ and choose $\mathcal{P}$ according to the hypothesis. Then,

$$L\left(f,\mathcal{P}\right) \le \underline{\int_a^b} f \le \overline{\int_a^b} f \le U\left(f,\mathcal{P}\right) < L\left(f,\mathcal{P}\right) + \epsilon.$$

It follows that $\left|\overline{\int_a^b} f - \underline{\int_a^b} f\right| < \epsilon$, and since ϵ was arbitrary, we have $\overline{\int_a^b} f = \underline{\int_a^b} f$. Thus, f is Darboux integrable. $\square$

2.4.2 *Equivalence of the Riemann and Darboux definitions*

In this section, we will prove the equivalence of the Riemann and Darboux definitions. To begin, we use Theorem 2.19 to prove a Cauchy-type characterization of Darboux integrability.

Theorem 2.20 *Let $f : [a,b] \to \mathbb{R}$ be a bounded function. Then, f is Darboux integrable if, and only if, given $\epsilon > 0$, there is a $\delta > 0$ so that $U\left(f,\mathcal{P}\right) - L\left(f,\mathcal{P}\right) < \epsilon$ for any partition $\mathcal{P}$ with $\mu\left(\mathcal{P}\right) < \delta$.*

Proof. Let M be a bound for $|f|$ on $[a,b]$. Suppose that f is Darboux integrable and fix $\epsilon > 0$. By Theorem 2.19, there is a partition $\mathcal{P}' = \{y_0, y_1, \ldots, y_m\}$ such that $U\left(f,\mathcal{P}'\right) - L\left(f,\mathcal{P}'\right) < \dfrac{\epsilon}{2}$. Set $\delta = \dfrac{\epsilon}{8Mm}$ and let $\mathcal{P} = \{x_0, x_1, \ldots, x_n\}$ be a partition of $[a,b]$ with $\mu\left(\mathcal{P}\right) < \delta$. Set

$$M_i = \sup\left\{f\left(x\right) : x_{i-1} \le x \le x_i\right\}$$

and

$$m_i = \inf\left\{f\left(x\right) : x_{i-1} \le x \le x_i\right\}.$$

Separate $\mathcal{P}$ into two classes. Let I be the set of indices of all subintervals $[x_{i-1}, x_i]$ which contain a point of $\mathcal{P}'$ and $J = \{0, 1, \ldots, n\} \setminus I$. Then,

$$\sum_{i \in I} \left(M_i - m_i\right)\left(x_i - x_{i-1}\right) \le 2M \sum_{i \in I} \left(x_i - x_{i-1}\right)$$

$$\le 4Mm\mu\left(\mathcal{P}\right) < 4Mm\delta < \frac{\epsilon}{2},$$

where the second inequality follows from the fact that a point of $\mathcal{P}'$ may be contained in two subintervals $[x_{i-1}, x_i]$. If $i \in J$, then there is a k such that $[x_{i-1}, x_i]$ is contained in $[y_{k-1}, y_k]$. It follows that

$$\sum_{i \in J} \left(M_i - m_i\right)\left(x_i - x_{i-1}\right) \le U\left(f,\mathcal{P}'\right) - L\left(f,\mathcal{P}'\right) < \frac{\epsilon}{2}.$$

Combining these estimate shows $U(f, \mathcal{P}) - L(f, \mathcal{P}) < \epsilon$. Another application of Theorem 2.19 shows the other implication and completes the proof of the theorem. $\qquad\square$

Theorem 2.21 *Let $f : [a, b] \to \mathbb{R}$. Then, f is Riemann integrable if, and only if, f is bounded and Darboux integrable.*

Proof. Suppose that f is bounded and Darboux integrable and let $A = \int_{\underline{a}}^b f = \overline{\int_a^b} f$. Fix $\epsilon > 0$ and choose δ by Theorem 2.20. Let $\mathcal{P}$ be a partition with mesh less than δ and let $\{t_i\}_{i=1}^n$ be a set of sampling points for $\mathcal{P}$. Then, by definition, $L(f, \mathcal{P}) \leq A \leq U(f, \mathcal{P})$ and $L(f, \mathcal{P}) \leq S(f, \mathcal{P}, \{t_i\}_{i=1}^n) \leq U(f, \mathcal{P})$, while by construction, $U(f, \mathcal{P}) - L(f, \mathcal{P}) < \epsilon$. Thus, $\mu(\mathcal{P}) < \delta$ implies $|S(f, \mathcal{P}, \{t_i\}_{i=1}^n) - A| < \epsilon$ for any set of sampling points $\{t_i\}_{i=1}^n$. Hence, f is Riemann integrable with Riemann integral equal to A.

Suppose f is Riemann integrable and $\epsilon > 0$. By Proposition 2.11, f is bounded. By Theorem 2.19, to show that f is Darboux integrable, it is enough to find a partition $\mathcal{P}$ such that $U(f, \mathcal{P}) - L(f, \mathcal{P}) < \epsilon$. Since f is Riemann integrable, there is a δ so that if $\mathcal{P}$ is a partition with mesh less than δ, then

$$\left| S(f, \mathcal{P}, \{t_i\}_{i=1}^n) - \int_a^b f \right| < \frac{\epsilon}{4}$$

for any set of sampling points $\{t_i\}_{i=1}^n$. Fix such a partition $\mathcal{P} = \{x_0, x_1, \ldots, x_n\}$. By the definition of M_i and m_i, there are points $T_i, t_i \in [x_{i-1}, x_i]$ such that $M_i < f(T_i) + \epsilon/4(b-a)$ and $f(t_i) - \epsilon/4(b-a) < m_i$, for $i = 1, \ldots, n$. Consequently,

$$U(f, \mathcal{P}) = \sum_{i=1}^n M_i (x_i - x_{i-1}) < \sum_{i=1}^n \left\{ f(T_i) + \frac{\epsilon}{4(b-a)} \right\} (x_i - x_{i-1})$$

$$= S(f, \mathcal{P}, \{T_i\}_{i=1}^n) + \frac{\epsilon}{4(b-a)} \sum_{i=1}^n (x_i - x_{i-1})$$

$$< \int_a^b f + \frac{\epsilon}{4} + \frac{\epsilon}{4}$$

$$= \int_a^b f + \frac{\epsilon}{2}.$$

Similarly, using $\{t_i\}_{i=1}^n$, we see that $L(f, \mathcal{P}) > \int_a^b f - \frac{\epsilon}{2}$. Thus, $U(f, \mathcal{P}) - L(f, \mathcal{P}) < \epsilon$ and f is Darboux integrable. $\qquad\square$

Consequently, we will refer to Darboux integrable functions as being Riemann integrable.

2.4.3 *Lattice properties*

Fix an interval $[a, b]$. We call a function $\varphi : [a, b] \to \mathbb{R}$ a *step function* if there is a partition $\mathcal{P} = \{x_0, x_1, \ldots, x_n\}$ of $[a, b]$ and scalars $\{a_1, \ldots, a_n\}$ such that $\varphi(x) = a_i$ for $x_{i-1} < x < x_i$, $i = 1, \ldots, n$. We are not concerned with the definition of φ at x_i; it could be a_i, a_{i+1} or any other value. Changing the value of φ at a finite number of points has no effect on the integral. See Exercise 2.2. Step functions are clearly bounded; they assume a finite number of values. By Exercise 2.1 and linearity, we see that step functions are Riemann integrable with integral $\int_a^b \varphi = \sum_{i=1}^n a_i (x_i - x_{i-1})$.

Let $f : [a, b] \to \mathbb{R}$ and let $\mathcal{P} = \{x_0, x_1, \ldots, x_n\}$, and define φ and ψ by

$$\varphi(x) = \sum_{i=1}^{n-1} m_i \chi_{[x_{i-1}, x_i)}(x) + m_n \chi_{[x_{n-1}, x_n]}(x)$$

and

$$\psi(x) = \sum_{i=1}^{n-1} M_i \chi_{[x_{i-1}, x_i)}(x) + M_n \chi_{[x_{n-1}, x_n]}(x).$$

Clearly, φ and ψ are step functions, $\varphi \leq f \leq \psi$, and $\int_a^b \varphi = L(f, \mathcal{P})$ and $\int_a^b \psi = U(f, \mathcal{P})$. As a consequence of Theorem 2.19, we have the first half of the following result.

Theorem 2.22 *Let $f : [a, b] \to \mathbb{R}$. Then, f is Riemann integrable if, and only if, for each $\epsilon > 0$ there are step functions φ and ψ such that $\varphi \leq f \leq \psi$ and*

$$\int_a^b (\psi - \varphi) < \epsilon.$$

Proof. We need only show that the existence of such step functions for each $\epsilon > 0$ implies that f is Riemann integrable. Fix $\epsilon > 0$ and choose φ and ψ such that $\int_a^b (\psi - \varphi) < \dfrac{\epsilon}{2}$. First, we partition $[a, b]$ as follows. Let $\mathcal{P}_\varphi$ and $\mathcal{P}_\psi$ be partitions defining φ and ψ, respectively, and set $\mathcal{P} = \mathcal{P}_\varphi \cup \mathcal{P}_\psi$. Next, we view φ and ψ as step functions defined by the partition $\mathcal{P}$, so that we can assume that φ and ψ are defined by the same partition.

Suppose that our fixed partition $\mathcal{P}$ equals $\{x_0, x_1, \ldots, x_n\}$. Since $\varphi \leq f \leq \psi$ and φ and ψ are bounded, there is a $B > 0$ such that $|f(x)| \leq B$

for all $x \in [a, b]$. Choose $y_0' \in (x_0, x_1)$ such that $|y_0' - x_0| < \dfrac{\epsilon}{8Bn}$ and, for $i = 1, \ldots, n - 1$, inductively choose $y_i \in (y_{i-1}', x_i)$ and $y_i' \in (x_i, x_{i+1})$ such that $|y_i' - y_i| < \dfrac{\epsilon}{8Bn}$. Finally, choose $y_n \in (y_{n-1}', x_n)$ such that $|x_n - y_n| < \dfrac{\epsilon}{8Bn}$. The partition

$$\mathcal{P}' = \left\{ x_0, y_0', y_1, x_1, y_1', y_2, \ldots y_{n-2}', y_{n-1}, x_{n-1}, y_{n-1}', y_n, x_n \right\}$$

is a refinement of $\mathcal{P}$, and we are done if we can show that $U(f, \mathcal{P}') - L(f, \mathcal{P}') < \epsilon$. We consider two types of intervals: those of the form $[y_{i-1}', y_i]$ and the ones with an x_i for an endpoint. Suppose I is a subinterval determined by $\mathcal{P}'$ with an x_i for an endpoint. Then,

$$(\sup\{f(x) : x \in I\} - \inf\{f(x) : x \in I\}) \ell(I) \le 2B\ell(I) < 2B\frac{\epsilon}{8Bn} = \frac{\epsilon}{4n}.$$

Since there are $2n$ such intervals, the sum of these terms contribute less than $\dfrac{\epsilon}{2}$ to the difference $U(f, \mathcal{P}') - L(f, \mathcal{P}')$.

Next, consider an interval of the form $J_i = [y_{i-1}', y_i]$. On such an interval, φ and ψ are constant, equal to a_i and b_i, say. Thus, since $\varphi \le f \le \psi$ on the interval,

$$(\sup\{f(x) : x \in J_i\} - \inf\{f(x) : x \in J_i\}) \ell(J_i) \le (b_i - a_i) \ell(J_i).$$

Summing over all such intervals, we get a contribution to $U(f, \mathcal{P}') - L(f, \mathcal{P}')$ that is less than

$$\sum_{i=1}^{n} (b_i - a_i) \ell(J_i) \le \int_a^b (\psi - \varphi) < \frac{\epsilon}{2}.$$

Combining these two estimates shows that $U(f, \mathcal{P}') - L(f, \mathcal{P}') < \epsilon$ and completes the proof. $\qquad\square$

It is easy to see that the sum and product of step functions are step functions. Given functions f and g, we define the *maximum* of f and g, denoted $f \vee g$, by $f \vee g(x) = \max\{f(x), g(x)\}$ and the *minimum* of f and g, $f \wedge g$, by $f \wedge g(x) = \min\{f(x), g(x)\}$. It follows that the maximum and the minimum of two step functions is also a step function. See Exercise 2.10.

Given a function f, we define the positive and negative parts of f, denoted by f^+ and f^- respectively, by $f^+ = \max\{f, 0\}$ and $f^- = \max\{-f, 0\}$. From these definitions, we see that $f = f^+ - f^-$, $|f| =$

$f^+ + f^-$, $f^+ = \dfrac{|f| + f}{2}$ and $f^- = \dfrac{|f| - f}{2}$. We will now use step functions to show that these operations preserve integrability.

Theorem 2.23 *If $f_1, f_2 : [a, b] \to \mathbb{R}$ are Riemann integrable, then $f_1 \vee f_2$ and $f_1 \wedge f_2$ are Riemann integrable.*

Proof. Fix $\epsilon > 0$. By Theorem 2.22, for $i = 1, 2$, there are step functions φ_i and ψ_i such that $\varphi_i \leq f_i \leq \psi_i$ and $\int_a^b (\psi_i - \varphi_i) < \dfrac{\epsilon}{2}$. Then $\varphi_1 \vee \varphi_2 \leq f_1 \vee f_2 \leq \psi_1 \vee \psi_2$. Since $\psi_1 \vee \psi_2 - \varphi_1 \vee \varphi_2 \leq \psi_1 + \psi_2 - \varphi_1 - \varphi_2$, which follows by checking various cases, we see that

$$\int_a^b (\psi_1 \vee \psi_2 - \varphi_1 \vee \varphi_2) \leq \int_a^b [(\psi_1 - \varphi_1) + (\psi_2 - \varphi_2)] < \epsilon.$$

Applying the corollary one more time, we have that $f_1 \vee f_2$ is Riemann integrable. Since $f_1 \wedge f_2 = f_1 + f_2 - f_1 \vee f_2$, it follows that $f_1 \wedge f_2$ is Riemann integrable. $\square$

A set of real-valued functions with a common domain is called a *vector space* if it contains all finite linear combinations of its elements. For example, by linearity, the set of Riemann integrable functions on $[a, b]$ is a vector space. A vector space S of real-valued functions is called a *vector lattice* if $f, g \in S$ implies that $f \vee g, f \wedge g \in S$. Thus, the set of Riemann integrable functions on $[a, b]$ is a vector lattice.

An immediate consequence of the previous theorem is the following corollary.

Corollary 2.24 *Suppose f is Riemann integrable on $[a, b]$. Then, f^+, f^- and $|f|$ are Riemann integrable on $[a, b]$ and*

$$\left| \int_a^b f \right| \leq \int_a^b |f|.$$

We leave the proof as an exercise. Note that $|f|$ may be Riemann integrable while f is not. See Exercises 2.11 and 2.12.

Another application of the use of step functions allows us to see that the product of Riemann integrable functions is Riemann integrable.

Corollary 2.25 *If $f_1, f_2 : [a, b] \to \mathbb{R}$ are Riemann integrable, then $f_1 f_2$ is Riemann integrable.*

Proof. By the previous corollary, we may assume that each $f_i \geq 0$. Choose $M > 0$ so that $f_i(x) \leq M$ for $i = 1, 2$ and $x \in [a, b]$. There are

step functions φ_i and ψ_i such that $\varphi_i \leq f_i \leq \psi_i$ and $\int_a^b (\psi_i - \varphi_i) < \dfrac{\epsilon}{2M}$.
Moreover, we may assume that $0 \leq \varphi_i$ and $\psi_i \leq M$. In fact, it is enough to
set $\varphi_i' = \max \{\varphi_i, 0\}$ and $\psi_i' = \min \{\psi_i, M\}$ and observe that $\varphi_i' \leq f_i \leq \psi_i'$
and $\int_a^b (\psi_i' - \varphi_i') \leq \int_a^b (\psi_i - \varphi_i)$. Hence, $\varphi_1 \varphi_2 \leq f_1 f_2 \leq \psi_1 \psi_2$ and

$$
\int_a^b (\psi_1 \psi_2 - \varphi_1 \varphi_2) = \int_a^b (\psi_1 \psi_2 - \psi_1 \varphi_2 + \psi_1 \varphi_2 - \varphi_1 \varphi_2)
$$

$$
\leq \int_a^b [M (\psi_2 - \varphi_2) + M (\psi_1 - \varphi_1)] < 2M \frac{\epsilon}{2M} = \epsilon.
$$

By Theorem 2.22, $f_1 f_2$ is Riemann integrable. $\square$

2.4.4   *Integrable functions*

The Darboux condition or, more correctly, the condition of Theorem 2.19
makes it easy to show that certain collections of functions are Riemann integrable. We now prove that monotone functions and continuous functions
are Riemann integrable.

Theorem 2.26   *Suppose that f is a monotone function on $[a, b]$. Then,
f is Riemann integrable on $[a, b]$.*

Proof.   Without loss of generality, we may assume that f is increasing.
Clearly, f is bounded by $\max \{|f(a)|, |f(b)|\}$. Fix $\epsilon > 0$. Let $\mathcal{P}$ be a
partition with mesh less than $\epsilon / (f(b) - f(a))$. (If $f(b) = f(a)$, then f
is constant and the result is a consequence of Example 2.5 and linearity.)
Since f is increasing, $M_i = f(x_i)$ and $m_i = f(x_{i-1})$. It follows that

$$
U(f, \mathcal{P}) - L(f, \mathcal{P}) = \sum_{i=1}^n \{M_i - m_i\} (x_i - x_{i-1})
$$

$$
= \sum_{i=1}^n \{f(x_i) - f(x_{i-1})\} (x_i - x_{i-1})
$$

$$
< \sum_{i=1}^n \{f(x_i) - f(x_{i-1})\} \frac{\epsilon}{f(b) - f(a)}
$$

$$
= (f(b) - f(a)) \frac{\epsilon}{f(b) - f(a)} = \epsilon,
$$

where the next to last equality uses the fact that $\sum_{i=1}^n \{f(x_i) - f(x_{i-1})\}$ is
a telescoping sum. By Theorems 2.19 and 2.21, f is Riemann integrable. $\square$

Suppose that f is a continuous function on $[a, b]$. Then, f is uniformly continuous. If $\mathcal{P}$ is a partition with sufficiently small mesh (depending on uniform continuity) and $\{t_i\}_{i=1}^n$ and $\{t_i'\}_{i=1}^n$ are sampling points for $\mathcal{P}$, then $S\left(f, \mathcal{P}, \{t_i\}_{i=1}^n\right) - S\left(f, \mathcal{P}, \{t_i'\}_{i=1}^n\right)$ can be made as small as desired. Thus, it seems likely that the Riemann sums for f will satisfy a Cauchy condition and f will be Riemann integrable. Unfortunately, the Cauchy condition must hold for Riemann sums defined by different partitions, which makes a proof along these lines complicated. Such problems can be avoided by using Theorem 2.19, and we have

Theorem 2.27 *Suppose that $f : [a, b] \to \mathbb{R}$ is continuous on $[a, b]$. Then, f is Riemann integrable on $[a, b]$.*

Proof. Since f is continuous on $[a, b]$, it is uniformly continuous there. Let $\epsilon > 0$ and choose a δ so that if $x, y \in [a, b]$ and $|x - y| < \delta$, then $|f(x) - f(y)| < \dfrac{\epsilon}{b-a}$. Let $\mathcal{P}$ be a partition of $[a, b]$ with mesh less than δ. Since f is continuous on the compact interval $[x_{i-1}, x_i]$, there are points $T_i, t_i \in [x_{i-1}, x_i]$ such that $M_i = f(T_i)$ and $m_i = f(t_i)$, for $i = 1, \ldots, n$. Since $|T_i - t_i| \leq \mu(\mathcal{P}) < \delta$,

$$M_i - m_i = |f(T_i) - f(t_i)| < \frac{\epsilon}{b-a}.$$

Thus,

$$U(f, \mathcal{P}) - L(f, \mathcal{P}) = \sum_{i=1}^n \{M_i - m_i\}(x_i - x_{i-1}) < \sum_{i=1}^n \frac{\epsilon}{b-a}(x_i - x_{i-1}) = \epsilon$$

and the proof is completed as in the previous theorem. $\square$

2.4.5 *Additivity of the integral over intervals*

We have observed that the integral is an operator, a function acting on functions. We can also view the integral as a function acting on sets. To do this, fix a function $f : [a, b] \to \mathbb{R}$, and let $E \subset [a, b]$. We say that f is *Riemann integrable over E* if the function $f\chi_E$ is Riemann integrable over $[a, b]$ and we define the Riemann integral of f over E to be

$$F(E) = \int_E f = \int_a^b f\chi_E.$$

Unfortunately, F may not be defined for many subsets of E. One of the recurring themes in developing an integration theory is to enlarge as much as possible the collection of sets that are allowable as inputs. For the Riemann integral, a natural collection of sets is the collection of finite unions of subintervals of $[a, b]$. As we will see below, if f is Riemann integrable on $[a, b]$, then f is Riemann integrable on every subinterval of $[a, b]$.

Proposition 2.28 *Suppose that $f : [a, b] \to \mathbb{R}$ is Riemann integrable and $c \in (a, b)$. Then, f is Riemann integrable on $[a, c]$ and $[c, b]$, and*

$$\int_a^c f + \int_c^b f = \int_a^b f.$$

Proof. We first claim that f is Riemann integrable on $[a, c]$ and $[c, b]$. Given $\epsilon > 0$, it is enough to show that there is a partition $\mathcal{P}_{[a,c]}$ of $[a, c]$ such that

$$U\left(f, \mathcal{P}_{[a,c]}\right) - L\left(f, \mathcal{P}_{[a,c]}\right) < \epsilon,$$

and a similar result for $[c, b]$. By Theorem 2.20, there is a $\delta > 0$ so that if $\mathcal{P}$ is a partition of $[a, b]$ with $\mu\left(\mathcal{P}\right) < \delta$, then $U\left(f, \mathcal{P}\right) - L\left(f, \mathcal{P}\right) < \epsilon$. Let $\mathcal{P}_{[a,c]}$ be any partition of $[a, c]$ with $\mu\left(\mathcal{P}_{[a,c]}\right) < \delta$, let $\mathcal{P}_{[c,b]}$ be any partition of $[c, b]$ with $\mu\left(\mathcal{P}_{[c,b]}\right) < \delta$, and set $\mathcal{P} = \mathcal{P}_{[a,c]} \cup \mathcal{P}_{[c,b]}$. Then, $\mu\left(\mathcal{P}\right) < \delta$ and

$$\left\{U\left(f, \mathcal{P}_{[a,c]}\right) - L\left(f, \mathcal{P}_{[a,c]}\right)\right\} + \left\{U\left(f, \mathcal{P}_{[c,b]}\right) - L\left(f, \mathcal{P}_{[c,b]}\right)\right\}$$
$$= U\left(f, \mathcal{P}\right) - L\left(f, \mathcal{P}\right) < \epsilon.$$

Since for any bounded function g and partition $\mathcal{P}$, $L\left(g, \mathcal{P}\right) \leq U\left(g, \mathcal{P}\right)$, it follows that

$$U\left(f, \mathcal{P}_{[a,c]}\right) - L\left(f, \mathcal{P}_{[a,c]}\right) < \epsilon$$

and

$$U\left(f, \mathcal{P}_{[c,b]}\right) - L\left(f, \mathcal{P}_{[c,b]}\right) < \epsilon$$

so that f is Riemann integrable on $[a, c]$ and $[c, b]$.

To see that $\int_a^c f + \int_c^b f = \int_a^b f$, we fix $\epsilon > 0$ and choose partitions $\mathcal{P}_{[a,c]}$ and $\mathcal{P}_{[c,b]}$ such that $U\left(f, \mathcal{P}_{[a,c]}\right) - \int_a^c f < \frac{\epsilon}{2}$ and $U\left(f, \mathcal{P}_{[c,b]}\right) - \int_c^b f < \frac{\epsilon}{2}$.

Set $\mathcal{P} = \mathcal{P}_{[a,c]} \cup \mathcal{P}_{[c,b]}$. Then,

$$\left| U(f, \mathcal{P}) - \left(\int_a^c f + \int_c^b f \right) \right| = \left| \left(U(f, \mathcal{P}_{[a,c]}) - \int_a^c f \right) \right.$$
$$\left. + \left(U(f, \mathcal{P}_{[c,b]}) - \int_c^b f \right) \right|$$
$$< \epsilon.$$

Since we can do this for any $\epsilon > 0$ and $\int_a^b f$ is the infimum of the $U(f, \mathcal{P})$, we see that $\int_a^c f + \int_c^b f = \int_a^b f$. $\qquad \square$

We leave it as an exercise for the reader to show that if f is Riemann integrable on $[a, c]$ and $[c, b]$ then f is Riemann integrable on $[a, b]$ (see Exercise 2.15).

Suppose f is Riemann integrable on $[a, b]$ and $[c, d] \subset [a, b]$. By applying the previous proposition twice, if necessary, we have

Corollary 2.29 *Suppose that $f : [a, b] \to \mathbb{R}$ is Riemann integrable and $[c, d] \subset [a, b]$. Then, f is Riemann integrable on $[c, d]$.*

Let I be an interval. We define the *interior* of I, denoted I^o, to be the set of $x \in I$ such that there is a $\delta > 0$ so that the δ-neighborhood of x is contained in I, $(x - \delta, x + \delta) \subset I$. Suppose that $f : [a, b] \to \mathbb{R}$ and $I, J \subset [a, b]$ are intervals with disjoint interiors, $I^o \cap J^o = \emptyset$. Then, if f is Riemann integrable on $[a, b]$, we have

$$\int_{I \cup J} f = \int_I f + \int_J f,$$

which is called an *additivity condition*. When I and J are contiguous intervals, then $I \cup J$ is an interval and this equality is an application of the previous proposition. When I and J are at a positive distance, then $I \cup J$ is no longer an interval. See Exercise 2.16.

2.5 Fundamental Theorem of Calculus

The Fundamental Theorem of Calculus consists of two parts which relate the processes of differentiation and integration and show that in some sense these two operations are inverses of one another. We begin by considering the integration of derivatives. Suppose that $f : [a, b] \to \mathbb{R}$ is differentiable on $[a, b]$ with derivative f'. The first part of the Fundamental Theorem of Calculus involves the familiar formula from calculus,

$$\int_a^b f' = f(b) - f(a). \tag{2.1}$$

Theorem 2.30 *(Fundamental Theorem of Calculus: Part I) Suppose that $f : [a, b] \to \mathbb{R}$ and f' is Riemann integrable on $[a, b]$. Then, (2.1) holds.*

Proof. Since f' is Riemann integrable, we are done if we can find a sequence of partitions $\{\mathcal{P}_k\}_{k=1}^\infty$ and corresponding sampling points $\left\{t_i^{(k)}\right\}_{i=1}^{n_k}$ such that $\mu(\mathcal{P}_k) \to 0$ as $k \to \infty$ and $S\left(f', \mathcal{P}_k, \left\{t_i^{(k)}\right\}_{i=1}^{n_k}\right) = f(b) - f(a)$ for all k. In fact, let $\mathcal{P} = \{x_0, x_1, \ldots, x_n\}$ be any partition of $[a, b]$. Since f is differentiable on (a, b) and continuous on $[a, b]$, we may apply the Mean Value Theorem to any subinterval of $[a, b]$. Hence, for $i = 1, \ldots, n$, there is a $y_i \in [x_{i-1}, x_i]$ such that $f(x_i) - f(x_{i-1}) = f'(y_i)(x_i - x_{i-1})$. Thus,

$$\sum_{i=1}^n f'(y_i)(x_i - x_{i-1}) = \sum_{i=1}^n [f(x_i) - f(x_{i-1})]$$

which is a telescoping sum equal to $f(x_n) - f(x_0) = f(b) - f(a)$. Thus, for any partition $\mathcal{P}$, there is a collection of sampling points $\{y_i\}_{i=1}^n$ such that

$$S(f', \mathcal{P}, \{y_i\}_{i=1}^n) = f(b) - f(a).$$

Taking any sequence of partitions with mesh approaching 0 and associating sampling points as above, we see that $\int_a^b f' = f(b) - f(a)$. $\qquad \square$

The key hypothesis in Theorem 2.30 is that f' is Riemann integrable. The following example shows that (2.1) does not hold in general for the Riemann integral.

Example 2.31 Define $f : [0, 1] \to \mathbb{R}$ by

$$f(x) = \begin{cases} x^2 \cos \dfrac{\pi}{x^2} & \text{if } 0 < x \le 1 \\ 0 & \text{if } x = 0 \end{cases}.$$

Then, f is differentiable on $[0, 1]$ with derivative

$$f'(x) = \begin{cases} 2x \cos \dfrac{\pi}{x^2} + \dfrac{2\pi}{x} \sin \dfrac{\pi}{x^2} & \text{if } 0 < x \le 1 \\ 0 & \text{if } x = 0 \end{cases}.$$

Since f' is not bounded on $[0, 1]$, f' is not Riemann integrable on $[0, 1]$.

There are also examples of bounded derivatives which are not Riemann integrable, but these are more difficult to construct. (See, for example, [Be, Section 1.3, page 20], [LV, Section 1.4.5] or [Sw1, Section 3.3, page 98].)

We will see later in Chapter 4 that the derivative f' in Example 2.31 is also not Lebesgue integrable so a general version of the Fundamental Theorem of Calculus for the Lebesgue integral also requires an integrability assumption on the derivative. In Chapter 4 we will construct an integral, called the gauge or Henstock-Kurzweil integral, for which the Fundamental Theorem of Calculus holds in full generality; that is, the Henstock-Kurzweil integral integrates all derivatives and (2.1) holds.

The second part of the Fundamental Theorem of Calculus concerns the differentiation of indefinite integrals. Suppose that $f : [a, b] \to \mathbb{R}$ is Riemann integrable on $[a, b]$. We define the *indefinite integral* of f at $x \in [a, b]$ by

$$F(x) = \int_a^x f(t)\, dt,$$

where $F(a) = \int_a^a f = 0$. If $a \leq x < y \leq b$, we define $\int_y^x f = -\int_x^y f$.

Let f be Riemann integrable on $[a, b]$. Choose $M > 0$ so that $|f(x)| \leq M$ for all $x \in [a, b]$. Let $x, y \in [a, b]$ and consider the difference $F(x) - F(y)$. Using the additivity of the Riemann integral, we have

$$|F(x) - F(y)| = \left| \int_a^x f(t)\, dt - \int_a^y f(t)\, dt \right|$$
$$= \left| \int_x^y f(t)\, dt \right| \leq \int_{\min(x,y)}^{\max(x,y)} |f(t)|\, dt \leq M\, |y - x|.$$

A function g satisfying an inequality of the form

$$|g(x) - g(y)| \leq C\, |x - y|$$

is said to satisfy a *Lipschitz condition* on $[a, b]$ with *Lipschitz constant* C. Thus, any indefinite integral satisfies a Lipschitz condition and any such function is uniformly continuous.

The second half of the Fundamental Theorem of Calculus concerns the differentiation of indefinite integrals.

Theorem 2.32 *(Fundamental Theorem of Calculus: Part II) Suppose that $f : [a, b] \to \mathbb{R}$ is Riemann integrable. Set $F(x) = \int_a^x f(t)\, dt$. Then, F* *is continuous on $[a, b]$. If f is continuous at $\xi \in [a, b]$, then F is differentiable at ξ and $F'(\xi) = f(\xi)$.*

Proof.   To see that F is continuous, we need only set $\delta = \epsilon/M$ in the Lipschitz estimate on F above. So, we only need show that the continuity of f implies the differentiability of F.

Suppose f is continuous at ξ and let $\epsilon > 0$. There is a $\delta > 0$ such that $|f(x) - f(\xi)| < \dfrac{\epsilon}{2}$ whenever $x \in [a, b]$ and $|x - \xi| < \delta$. Thus, if $0 < |x - \xi| < \delta$, then

$$\left| \frac{F(x) - F(\xi)}{x - \xi} - f(\xi) \right| = \left| \frac{1}{x - \xi} \int_\xi^x f(t)\, dt - f(\xi) \right|$$

$$= \frac{1}{|x - \xi|} \left| \int_\xi^x \{f(t) - f(\xi)\}\, dt \right|$$

$$\leq \frac{1}{|x - \xi|} \int_\xi^x |f(t) - f(\xi)|\, dt$$

$$\leq \frac{1}{|x - \xi|} \frac{\epsilon}{2} |x - \xi| = \frac{\epsilon}{2} < \epsilon.$$

That is, F is differentiable at ξ and $F'(\xi) = f(\xi)$.   □

The theorem tells us that F must be differentiable at points where f is continuous. If f is not continuous at a point, F may or may not be differentiable.

Example 2.33   Define the *signum function* sgn: $\mathbb{R} \to \mathbb{R}$ by

$$sgn\, x = \begin{cases} \dfrac{x}{|x|} & \text{if } x \neq 0 \\ 0 & \text{if } x = 0 \end{cases}.$$

The function *sgn* is continuous for $x \neq 0$ and is not continuous at 0. The indefinite integral of *sgn* is $F(x) = |x|$, which is continuous everywhere and differentiable except at 0. Here, the indefinite integral is not differentiable at the point where the function is not continuous.

Next, consider $g(x) = \begin{cases} 0 & \text{if } x \neq 0 \\ 1 & \text{if } x = 0 \end{cases}$, which is continuous at every x except 0. In this case, $F(x) = 0$ for all x is differentiable at 0, even though f is not continuous there.

This theorem only guarantees that F is differentiable at points at which f is continuous. In fact, F is differentiable at "most" points. We will discuss this in Chapter 4.

2.5.1 *Integration by parts and substitution*

Two of the most familiar results from the calculus, integration by parts and by substitution, are consequences of the Fundamental Theorem of Calculus. Integration by parts, which follows from the product rule for differentiation, is a kind of product rule for integration.

Theorem 2.34 *(Integration by parts) Suppose that $f, g : [a, b] \to \mathbb{R}$ and f' and g' are Riemann integrable on $[a, b]$. Then, fg' and $f'g$ are Riemann integrable on $[a, b]$ and*

$$\int_a^b f(x) g'(x) \, dx = f(b) g(b) - f(a) g(a) - \int_a^b f'(x) g(x) \, dx.$$

Proof. Note that f and g are continuous and hence Riemann integrable by Theorem 2.27. By Corollary 2.25, fg' and $f'g$ are Riemann integrable. Thus, $(fg)' = fg' + f'g$ is Riemann integrable and, by Theorem 2.30,

$$\int_a^b f(x) g'(x) \, dx + \int_a^b f'(x) g(x) \, dx = \int_a^b (fg)'(x) \, dx$$
$$= f(b) g(b) - f(a) g(a).$$

The result now follows. $\qquad\qquad\qquad\qquad\qquad\qquad\qquad\qquad\square$

We now consider integration by substitution, or change of variables.

Theorem 2.35 *(Change of variables) Let $\phi : [a, b] \to \mathbb{R}$ be continuously differentiable. Assume $\phi([a, b]) = [c, d]$ with $\phi(a) = c$ and $\phi(b) = d$. If $f : [c, d] \to \mathbb{R}$ is continuous, then $f(\phi) \phi'$ is Riemann integrable on $[a, b]$ and*

$$\int_a^b f(\phi(t)) \phi'(t) \, dt = \int_c^d f.$$

Proof. Define F and H by $F(x) = \int_c^x f(t) \, dt$ and $H(y) = \int_a^y f(\phi(t)) \phi'(t) \, dt$. By hypothesis, both these integrands are continuous so that, by Theorem 2.32, F and H are differentiable. Consequently, $F \circ \phi$ is differentiable on $[a, b]$ and by the Chain Rule,

$$(F \circ \phi)'(y) = F'(\phi(y)) \phi'(y) = f(\phi(y)) \phi'(y) = H'(y),$$

so that $(F \circ \phi)(y) = H(y) + C$. Since $F(c) = 0$, $H(a) = 0$ and $F(c) = F(\phi(a)) = H(a) + C$, $C = 0$. We now have

$$\int_c^d f = F(d) = F(\phi(b)) = H(b) = \int_a^b f(\phi(t)) \phi'(t) \, dt$$

as we wished to show. $\square$

2.6   Characterizations of integrability

We now characterize Riemann integrability in terms of the local behavior of the function. Let $f : [a, b] \to \mathbb{R}$ be bounded and let S be a nonempty subset of $[a, b]$. The *oscillation of* f *over* S is defined to be

$$\omega (f, S) = \sup \{f (t) : t \in S\} - \inf \{f (t) : t \in S\}.$$

It follows immediately that if $S \subset T$ then $\omega (f, S) \leq \omega (f, T)$. Let $x \in [a, b]$ and, for $\delta > 0$, set $U_\delta (x) = \{t \in [a, b] : |t - x| < \delta\}$. The *oscillation of* f *at* x is defined to be

$$\omega (f, x) = \lim_{\delta \to 0^+} \omega (f, U_\delta).$$

Note that the limit exists since $\omega (f, U_\delta)$ is a decreasing function of δ. It is easy to see that f is continuous at x if, and only if, $\omega (f, x) = 0$. (See Exercise 2.30.)

Let S be a subset of $\mathbb{R}$. The *closure* of S, denoted $\overline{S}$, is the set of all $x \in \mathbb{R}$ for which there is a sequence $\{s_n\}_{n=1}^\infty \subset S$ that converges to x. Note, in particular, that $S \subset \overline{S}$. If S is a bounded interval, then $\overline{S}$ is the union of S with the set of its endpoints. Our first characterization of Riemann integrability is in terms of the oscillation of the function f. We begin with a lemma.

Lemma 2.36   *Suppose that* $\omega (f, x) < \epsilon$ *for every* $x \in [a, b]$. *Then, there is a partition* $\mathcal{P} = \{x_0, x_1, \ldots, x_n\}$ *of* $[a, b]$ *such that* $\omega (f, [x_{i-1}, x_i]) < \epsilon$ *for* $i = 1, \ldots, n$.

Proof.   For each $t \in [a, b]$, there is an open interval I_t centered at t such that $\omega \left(f, \overline{I_t} \cap [a, b]\right) < \epsilon$. Since $\{I_t : t \in [a, b]\}$ is an open cover of $[a, b]$, there is a finite subcover $\{I_{t_1}, I_{t_2}, \ldots, I_{t_n}\}$. The set of endpoints of these intervals that lie in (a, b) along with the points a and b yield a partition $\{x_0, x_1, \ldots, x_n\}$ of $[a, b]$ such that for each $i = 1, \ldots, n$, there is a k so that $[x_{i-1}, x_i] \subset \overline{I_{t_k}}$. Hence, $\omega (f, [x_{i-1}, x_i]) \leq \omega \left(f, \overline{I_{t_k}} \cap [a, b]\right) < \epsilon$. $\square$

For our first characterization, we require the notion of the *outer Jordan content* of a subset S of $[a, b]$. Let $\mathcal{P} = \{x_0, x_1, \ldots, x_n\}$ be a partition of $[a, b]$ and let $J (S, \mathcal{P})$ be the sum of the lengths of the closed intervals

$[x_{i-1}, x_i]$ which contain points of the closure of S. The outer Jordan content of S, denoted $\bar{c}(S)$, is defined to be the infimum of $J(S, \mathcal{P})$ as $\mathcal{P}$ runs through all partitions of $[a, b]$. Note that $J(S, \mathcal{P}) = U(\chi_{\bar{S}}, \mathcal{P})$ and, consequently, $\bar{c}(S) = \overline{\int}_a^b \chi_{\bar{S}}$. (For a discussion of Jordan content, see [Bar].)

A finite subset of $[a, b]$ obviously has outer Jordan content 0, but an infinite set can also have outer Jordan content 0. (See Exercise 2.26.) The set function $\bar{c}$ is monotone in the sense that if $S \subset T$ then $\bar{c}(S) \leq \bar{c}(T)$ and is also subadditive in the sense that if $S, T \subset [a, b]$, then $\bar{c}(S \cup T) \leq \bar{c}(S) + \bar{c}(T)$. (See Exercise 2.27.)

For $\epsilon > 0$, set $D_\epsilon(f) = \{x \in [a, b] : \omega(f, x) \geq \epsilon\}$. We characterize Riemann integrability in terms of the outer Jordan content of the sets $D_\epsilon(f)$.

Theorem 2.37 *Let $f : [a, b] \to \mathbb{R}$. Then, f is Riemann integrable over $[a, b]$ if, and only if, f is bounded and for every $\epsilon > 0$, the set $D_\epsilon(f)$ has outer Jordan content 0.*

Proof. Suppose first that f is bounded and for every $\epsilon > 0$, the set $D_\epsilon(f)$ has outer Jordan content 0. Choose $M > 0$ such that $|f(t)| \leq M$ for $a \leq t \leq b$ and let $\epsilon > 0$. Let $\mathcal{P}$ be the partition of $[a, b]$ such that the sum of the lengths of the subintervals determined by $\mathcal{P}$ that contain points of $D_{\epsilon/2(b-a)}$ is less than $\frac{\epsilon}{4M}$. Let these subintervals be labeled $\{I_1, I_2, \ldots, I_k\}$ and label the remaining subintervals determined by $\mathcal{P}$ by $\{J_1, J_2, \ldots, J_l\}$. Applying the previous lemma to each J_j, we may assume that $\omega(f, J_j) < \frac{\epsilon}{2(b-a)}$ for $j = 1, \ldots, l$. We then have

$$U(f, \mathcal{P}) - L(f, \mathcal{P}) \leq \sum_{i=1}^k \omega(f, I_i)\ell(I_i) + \sum_{j=1}^l \omega(f, J_j)\ell(J_j)$$

$$< 2M\frac{\epsilon}{4M} + \frac{\epsilon}{2(b-a)}(b-a) = \epsilon$$

so that f is Riemann integrable by Theorem 2.19.

For the converse, assume that there is an $\epsilon > 0$ such that $\bar{c}(D_\epsilon(f)) = c > 0$. We will use Theorem 2.19 to show that f is not Riemann integrable. Let $\mathcal{P} = \{x_0, x_1, \ldots, x_n\}$ be a partition of $[a, b]$ and let I be the set of all indices i such that the intersection of $[x_{i-1}, x_i]$ and $D_\epsilon(f)$ is nonempty. Let $I' \subset I$ be the set of indices i such that $(x_{i-1}, x_i) \cap D_\epsilon(f) \neq \emptyset$. By Exercise 2.31, for $i \in I'$, $\omega(f, [x_{i-1}, x_i]) \geq \epsilon$. Let $\eta > 0$. Suppose $i \in I \setminus I'$. Then, at least one of the endpoints of $[x_{i-1}, x_i]$ is in $D_\epsilon(f)$. Refine $\mathcal{P}$ by adding $y_i, y_i' \in (x_{i-1}, x_i)$ such that $y_i < y_i'$, $|y_i - x_{i-1}| < \frac{\eta}{2n}$ and $|y_i' - x_i| < \frac{\eta}{2n}$.

For $i \in I \setminus I'$, label the intervals $[x_{i-1}, y_i]$ and $[y_i', x_i]$ by $J_1, \ldots, J_m$ where $m \leq 2n$. Note that $[y_i, y_i'] \cap D_\epsilon(f) = \emptyset$ so that

$$c \leq \sum_{i \in I'} (x_i - x_{i-1}) + \sum_{k=1}^{m} \ell(J_k)$$

$$\leq \sum_{i \in I'} (x_i - x_{i-1}) + 2n \frac{\eta}{2n} = \sum_{i \in I'} (x_i - x_{i-1}) + \eta.$$

Since $\eta > 0$ is arbitrary, it follows that

$$c \leq \sum_{i \in I'} (x_i - x_{i-1}).$$

Hence,

$$U(f, \mathcal{P}) - L(f, \mathcal{P}) = \sum_{i=1}^{n} \omega(f, [x_{i-1}, x_i])(x_i - x_{i-1})$$

$$\geq \sum_{i \in I'} \omega(f, [x_{i-1}, x_i])(x_i - x_{i-1}) \geq \epsilon c.$$

Since this is true for any partition $\mathcal{P}$, it follows that f is not Riemann integrable. $\square$

Remark 2.38 *If S is a subset of $[a, b]$ with outer Jordan content 0, then for $\delta > 0$, there is a finite number of non-overlapping, closed intervals $\{I_1, \ldots, I_n\}$ such that $S \subset \cup_{i=1}^{n} I_i$ and $\sum_{i=1}^{n} \ell(I_i) < \delta$. Set $\eta = \delta - \sum_{i=1}^{n} \ell(I_i) > 0$. If $I_i = [a_i, b_i]$, set $J_i = \left(a_i - \frac{\eta}{4n}, b_i + \frac{\eta}{4n}\right)$. Then, $S \subset \cup_{i=1}^{n} I_i \subset \cup_{i=1}^{n} J_i$ and*

$$\sum_{i=1}^{n} \ell(J_i) = \sum_{i=1}^{n} \left\{ \ell(I_i) + \frac{\eta}{2n} \right\} = \sum_{i=1}^{n} \ell(I_i) + \frac{\eta}{2} < \delta.$$

If two intervals in the set $\{J_i\}_{i=1}^{n}$ have a nonempty intersection, we can replace them by their union. This will not change the union of the intervals and will decrease the sum of their lengths. Thus, we can cover S by non-overlapping, open intervals, the sum of whose lengths is less than δ.

While this theorem gives a characterization of Riemann integrability, the test involves an infinite number of conditions and, consequently, is not practical to employ. However, if $\epsilon < \epsilon'$ then $D_{\epsilon'}(f) \subset D_\epsilon(f)$, so that $D(f) = \cup_{\epsilon > 0} D_\epsilon(f)$ is a kind of limit of $D_\epsilon(f)$ as ϵ decreases to 0. As a consequence of Exercise 2.30, we see that $D(f) = \{t \in [a, b] : f \text{ is discontinuous at } t\}$. Our second characterization, due to Lebesgue, gives a characterization of

Riemann integrability in terms of the single set $D(f)$. We will use the following lemma.

Lemma 2.39 *For each $\epsilon > 0$, the set $D_\epsilon(f)$ is closed in $[a, b]$.*

Proof. Let $x \in [a, b] \setminus D_\epsilon(f)$ and set $\eta = \omega(f, x)$. Since $\eta < \epsilon$, there is a neighborhood $U_\delta(x)$ of x such that $\omega(f, U_\delta(x)) < \dfrac{\eta + \epsilon}{2} < \epsilon$, so if $x_1, x_2 \in U_\delta(x)$, then $|f(x_1) - f(x_2)| < \dfrac{\eta + \epsilon}{2}$. Thus, $U_\delta(x) \cap D_\epsilon(f) = \emptyset$, so that the complement of $D_\epsilon(f)$ is open in $[a, b]$. It follows that $D_\epsilon(f)$ is closed in $[a, b]$. $\qquad\square$

Theorem 2.40 *Let $f : [a, b] \to \mathbb{R}$. Then, f is Riemann integrable if, and only if, f is bounded and, for every $\delta > 0$, $D(f)$ can be covered by a countable number of open intervals, the sum of whose lengths is less than δ.*

Proof. Suppose f is Riemann integrable. By our first characterization and the previous remark, for each n, $D_{1/n}(f)$ can be covered by a finite number of open intervals, the sum of whose lengths is less than $\delta 2^{-n}$. By Exercise 2.32, $D(f) = \cup \{D_{1/n}(f) : n \in \mathbb{N}\}$, so that $D(f)$ can be covered by a countable number of intervals, the sum of whose lengths is less than $\sum_{n=1}^{\infty} \delta 2^{-n} = \delta$.

Next, let $\epsilon, \delta > 0$ and assume that there exist open intervals $\{I_i\}_{i=1}^{\infty}$ covering $D(f)$ such that $\sum_{i=1}^{\infty} \ell(I_i) < \delta$. By the previous lemma, $D_\epsilon(f)$ is closed in $[a, b]$ and, since $D_\epsilon(f) \subset D(f)$, there exist a finite number of open intervals $\{I_1, I_2, \ldots, I_n\}$ which cover $D_\epsilon(f)$. The endpoints of $\{I_1, I_2, \ldots, I_n\}$ in $[a, b]$ along with a and b comprise a partition of $[a, b]$ such that the sum of the lengths of the intervals, determined by the partition, which intersect $D_\epsilon(f)$ is less than δ. Since this is true for every δ, $D_\epsilon(f)$ has outer Jordan content 0. Since ϵ is arbitrary, by the previous theorem, f is Riemann integrable. $\qquad\square$

2.6.1 *Lebesgue measure zero*

We can use one of the basic ideas of Lebesgue measure to give a restatement of Theorem 2.40 in other terms. A subset $E \subset \mathbb{R}$ is said to have *Lebesgue measure* 0 or is called a *null set* if, for every $\delta > 0$, E can be covered by a countable number of open intervals the sum of whose length is less than δ. The following example shows that a countable set has measure zero.

Example 2.41 Let $C \subset \mathbb{R}$ be a countable set. Then, we can write $C = \{c_i\}_{i=1}^{\infty}$. Fix $\delta > 0$ and let $I_i = \left(c_i - \delta 2^{-i-2}, c_i + \delta 2^{-i-2}\right)$. Then, I_i is an open interval containing c_i and having length $\ell(I_i) = \delta 2^{-i-1}$. It follows that $C \subset \cup_{i=1}^{\infty} I_i$ and $\sum_{i=1}^{\infty} \ell(I_i) = \sum_{i=1}^{\infty} \delta 2^{-i-1} = \delta/2 < \delta$. Thus, C is a null set.

Thus, every countable set is null. In Chapter 3, we will give an example of an uncountable set that is null.

A statement about the points of a set E is said to hold *almost everywhere* (a.e.) in E if the points in E for which the statement fails to hold has Lebesgue measure 0. For example, a function $g : [a, b] \to \mathbb{R}$ is equal to 0 a.e. in $[a, b]$ means that the set $\{t \in [a, b] : g(t) \neq 0\}$ has Lebesgue measure 0. The following corollary, due to Lebesgue, restates the previous theorem in terms of null sets.

Corollary 2.42 *A bounded function $f : [a, b] \to \mathbb{R}$ is Riemann integrable if, and only if, f is continuous a.e. in $[a, b]$.*

2.7 Improper integrals

Since the Riemann integral is restricted to bounded functions defined on bounded intervals, it is necessary to make special definitions in order to allow unbounded functions or unbounded intervals. These extensions, sometimes called *improper integrals*, were first carried out by Cauchy and we will refer to the extensions as *Cauchy-Riemann integrals*. (See [C, (2) 4, pages 140-150].) First, we consider the case of an unbounded function defined on a bounded interval.

Let $f : [a, b] \to \mathbb{R}$ and assume that f is Riemann integrable on every subinterval $[c, b]$, $a < c < b$. Note that this guarantees that f is bounded on $[c, b]$ for $c \in (a, b)$ but not necessarily on all of $[a, b]$.

Definition 2.43 Let $f : [a, b] \to \mathbb{R}$ be as above. We say that f is *Cauchy-Riemann integrable* over $[a, b]$ if $\lim_{c \to a+} \int_c^b f$ exists, and we define the Cauchy-Riemann integral of f over $[a, b]$ to be

$$\int_a^b f = \lim_{c \to a+} \int_c^b f.$$

When the limit exists, we say that the Cauchy-Riemann integral of f *converges*; if the limit fails to exist, we say the integral *diverges*.

By Exercise 2.35 we see that if f is Riemann integrable over $[a, b]$, then this definition agrees with the original definition of the Riemann integral and, thus, gives an extension of the Riemann integral.

Example 2.44 Let $p \in \mathbb{R}$ and define $f : [0, 1] \to \mathbb{R}$ by $f(t) = t^p$, for $0 < t \le 1$ and $f(0) = 0$. For $p \ne -1$, $\int_c^1 t^p dt = \dfrac{1 - c^{p+1}}{p + 1}$ so $\lim_{c \to 0^+} \int_c^1 t^p dt$ exists and equals $\dfrac{1}{p + 1}$ if, and only if, $p > -1$, and then $\int_0^1 t^p dt = \dfrac{1}{p + 1}$. If $p = -1$, $\int_c^1 \dfrac{1}{t} dt = -\ln c$ which does not have a finite limit as $c \to 0^+$. Thus, t^p is integrable if, and only if, $p > -1$.

Similarly, if f is Riemann integrable over every subinterval $[a, c]$, $a < c < b$, then f is said to be Cauchy-Riemann integrable over $[a, b]$ if $\int_a^b f = \lim_{c \to b^-} \int_a^c f$ exists. This definition follows by applying the previous definition to the function $g(x) = f(a + b - x)$.

If a function $f : [a, b] \to \mathbb{R}$ has a singularity or becomes unbounded at an interior point c of $[a, b]$, then f is defined to be Cauchy-Riemann integrable over $[a, b]$ if f is Cauchy-Riemann integrable over both $[a, c]$ and $[c, b]$ and the integral over $[a, b]$ is defined to be

$$\int_a^b f = \int_a^c f + \int_c^b f.$$

Note that if f is Cauchy-Riemann integrable over $[a, b]$, then

$$\lim_{\epsilon \to 0^+} \left(\int_a^{c-\epsilon} f + \int_{c+\epsilon}^b f \right) \tag{2.2}$$

exists and equals $\int_a^b f$. However, the limit in (2.2) may exist and f may fail to be Cauchy-Riemann integrable over $[a, b]$, as the following example shows.

Example 2.45 Let $f(t) = t^{-3}$ for $0 < |t| \le 1$ and $f(0) = 0$. Then, since f is an odd function (see Exercise 2.6),

$$\lim_{\epsilon \to 0^+} \left(\int_{-1}^{-\epsilon} f + \int_\epsilon^1 f \right)$$

exists (and equals 0), but f is not Cauchy-Riemann integrable over $[-1, 1]$ since f is not Cauchy-Riemann integrable over $[0, 1]$ by Example 2.44.

If f is Riemann integrable over $[a, c - \epsilon]$ and $[c + \epsilon, b]$ for every small $\epsilon > 0$, the limit in (2.2) is called the *Cauchy principal value* of f over $[a, b]$ and is often denoted by $pv \int_a^b f$.

Suppose now that f is defined on an unbounded interval such as $[a, \infty)$. We next define the Cauchy-Riemann integral for such functions.

Definition 2.46 Let $f : [a, \infty) \to \mathbb{R}$. We say that f is *Cauchy-Riemann integrable* over $[a, \infty)$ if f is Riemann integrable over $[a, b]$ for every $b > a$ and $\lim_{b \to \infty} \int_a^b f$ exists. We define the Cauchy-Riemann integral of f over $[a, \infty)$ to be

$$\int_a^\infty f = \lim_{b \to \infty} \int_a^b f.$$

If the limit exists, we say that the Cauchy-Riemann integral of f *converges*; if the limit fails to exist, we say the integral *diverges*.

A similar definition is made for functions defined on intervals of the form $(-\infty, b]$.

Example 2.47 Let $p \in \mathbb{R}$ and let $f(t) = t^p$, for $t \geq 1$. For $p \neq -1$, $\int_1^b t^p dt = \dfrac{b^{p+1} - 1}{p + 1}$ so $\lim_{b \to \infty} \int_1^b t^p dt$ exists and equals $\dfrac{-1}{p + 1}$ if, and only if, $p < -1$. If $p = -1$, $\int_1^b \dfrac{1}{t} dt = \ln b$ which does not have a finite limit as $b \to \infty$. Thus, t^p is Cauchy-Riemann integrable over $[1, \infty)$ if, and only if, $p < -1$ and, then, $\int_1^\infty t^p dt = \dfrac{-1}{p + 1}$.

If $f : (-\infty, \infty) \to \mathbb{R}$, then f is Cauchy-Riemann integrable over $(-\infty, \infty)$ if, and only if, $\int_{-\infty}^a f$ and $\int_a^\infty f$ both exist for some a and the Cauchy-Riemann integral of f over $(-\infty, \infty)$ is defined to be

$$\int_{-\infty}^\infty f = \int_{-\infty}^a f + \int_a^\infty f.$$

Exercise 2.38 shows that the value of the integral is independent of the choice of a.

As in the case of integrals over bounded intervals, if $f : (-\infty, \infty) \to \mathbb{R}$ is Cauchy-Riemann integrable over $(-\infty, \infty)$, then the limit

$$\lim_{a \to \infty} \int_{-a}^a f \qquad (2.3)$$

exists. However, the limit in (2.3) may exist and f may fail to be Cauchy-Riemann integrable over $(-\infty, \infty)$. See Exercise 2.39. The limit in (2.3), if it exists, is called the *Cauchy principal value* of f over $(-\infty, \infty)$ and is often denoted $pv \int_{-\infty}^{\infty} f$.

We saw in Corollary 2.24 that if a function $f : [a, b] \to \mathbb{R}$ is Riemann integrable over $[a, b]$, then $|f|$ is Riemann integrable over $[a, b]$. We show in the next example that this property does not hold for the Cauchy-Riemann integral. First, we establish a preliminary result called a *comparison test*.

Proposition 2.48 *(Comparison Test) Let $f, g : [a, \infty) \to \mathbb{R}$ and suppose that $|f(t)| \leq g(t)$ for $t \geq a$. Assume that f is Riemann integrable over $[a, b]$ for every $b > a$ and that g is Cauchy-Riemann integrable over $[a, \infty)$. Then, f (and $|f|$) is Cauchy-Riemann integrable over $[a, \infty)$.*

Proof. To show that $\lim_{b \to \infty} \int_a^b f$ exists, it suffices to show that the Cauchy condition is satisfied for this limit. However, if $c > b > a$, then

$$\left| \int_a^b f - \int_a^c f \right| = \left| \int_b^c f \right| \leq \int_b^c |f| \leq \int_b^c g \to 0$$

as $b, c \to \infty$, since, by assumption, $\lim_{b \to \infty} \int_a^b g$ exists and so its terms satisfy a Cauchy condition. $\square$

Example 2.49 The function $\dfrac{\sin x}{x}$ is Cauchy-Riemann integrable over $[\pi, \infty)$ but $\left| \dfrac{\sin x}{x} \right|$ is not. First, we show that $\int_\pi^\infty \dfrac{\sin x}{x} dx$ exists. Integration by parts gives

$$\int_\pi^b \frac{\sin x}{x} dx = -\frac{\sin b}{b^2} - \int_\pi^b \frac{\cos x}{x^2} dx.$$

Now, $\lim_{b \to \infty} \dfrac{\sin b}{b^2} = 0$ and $\int_\pi^\infty \dfrac{\cos x}{x^2} dx$ exists by Proposition 2.48 and Example 2.47 since $\left| \dfrac{\cos x}{x^2} \right| \leq \dfrac{1}{x^2}$.

Next, we consider $\int_\pi^\infty \left| \dfrac{\sin x}{x} \right| dx$. To see that this integral does not exist, note that

$$\int_\pi^{k\pi} \left| \frac{\sin x}{x} \right| dx = \sum_{j=1}^{k-1} \int_{j\pi}^{(j+1)\pi} \left| \frac{\sin x}{x} \right| dx$$

$$\geq \sum_{j=1}^{k-1} \frac{1}{(j+1)\pi} \int_{j\pi}^{(j+1)\pi} |\sin x| \, dx = \sum_{j=1}^{k-1} \frac{2}{(j+1)\pi}$$

which diverges to ∞ as $k \to \infty$.

A function f defined on an interval I is said to be *absolutely integrable* over I if both f and $|f|$ are integrable over I. If f is integrable over I but $|f|$ is not integrable over I, f is said to be *conditionally integrable* over I. The previous example shows that the Cauchy-Riemann integral admits conditionally integrable functions whereas Corollary 2.24 shows that there are no such functions for the Riemann integral. Note that the comparison test in Proposition 2.48 is a test for absolute integrability.

We will see later that the Henstock-Kurzweil integral admits conditionally integrable functions whereas the Lebesgue integral does not.

Let S be the set of Cauchy-Riemann integrable functions. It follows from standard limit theorems that S is a vector space of functions. However, the last example shows that, in contrast with the space of Riemann integrable functions, S is not a vector lattice of functions. From the fact that $f(x) = \dfrac{\sin x}{x}$ is conditionally integrable over $[\pi, \infty)$, it follows that neither $f^{+} = f \vee 0$ nor $f^{-} = f \wedge 0$ is Cauchy-Riemann integrable over $[\pi, \infty)$. For a more thorough discussion of the Cauchy-Riemann integral, see [Br], [CS], [Fi] and [Fl].

2.8   Exercises

Riemann's definition

Exercise 2.1   In Example 2.5, we assume that I is a closed interval. Suppose that I is any interval with endpoints c and d; that is, suppose I has one of the forms (c, d), $(c, d]$, or $[c, d)$. Prove that $\int_a^b \chi_I = d - c$.

Exercise 2.2   Suppose $f : [a, b] \to \mathbb{R}$ is Riemann integrable. Show that if f is altered at a finite number of points, then the altered function is Riemann integrable and that the value of the integral is unchanged. Can this statement be changed to a countable number of points?

Exercise 2.3   Suppose that $f, h : [a, b] \to \mathbb{R}$ are Riemann integrable with $\int_a^b f = \int_a^b h$. Suppose that $f \le g \le h$. Prove that g is Riemann integrable.

Exercise 2.4   If $f : [a, b] \to \mathbb{R}$ is continuous, nonnegative and $\int_a^b f = 0$, prove that $f \equiv 0$. Is continuity important? Is positivity? In each case, either prove the result or give a counterexample.

Basic properties

Exercise 2.5 Suppose that f is continuous and nonnegative on $[a, b]$. If there is a $c \in [a, b]$ such that $f(c) > 0$, prove that $\int_a^b f > 0$.

Exercise 2.6 Let $f : [-a, a] \to \mathbb{R}$ be Riemann integrable. We say that f is an *odd function* if $f(-x) = -f(x)$ for all $x \in [-a, a]$ and we say that f is an *even function* if $f(-x) = f(x)$ for all $x \in [-a, a]$.

(1) If f is an odd function, prove that $\int_{-a}^{a} f = 0$.
(2) If f is an even function, prove that $\int_{-a}^{a} f = 2 \int_0^a f$.

Darboux's definition

Exercise 2.7 Let $f : [a, b] \to \mathbb{R}$. Suppose there are partitions $\mathcal{P}$ and $\mathcal{P}'$ such that $L(f, \mathcal{P}) = U(f, \mathcal{P}')$. Prove that f is Darboux integrable.

Exercise 2.8 Suppose that $f, g : [a, b] \to \mathbb{R}$, $a < c < b$, and $\alpha \geq 0$.

(1) Prove the following results for upper and lower integrals:

    (a) $\overline{\int_a^b} (f + g) \leq \overline{\int_a^b} f + \overline{\int_a^b} g$ and $\underline{\int_a^b} f + \underline{\int_a^b} g \leq \underline{\int_a^b} (f + g)$;

    (b) $\overline{\int_a^b} \alpha f = \alpha \overline{\int_a^b} f$ and $\underline{\int_a^b} \alpha f = \alpha \underline{\int_a^b} f$;

    (c) $\overline{\int_a^b} f = \overline{\int_a^c} f + \overline{\int_c^b} f$ and $\underline{\int_a^b} f = \underline{\int_a^c} f + \underline{\int_c^b} f$.

(2) Give examples to show that strict inequalities can occur in part *(1.a)*.

Exercise 2.9 Let $f : [a, b] \to \mathbb{R}$ be bounded. Define the *upper* and *lower indefinite integrals* of f by $\overline{F}(x) = \overline{\int_a^x} f(t)\, dt$ and $\underline{F}(x) = \underline{\int_a^x} f(t)\, dt$. Prove that $\overline{F}$ and $\underline{F}$ satisfy Lipschitz conditions. Suppose that f is continuous at x. Show that the upper and lower indefinite integrals are differentiable at x with derivatives equal to $f(x)$.

Exercise 2.10 Let φ and ψ be step functions and $\alpha \in \mathbb{R}$. Prove that $\alpha\varphi$, $\varphi + \psi$, $\varphi\psi$, $\varphi \vee \psi$, and $\varphi \wedge \psi$ are step functions.

Exercise 2.11 Prove Corollary 2.24.

Exercise 2.12 Give an example of a function $f : [0, 1] \to \mathbb{R}$ such that $|f|$ is Riemann integrable but f is not Riemann integrable.

Exercise 2.13 Suppose that $f : [0, 1] \to \mathbb{R}$ is continuous. Show that

$$\lim_{n \to \infty} \int_0^1 f(x^n)\, dx = f(0).$$

Exercise 2.14   Suppose that $f : [a, b] \to \mathbb{R}$ is continuous and nonnegative and set $M = \sup\{f(t) : a \le t \le b\}$. Show

$$\lim_{n \to \infty} \left(\int_a^b f^n \right)^{1/n} = M.$$

Exercise 2.15   Let $f : [a, b] \to \mathbb{R}$ and suppose that f is Riemann integrable on $[a, c]$ and $[c, b]$. Prove that f is Riemann integrable on $[a, b]$.

Exercise 2.16   Suppose $f : [a, b] \to \mathbb{R}$ and $I, J \subset [a, b]$ are intervals with disjoint interiors. Prove that

$$\int_{I \cup J} f = \int_I f + \int_J f$$

Fundamental Theorem of Calculus

Exercise 2.17   Suppose that $f : [a, b] \to \mathbb{R}$ is Riemann integrable and $m \le f(x) \le M$ for all $x \in [a, b]$. Prove that

$$m \le \frac{1}{b - a} \int_a^b f \le M.$$

Exercise 2.18   Prove the *Mean Value Theorem*. If $f : [a, b] \to \mathbb{R}$ is continuous, prove there is a $c \in [a, b]$ such that $\int_a^b f = f(c)(b - a)$.

   If f is also nonnegative, give a geometric interpretation of this result.

Exercise 2.19   Prove the following version of the *Mean Value Theorem*. If $f : [a, b] \to \mathbb{R}$ is continuous and $g : [a, b] \to \mathbb{R}$ is nonnegative and Riemann integrable on $[a, b]$, then there is a $c \in [a, b]$ such that

$$\int_a^b f(x) g(x) \, dx = f(c) \int_a^b g(x) \, dx.$$

Exercise 2.20   Suppose that f is continuous and strictly increasing on $[0, a]$, differentiable on $(0, a)$, and $f(0) = 0$. Define g by

$$g(x) = \int_0^x f(t) \, dt + \int_0^{f(x)} f^{-1}(t) \, dt - x f(x)$$

for $x \in [0, a]$.

(1) Prove that $g \equiv 0$ on $[0, a]$.

(2) Use this result to prove *Young's inequality*: for $0 < b \leq f(a)$,

$$ab \leq \int_0^a f(x)\, dx + \int_0^b f^{-1}(x)\, dx.$$

(3) Deduce *Hölder's inequality*: If $a, b \geq 0$, then

$$ab \leq \frac{a^p}{p} + \frac{b^{p'}}{p'},$$

where $p > 1$ and $\dfrac{1}{p} + \dfrac{1}{p'} = 1$.

Exercise 2.21   Evaluate $\int_0^\pi \cos 2\theta \sin 3\theta\, d\theta$ and $\int_0^2 x^2 e^x dx$.

Exercise 2.22   Evaluate $\int_0^4 x^2 \left(2x^3 + 16\right)^{1/2} dx$.

Characterizations of integrability

Exercise 2.23   Let $f : [a, b] \to \mathbb{R}$ be Riemann integrable. Let $p, c > 0$. Prove the following two statements.

(1) If $f \geq 0$, then f^p is Riemann integrable.

(2) If $|f| \geq c > 0$, then $\dfrac{1}{f}$ is Riemann integrable.

Exercise 2.24   Let $f : [a, b] \to \mathbb{R}$ be Riemann integrable and suppose $m \leq f(x) \leq M$ for all $x \in [a, b]$. Suppose $\varphi : [m, M] \to \mathbb{R}$ is continuous. Prove that $\varphi \circ f$ is Riemann integrable.

Exercise 2.25   Show that the composition of Riemann integrable functions is not necessarily Riemann integrable. [Hint: define f and φ on $[0, 1]$ by

$$f(x) = \begin{cases} 0 & \text{if} \quad x \text{ is irrational} \\ \dfrac{m}{n} & \text{if } x = \dfrac{m}{n} \in \mathbb{Q} \text{ and } (m, n) = 1 \end{cases} \quad \text{and} \quad \varphi(x) = \begin{cases} 0 \text{ if} & x = 0 \\ 1 \text{ if } 0 < x \leq 1 \end{cases}.$$

Note that f is continuous a.e. and φ is Riemann integrable.]

Exercise 2.26   Show that a finite set has outer Jordan content 0. Show that $S = \left\{ \dfrac{1}{k} : k \in \mathbb{N} \right\} \subset [0, 1]$ has outer Jordan content 0. Give an example of a countable subset of $[0, 1]$ with positive outer Jordan content.

Exercise 2.27   If $S \subset T \subset [a, b]$, show $\bar{c}(S) \leq \bar{c}(T)$. If $S, T \subset [a, b]$, show that $\bar{c}(S \cup T) \leq \bar{c}(S) + \bar{c}(T)$. If $T \subset [a, b]$ has outer Jordan content 0 and $S \subset [a, b]$, show $\bar{c}(S \cup T) = \bar{c}(S)$.

Exercise 2.28   Let $f : [a, b] \to \mathbb{R}$ be a bounded function. Suppose that $f = 0$ except on a set of outer Jordan content 0. Prove that f is Riemann integrable and $\int_a^b f = 0$.

Exercise 2.29   Suppose that $f, g : [a, b] \to \mathbb{R}$ are bounded and f is Riemann integrable. If $f = g$ except on a set of outer Jordan content 0, prove that g is Riemann integrable and $\int_a^b g = \int_a^b f$.

Exercise 2.30   Let $f : [a, b] \to \mathbb{R}$ be bounded. Prove that f is continuous at $x \in [a, b]$ if, and only if, $\omega(f, x) = 0$.

Exercise 2.31   Let $f : [a, b] \to \mathbb{R}$, $D_\epsilon(f) = \{x \in [a, b] : \omega(f, x) \geq \epsilon\}$ and let $(c, d) \subset [a, b]$ be an interval. Suppose that $D_\epsilon(f) \cap (c, d) \neq \emptyset$. Prove that $\omega(f, (c, d)) \geq \epsilon$. Show by example that we cannot replace the open interval (c, d) by the closed interval $[c, d]$.

Exercise 2.32   Let $f : [a, b] \to \mathbb{R}$ and let

$$D(f) = \{t \in [a, b] : f \text{ is discontinuous at } t\}.$$

Prove that $D(f) = \cup \{D_\epsilon(f) : \epsilon > 0\} = \cup \{D_{1/n}(f) : n \in \mathbb{N}\}$.

Exercise 2.33   Prove that every subset of a null set is a null set. Prove that a countable union of null sets is a null set.

Improper integrals

Exercise 2.34   Determine whether the following improper integrals converge or diverge:

(1) $\int_1^4 \dfrac{dx}{\sqrt{x-1}}$

(2) $\int_0^1 \dfrac{x^2+1}{x\sqrt{1-x}} dx$

(3) $\int_0^1 \ln x\, dx$

(4) $\int_0^{\pi/2} \tan x\, dx$

(5) $\int_0^\infty \dfrac{x\, dx}{(x+2)^2 (x+1)}$

(6) $\int_0^\infty \dfrac{dx}{x\sqrt{x+1}}$

(7) $\int_0^\infty e^{-x} dx$

(8) $\int_2^\infty \dfrac{dx}{(x-1)^{3/4}}$

(9) $\int_{-\infty}^\infty \dfrac{dx}{1+x^2}$

(10) $\int_{-\infty}^\infty \dfrac{dx}{1-x^2}$

(11) $\int_1^\infty \dfrac{dx}{x \ln^2 x}$

(12) $\int_0^\infty \dfrac{e^x}{\sqrt{x}} dx$

Exercise 2.35 Suppose that $f : [a, b] \to \mathbb{R}$ is Riemann integrable over $[a, b]$. Prove that

$$\int_a^b f = \lim_{c \to a^+} \int_c^b f.$$

Exercise 2.36 Formulate and prove an analogue of the Comparison Test, Proposition 2.48, for improper integrals over $[a, b]$.

Exercise 2.37 Define the *gamma function* for $x > 0$ by

$$\Gamma(x) = \int_0^x e^{-t} t^{x-1} dt.$$

Prove the following results for Γ:

(1) The improper integral defining Γ converges.
(2) $\Gamma(x + 1) = x\Gamma(x)$.
(3) For $n \in \mathbb{N}$, $\Gamma(n) = (n - 1)!$.

Exercise 2.38 Suppose that f is Cauchy-Riemann integrable over $(-\infty, \infty)$. Prove that for any $a, b \in \mathbb{R}$,

$$\int_{-\infty}^a f + \int_a^\infty f = \int_{-\infty}^b f + \int_a^b f.$$

Hence, the Cauchy-Riemann integral of f is independent of the cutoff point a.

Exercise 2.39 Give an example of a function f defined on $(-\infty, \infty)$ which is not Cauchy-Riemann integrable but such that the Cauchy principal value integral of f over $(-\infty, \infty)$ exists.

Chapter 3

Convergence theorems and the Lebesgue integral

While the Riemann integral enjoys many desirable properties, it also has several shortcomings. As was pointed out in Chapter 2, one of these shortcomings concerns the fact that a general form of the Fundamental Theorem of Calculus does not hold for Riemann integrable functions. Another serious drawback which we will address in this chapter is the lack of "good" convergence theorems for the Riemann integral. A convergence theorem for an integral concerns a sequence of integrable functions $\{f_k\}_{k=1}^{\infty}$ which converge in some sense, such a pointwise, to a limit function f and involves sufficient conditions for interchanging the limit and the integral, that is to guarantee $\lim_k \int f_k = \int \lim_k f_k$.

In modern integration theories, the standard convergence theorems are the Monotone Convergence Theorem, in which the functions converge monotonically, and the Bounded Convergence Theorem, in which the functions are uniformly bounded. We begin the chapter by establishing a convergence theorem for the Riemann integral and then presenting an example that points out the deficiencies of the Riemann integral with respect to desirable convergence theorems. This example is used to motivate the presentation of Lebesgue's descriptive definition of the integral that bears his name. This leads to a discussion of outer measure, measure and measurable functions. The definition and derivation of the important properties of the Lebesgue integral on the real line, including the Monotone and Dominated Convergence Theorems, are then carried out. The Lebesgue integral on n-dimensional Euclidean space is discussed and versions of the Fubini and Tonelli Theorems on the equality of multiple and iterated integrals are established.

For the Riemann integral, we have the following basic convergence result.

Theorem 3.1   *Let $f, f_k : [a, b] \to \mathbb{R}$ for $k \in \mathbb{N}$. Suppose that each f_k is Riemann integrable and that the sequence $\{f_k\}_{k=1}^{\infty}$ converges to f uniformly on $[a, b]$. Then, f is Riemann integrable over $[a, b]$ and*

$$\lim_k \int_a^b f_k = \int_a^b f = \int_a^b \lim_k f_k. \tag{3.1}$$

Proof.   To prove that f is Riemann integrable, it is enough to show that the partial sums for f satisfy the Cauchy criterion. Fix $\epsilon > 0$ and choose an $N \in \mathbb{N}$ such that $|f(x) - f_k(x)| < \dfrac{\epsilon}{3(b-a)}$ for $k > N$ and all $x \in [a, b]$. Fix a $K > N$. Since f_K is Riemann integrable, the partial sums for f_K satisfy the Cauchy criterion, so that there is a $\delta > 0$ so that if $\mathcal{P}_j$, $j = 1, 2$, are partitions of $[a, b]$ with $\mu(\mathcal{P}_j) < \delta$ and $\left\{t_i^{(j)}\right\}_{i=1}^{n_j}$ are sets of sampling points relative to $\mathcal{P}_j$, then

$$\left| S\left(f_K, \mathcal{P}_1, \left\{t_i^{(1)}\right\}_{i=1}^{n_1}\right) - S\left(f_K, \mathcal{P}_2, \left\{t_i^{(2)}\right\}_{i=1}^{n_2}\right) \right| < \frac{\epsilon}{3}.$$

Let $\mathcal{P}_1$ and $\mathcal{P}_2$ be partitions of $[a, b]$ with mesh less than δ and let $\left\{t_i^{(j)}\right\}_{i=1}^{n_j}$ be corresponding sets of sampling points. Set $S_j(g) = S\left(g, \mathcal{P}_j, \left\{t_i^{(j)}\right\}_{i=1}^{n_j}\right)$. Then,

$$\left| S\left(f, \mathcal{P}_1, \left\{t_i^{(1)}\right\}_{i=1}^{n_1}\right) - S\left(f, \mathcal{P}_2, \left\{t_i^{(2)}\right\}_{i=1}^{n_2}\right) \right| = |S_1(f) - S_2(f)|$$

$$= |S_1(f) - S_1(f_K) + S_1(f_K) - S_2(f_K) + S_2(f_K) - S_2(f)|$$

$$\leq |S_1(f) - S_1(f_K)| + |S_1(f_K) - S_2(f_K)| + |S_2(f_K) - S_2(f)|.$$

For the first and third terms, by the uniform convergence, we have

$$|S_j(f) - S_j(f_K)| \leq \sum_{i=1}^{n_j} \left| f\left(t_i^{(j)}\right) - f_K\left(t_i^{(j)}\right) \right| \left(x_i^{(j)} - x_{i-1}^{(j)}\right)$$

$$< \frac{\epsilon}{3(b-a)} \sum_{i=1}^{n_j} \left(x_i^{(j)} - x_{i-1}^{(j)}\right) = \frac{\epsilon}{3},$$

while the middle term is less that $\dfrac{\epsilon}{3}$ by the choice of K. Thus,

$$\left| S\left(f, \mathcal{P}_1, \left\{t_i^{(1)}\right\}_{i=1}^{n_1}\right) - S\left(f, \mathcal{P}_2, \left\{t_i^{(2)}\right\}_{i=1}^{n_2}\right) \right| < \epsilon$$

so that f is Riemann integrable.

To see that $\int_a^b f = \lim_k \int_a^b f_k$, fix $\epsilon > 0$ and, by uniform convergence, choose $N \in \mathbb{N}$ such that $|f(x) - f_k(x)| < \dfrac{\epsilon}{b-a}$ for $k > N$ and $x \in [a, b]$. Then,

$$\int_a^b f_k - \epsilon < \int_a^b f < \int_a^b f_k + \epsilon$$

for all $k > N$. Thus, $\int_a^b f = \lim_k \int_a^b f_k$. $\qquad\square$

The uniform convergence assumption in Theorem 3.1 is quite strong, and it would be desirable to replace this assumption with a weaker hypothesis. However, it should be noted that, in general, pointwise convergence will not suffice for (3.1) to hold.

Example 3.2 Define $f_k : [0, 1] \to \mathbb{R}$ by $f_k(x) = k\chi_{(0,1/k]}(x)$. Then, $\{f_k\}_{k=1}^\infty$ converges pointwise to 0 but $\int_0^1 f_k = 1$ for every k, so (3.1) fails to hold.

In addition to the assumption of pointwise convergence, there are two natural assumptions which can be imposed on a sequence of integrable functions as in Theorem 3.1. The first is a uniform boundedness condition in which it is assumed that there exists an $M > 0$ such that $|f_k(x)| \le M$ for all k and x; a theorem with this hypothesis is referred to as a Bounded Convergence Theorem. The second assumption is to require that for each x, the sequence $\{f_k(x)\}_{k=1}^\infty$ converges monotonically to $f(x)$; a theorem with this hypothesis is referred to as a Monotone Convergence Theorem. Note that the sequence in the previous example does not satisfy either of these hypotheses. The following example shows that neither the Bounded nor Monotone Convergence Theorem holds for the Riemann integral.

Example 3.3 Let $\{r_n\}_{n=1}^\infty$ be an enumeration of the rational numbers in $[0, 1]$. For each $k \in \mathbb{N}$, define $f_k : [0, 1] \to \mathbb{R}$ by $f_k(r_n) = 1$ for $1 \le n \le k$ and $f_k(x) = 0$ otherwise. By Corollary 2.42, each f_k is Riemann integrable. For each $x \in [0, 1]$, the sequence $\{f_k(x)\}_{k=1}^\infty$ is increasing and bounded by 1. The sequence $\{f_k\}_{k=1}^\infty$ converges to the Dirichlet function defined in Example 2.7 which is not Riemann integrable.

We will see later in this chapter that both the Monotone and Bounded Convergence Theorems are valid for the Lebesgue integral. We will show in Chapter 4 that both theorems are also valid for the Henstock-Kurzweil integral.

It should be pointed out that there are versions of the Monotone and Bounded Convergence Theorems for the Riemann integral, but both of them require one assume the Riemann integrability of the limit function. It is desirable that the integrability of the limit function be part of the conclusion of these results. See [Lew1] for the Bounded Convergence Theorem and [Th] for the Monotone Convergence Theorem.

In the remainder of this chapter, we will construct and describe the fundamental properties of the Lebesgue integral. We begin by considering Lebesgue's descriptive definition of the Lebesgue integral.

3.1   Lebesgue's descriptive definition of the integral

H. Lebesgue (1875-1941) defined *le problème d'intégration* (the problem of integration) as follows. (See [Leb, Vol. II, page 114].) He wished to assign to each bounded function f defined on a finite interval $[a, b]$ a number, denoted by $\int_a^b f(x)\, dx$, that satisfied six conditions. Suppose that $a, b, c, h \in \mathbb{R}$. Then:

(1) $\int_a^b f(x)\, dx = \int_{a+h}^{b+h} f(x - h)\, dx.$

(2) $\int_a^b f(x)\, dx + \int_b^c f(x)\, dx + \int_c^a f(x)\, dx = 0.$

(3) $\int_a^b [f(x) + \varphi(x)]\, dx = \int_a^b f(x)\, dx + \int_a^b \varphi(x)\, dx.$

(4) If $f \geq 0$ and $b > a$ then $\int_a^b f(x)\, dx \geq 0.$

(5) $\int_0^1 1\, dx = 1.$

(6) If $\{f_k\}_{k=1}^\infty$ increases pointwise to f then $\int_a^b f_k(x)\, dx \to \int_a^b f(x)\, dx.$

In other words, he described the properties he wanted this "integral" to possess and then attempted to deduce a definition for this integral from these properties. He called this definition *descriptive*, to contrast with the constructive definitions, like Riemann's, in which an object is defined and then its properties are deduced from the definition.

Assuming these six conditions, we wish to determine other properties of this integral. To begin, notice that setting $\varphi = -f$ in (3) shows that $\int_a^b (-f)(x)\, dx = -\int_a^b f(x)\, dx$. If $f \geq g$, then (3) and (4) imply that

$$\int_a^b f(x)\, dx - \int_a^b g(x)\, dx = \int_a^b [f(x) - g(x)]\, dx \geq 0$$

so that $\int_a^b f(x)\, dx \geq \int_a^b g(x)\, dx$. Hence, this integral satisfies a monotonicity condition.

We next show that $\int_a^b 1dx = b - a$ for all $a, b \in \mathbb{R}$. From (1), we see that $b - a = d - c$ implies $\int_a^b 1dx = \int_c^d 1dx$. Thus, from (5), for any interval $[a, b]$ of length 1,

$$\int_a^b 1dx = 1.$$

From (2), by setting $c = b = a$, we see that $\int_a^a f(x)\, dx = 0$; then, setting $c = b$, we get $\int_a^c f(x)\, dx = -\int_c^a f(x)\, dx$, so that

$$\int_a^b 1dx + \int_b^c 1dx = \int_a^c 1dx.$$

Iterating this result shows

$$\int_{a_0}^{a_1} 1dx + \int_{a_1}^{a_2} 1dx + \cdots + \int_{a_{n-1}}^{a_n} 1dx = \int_{a_0}^{a_n} 1dx.$$

Setting $a_i = i$ yields $\int_0^n 1dx = n$, while setting $a_i = \dfrac{i}{n}$ shows that $\int_0^{1/n} 1dx = 1/n$. Again, by iteration, we see that

$$\int_0^q 1dx = q$$

for any rational number q. Finally, if $r \in \mathbb{R}$, let p and q be rational numbers such that $p < r < q$. Then, since $\chi_{[0,p]} \le \chi_{[0,r]} \le \chi_{[0,q]}$, by monotonicity,

$$0 \le \int_0^q 1dx - \int_0^r 1dx \le \int_0^q 1dx - \int_0^p 1dx = q - p,$$

which implies

$$0 \le q - \int_0^r 1dx \le q - p.$$

Letting p and q approach r, we conclude that for all real numbers r,

$$\int_0^r 1dx = r,$$

so that $\int_a^b 1dx = b - a$ for all $a, b \in \mathbb{R}$.

Setting $\varphi = f$ in (3), by iteration, we see that

$$\int_a^b nf(x)\, dx = n \int_a^b f(x)\, dx$$

for every natural number n. Setting $f = \varphi = 0$ shows that $\int_a^b 0\,dx = 0$, which in turn implies that this equality holds for any integer n. Since

$$\int_a^b f(x)\,dx = \int_a^b \left(n\frac{1}{n}\right) f(x)\,dx = n\int_a^b \frac{1}{n}f(x)\,dx,$$

it follows that

$$\int_a^b qf(x)\,dx = q\int_a^b f(x)\,dx$$

for any rational number q.

To see that this equality holds for any real number, note that since both f and $-f$ are bounded by $|f|$, monotonicity implies that $\left|\int_a^b f(x)\,dx\right| \le \int_a^b |f(x)|\,dx$. Now, fix a real number r. Let $M = \sup\{|f(x)| : x \in [a,b]\}$. Let $q \in \mathbb{Q}$ and choose a real number $p = p_q \in (0,1)$ such that $|r - q|(M + p) \in \mathbb{Q}$. Then,

$$\left|\int_a^b rf(x)\,dx - q\int_a^b f(x)\,dx\right| \le \int_a^b |r - q|\,|f(x)|\,dx$$

$$\le |r - q|(M + p)\int_a^b 1\,dx.$$

Letting q approach r, we see that $\int_a^b rf(x)\,dx = r\int_a^b f(x)\,dx$. Hence, from properties (3) and (4), we see that this integral must be linear.

By using properties (1) through (5), we have shown that $\int_a^b 1\,dx = b - a$ and the integral is linear. We have not made use of the crucial property (6).

Suppose we have an integral satisfying properties (1) through (5) and let f be Riemann integrable on $[a,b]$. Let $\mathcal{P} = \{x_0, x_1, \ldots, x_n\}$ be a partition of $[a,b]$. Recalling the definitions

$$m_i = \inf\{f(x) : x_{i-1} \le x \le x_i\}$$

and

$$M_i = \sup\{f(x) : x_{i-1} \le x \le x_i\},$$

we see

$$\sum_{i=1}^n m_i \ell\left([x_{i-1}, x_i]\right) \le \sum_{i=1}^n \int_{x_{i-1}}^{x_i} f = \int_a^b f \le \sum_{i=1}^n M_i \ell\left([x_{i-1}, x_i]\right),$$

which implies that

$$\int_{\underline{a}}^{b} f \le \int_{a}^{b} f \le \overline{\int_{a}^{b}} f.$$

Thus, if f is Riemann integrable on $[a, b]$, then $\int_{\underline{a}}^{b} f = \overline{\int_{a}^{b}} f$ and the middle integral must equal the Riemann integral. Thus, any integral that satisfies properties (1) through (5) must agree with the Riemann integral for Riemann integrable functions.

We now investigate property (6). Suppose that $f : [a, b] \to \mathbb{R}$ is bounded. Fix l and L such that $l \le f < L$. Given a partition $\mathcal{P} = \{l_0, l_1, \ldots, l_n\}$ of the interval $[l, L]$ with $l_0 = l$, $l_n = L$ and $l_i < l_{i+1}$ for $i = 1, \ldots, n$, let $E_i = \{x \in [a, b] : l_{i-1} \le f(x) < l_i\}$ for $i = 1, \ldots, n$, and consider the *simple function* φ defined by

$$\varphi(x) = \sum_{i=1}^{n} l_{i-1} \chi_{E_i}(x).$$

It then follows that $\varphi \le f$ on $[a, b]$ and, by the linearity of the integral, $\int_{a}^{b} \varphi(x)\, dx = \sum_{i=1}^{n} l_{i-1} \int_{a}^{b} \chi_{E_i}(x)\, dx$.

Now, fix a partition $\mathcal{P}_0$ and define a sequence of partitions $\{\mathcal{P}_k\}_{k=1}^{\infty}$ such that:

(1) $\mathcal{P}_k$ is a refinement of $\mathcal{P}_{k-1}$ for $k = 1, 2, \ldots$;
(2) $\mu(\mathcal{P}_k) \le \frac{1}{2}\mu(\mathcal{P}_{k-1})$ for $k = 1, 2, \ldots$.

Let φ_k be the function associated to $\mathcal{P}_k$ as above. Then, $\{\varphi_k\}_{k=1}^{\infty}$ is a sequence of simple functions that increase monotonically to f. In fact, by construction, $0 \le f - \varphi_k < \mu(\mathcal{P}_k)$ and $\mu(\mathcal{P}_k) \to 0$, so that $\{\varphi_k\}_{k=1}^{\infty}$ converges to f uniformly on $[a, b]$. Consequently, by (6)

$$\int_{a}^{b} \varphi_k \to \int_{a}^{b} f.$$

Thus, to evaluate the integral of f, it is enough to be able to integrate the functions φ_k, which in turn depends on integrals of the form $\int_{a}^{b} \chi_E(x)\, dx$. As Lebesgue said, "To know how to calculate the integral of any function, it suffices to know how to calculate the integrals of functions ψ which take only the values 0 and 1" [Leb, Vol. II, page 118]. If $E = [c, d]$ is an interval in $[a, b]$, then $\int_{a}^{b} \chi_E(x)\, dx = \int_{c}^{d} 1\, dx$, which is the length on the interval $[c, d]$. Thus, Lebesgue reduced the problem of integration to that of extending the definition of length from intervals in $\mathbb{R}$ to arbitrary subsets of $\mathbb{R}$, that is, to *le problème de la mesure des ensembles* (the problem of the measure of sets). His goal was to assign to each bounded set $E \subset \mathbb{R}$ a nonnegative number $m(E)$ satisfying the following conditions:

(1) congruent sets (that is, translations of a single set) have equal measure;
(2) the measure of a finite or countably infinite union of pairwise disjoint sets is equal to the sum of the measures of the individual sets (*countable additivity*); and,
(3) the measure of the set $[0,1]$ is 1.

As we shall see below in Remark 3.10, this problem has no solution.

3.2   Measure

Our goal is to extend the concept of length to sets other than intervals, with a function that preserves properties (1) through (3) of the problem of measure.

3.2.1   *Outer measure*

We first extend the length function by defining outer measure.

Definition 3.4   Let $E \subset \mathbb{R}$. We define the *(Lebesgue) outer measure* of E, $m^*(E)$, by

$$m^*(E) = \inf \left\{ \sum_{j \in \sigma} \ell(I_j) \right\},$$

where the infimum is taken over all countable collections of open intervals $\{I_j\}_{j \in \sigma}$ such that $E \subset \cup_{j \in \sigma} I_j$.

Notation 3.5   *Here and below, we use σ to represent a countable set, which may be finite or countably infinite.*

It follows immediately from the definition that $m^*(\emptyset) = 0$. Since $\ell(I) > 0$ for every open interval I, we see $m^*(E) \geq 0$. Since $\emptyset \subset (0, \epsilon)$ for every $\epsilon > 0$, $0 \leq m^*(\emptyset) \leq \epsilon$ for all $\epsilon > 0$. It follows that $m^*(\emptyset) = 0$.

We show that m^* extends the length function and establish the basic properties of outer measure. Given a set $E \subset \mathbb{R}$ and $h \in \mathbb{R}$, we define the *translation of E by h* to be the set

$$E + h = \{x \in \mathbb{R} : x = y + h \text{ for some } y \in E\}.$$

We say a set function F is *translation invariant* if $F(E) = F(E+h)$ whenever either side is defined.

Theorem 3.6 *The outer measure m^* satisfies the following properties:*

(1) m^* is monotone; that is, if $F \subset E \subset \mathbb{R}$ then $m^*(F) \le m^*(E)$;
(2) m^* is translation invariant;
(3) if I is an interval then $m^*(I) = \ell(I)$;
(4) m^* is countably subadditive; that is, if σ is a countable set and $E_i \subset \mathbb{R}$ for all $i \in \sigma$, then $m^*(\cup_{i \in \sigma} E_i) \le \sum_{i \in \sigma} m^*(E_i)$.

Proof. If $F \subset E$, then every cover of E by a countable collection of open intervals is a cover of F, which implies (1). We leave (2) as an exercise. See Exercise 3.1.

To prove (3), let $I \subset \mathbb{R}$ be an interval with endpoints a and b. For any $\epsilon > 0$, $(a - \epsilon, b + \epsilon)$ is an open interval containing I so that $m^*(I) \le b - a + 2\epsilon$. Hence, $m^*(I) \le b - a$.

Now, suppose that I is a bounded, closed interval. Let $\{I_j : j \in \sigma\}$ be a countable cover of I by open intervals. We claim that $\sum_{j \in \sigma} \ell(I_j) \ge b - a$ which will establish that $m^*(I) = b - a$. Since I is compact, a finite number of intervals from $\{I_j : j \in \sigma\}$ cover I; call this set $\{J_i : i = 1, \ldots, m\}$. (See [BS, pages 319-322].) It suffices to show that $\sum_{i=1}^{m} \ell(J_i) \ge b - a$. Since $I \subset \cup_{i=1}^{m} J_i$, there is an i_1 such that $J_{i_1} = (a_1, b_1)$ with $a_1 < a < b_1$. If $b_1 > b$, then $[a, b] \subset J_{i_1}$ and since $\sum_{i=1}^{m} \ell(J_i) \ge \ell(J_{i_1}) \ge b - a$, we are done. If $b_1 \le b$, there is an i_2 such that $J_{i_2} = (a_2, b_2)$ and $a_2 < b_1 < b_2$. Continuing this construction produces a finite number of intervals $\{J_{i_k} = (a_k, b_k) : k = 1, \ldots, n\}$ from $\{J_i : i = 1, \ldots, m\}$ such that $a_1 < a$, $a_i < b_{i-1} < b_i$ and $b_n > b$. Thus,

$$\sum_{i=1}^{m} \ell(J_i) \ge \sum_{k=1}^{n} (b_k - a_k) = b_n + \sum_{k=2}^{n} (b_{k-1} - a_k) - a_1 > b_n - a_1 > b - a,$$

as we wished to show. Thus, $m^*(I) = \ell(I)$.

If I is a bounded interval, then for any $\epsilon > 0$ there is a closed interval $J \subset I$ with $\ell(I) < \ell(J) + \epsilon$. Then, $m^*(I) \ge m^*(J) = \ell(J) > \ell(I) - \epsilon$. Thus, $m^*(I) \ge \ell(I)$, and by the remark above, the two are equal.

Finally, if I is an unbounded interval, then for every $r > 0$, there is a closed $J \subset I$ with $m^*(J) = \ell(J) = r$. Hence, $m^*(I) \geq r$ for every $r > 0$ which implies $m^*(I) = \infty$.

It remains to prove (4). Assume first that $m^*(E_i) < \infty$ for all $i \in \sigma$ and let $\epsilon > 0$. For each i, choose a countable collection of open interval $\{I_{i,n}\}_{n \in \sigma_i}$ such that $\sum_{n \in \sigma_i} \ell(I_{i,n}) < m^*(E_i) + 2^{-i}\epsilon$. Then, $\cup_{i \in \sigma} \{I_{i,n}\}_{n \in \sigma_i}$ is a countable collection of open intervals whose union contains $\cup_{i \in \sigma} E_i$. Thus, by Exercise 3.2,

$$m^*(\cup_i E_i) \leq \sum_{i,n} \ell(I_{i,n}) = \sum_{i \in \sigma} \sum_{n \in \sigma_i} \ell(I_{i,n})$$
$$< \sum_{i \in \sigma} \left(m^*(E_i) + 2^{-i}\epsilon \right) = \sum_{i \in \sigma} m^*(E_i) + \epsilon.$$

Since ϵ was arbitrary, it follows that $m^*(\cup_{i \in \sigma} E_i) \leq \sum_{i \in \sigma} m^*(E_i)$. Finally, if $m^*(E_i) = \infty$ for some i, then $\sum_{i \in \sigma} m^*(E_i) = \infty$ and the inequality follows. $\square$

Since $x \in (x - \epsilon, x + \epsilon)$ for every $\epsilon > 0$, we see that $m^*(\{x\}) = 0$ for all $x \in \mathbb{R}$. By the subadditivity of outer measure, we get

Corollary 3.7 *If $E \subset \mathbb{R}$ is a countable set, then $m^*(E) = 0$.*

We shall show in Section 3.2.3 that the converse of Corollary 3.7 is false. As a consequence of this corollary, we obtain

Corollary 3.8 *If I is a non-degenerate interval, then I is uncountable.*

The outer measure we have defined above is defined for every (bounded) set $E \subset \mathbb{R}$ and satisfies conditions (1) and (3) listed under the problem of measure. Unfortunately, outer measure is countably subadditive, but not countably additive, as the following example shows.

Example 3.9 We begin by defining an equivalence relation on $[0, 1]$. Let $x, y \in [0, 1]$. We say that $x \sim y$ if $x - y \in \mathbb{Q}$. By the Axiom of Choice, we choose a set $P \subset [0, 1]$ which contains exactly one point from each equivalence class determined by $\sim$. We need to make two observations about P:

(1) if $q, r \in \mathbb{Q}$ and $q \neq r$ then $(P + q) \cap (P + r) = \emptyset$;
(2) $[0, 1] \subset \cup (P + r)$, where the union is taken over all $r \in \mathbb{Q}_0 = \mathbb{Q} \cap [-1, 1]$.

To see (1), suppose that $x \in (P+q) \cap (P+r)$. Then, there exist $s, t \in P$ such that $x = q + s = r + t$. This implies that $s - t = r - q \neq 0$, and since $r - q \in \mathbb{Q}$, $s \sim t$. Since $s, t \in P$, this violates the definition of P, proving (1). For (2), let $x \in [0, 1]$. Then, x is in one of the equivalence classes determined by $\sim$, so there is an $s \in P$ such that $x \sim s$. Thus, $x - s = r \in \mathbb{Q}$ and since $x, s \in [0, 1]$, $r \in [-1, 1]$ and $x \in P + r$.

Note that $\cup_{r \in \mathbb{Q}_0} (P + r) \subset [-1, 2]$, so by monotonicity, translation invariance and countable subadditivity,

$$1 = m^* ([0, 1]) \leq m^* (\cup_{r \in \mathbb{Q}_0} (P + r)) \leq m^* ([-1, 2]) = 3$$

and $0 < m^* (\cup_{r \in \mathbb{Q}_0} (P + r)) < \infty$. On the other hand, by translation invariance, $m^* (P + r) = m^* (P)$ for any $r \in \mathbb{R}$, which implies

$$\sum_{r \in \mathbb{Q}_0} m^* (P + r) = \sum_{r \in \mathbb{Q}_0} m^* P$$

so that the sum is either 0, if $m^* (P) = 0$, or infinity, if $m^* (P) > 0$. In either case,

$$m^* (\cup_{r \in \mathbb{Q}_0} (P + r)) \neq \sum_{r \in \mathbb{Q}_0} m^* (P + r)$$

so that outer measure is not countably additive.

We will return to this example below.

Remark 3.10 *This example shows that there is no solution to Lebesgue's problem of measure. In the previous construction we have used the following facts to show that outer measure is not countably additive:*

(1) $m^* (P + r) = m^* (P)$;
(2) $0 < m^* (\cup_{r \in \mathbb{Q}_0} (P + r)) < \infty$.

The first follows from translation invariance. The second uses $m^* ([0, 1]) = 1$, monotonicity, and finite subadditivity (to show $m^* ([-1, 2]) \leq 3$). Since monotonicity is a consequence of finite subadditivity, the only properties we used were translation invariance, finite subadditivity, and $m^* ([0, 1]) = 1$. Thus, this example applies to any function satisfying these three properties. So, there is no function defined on all subsets of $\mathbb{R}$ that is translation invariant, countably additive and equals 1 on $[0, 1]$.

3.2.2 *Lebesgue measure*

Example 3.9 shows that m^* is not countably additive on the power set of $\mathbb{R}$. In order to obtain a countably additive set function which extends the length function, we restrict the domain of m^* to a suitable subset of the power set of $\mathbb{R}$. The members of this subset were called *measurable subsets* by Lebesgue. Lebesgue worked on a closed, bounded interval $I = [a, b]$, and for $E \subset I$, he defined the *inner measure* of E to be $m_* (E) = (b - a) - m^* (I \setminus E)$; that is, the inner measure of E is the length of I minus the outer measure of the complement of E in I. Lebesgue defined a subset $E \subset I$ to be measurable if $m^* (E) = m_* (E)$. Using the definition of inner measure and the fact that the outer measure of an interval is its length, Lebesgue's condition is equivalent to

$$m^* (I) = m^* (E) + m^* (I \setminus E).$$

Unfortunately, this procedure is not meaningful if we want to consider arbitrary subsets of $\mathbb{R}$ since the length of $\mathbb{R}$ is infinite. However, there is a characterization of Lebesgue measurable subsets of an interval I due to Constantin Carathéodory (1873-1950) that generalizes very nicely to arbitrary subsets of $\mathbb{R}$.

In the above equality, we assume that $E \subset I$, so that $E = E \cap I$. Carathéodory's idea was to test E with every subset of $\mathbb{R}$, instead of just an interval containing E. Thus, he was led to consider the condition

$$m^* (A) = m^* (A \cap E) + m^* (A \setminus E)$$

for every subset $A \subset \mathbb{R}$; A need not even be a measurable set! We now show that the two conditions are equivalent.

Theorem 3.11   *Let $I \subset \mathbb{R}$ be a bounded interval. If $E \subset I$, the following are equivalent:*

(1) $m^* (I) = m^* (E) + m^* (I \setminus E)$;
(2) $m^* (A) = m^* (A \cap E) + m^* (A \setminus E)$, for all $A \subset I$.

The proof will be based on several preliminary results. Given intervals $I, J \subset \mathbb{R}$, we define the *distance from I to J* by

$$d (I, J) = \inf \{|x - y| : x \in I, y \in J\}.$$

We begin by proving that outer measure is additive over intervals that are at a positive distance.

Lemma 3.12 *Let $I, J \subset \mathbb{R}$ be bounded intervals such that the distance from I to J is positive. Then,*

$$m^*(I \cup J) = m^*(I) + m^*(J).$$

Proof. By subadditivity, $m^*(I \cup J) \leq m^*(I) + m^*(J)$. To show the opposite inequality, fix $\epsilon > 0$ and choose a countable collection of open intervals $\{I_i\}_{i \in \sigma}$ such that $I \cup J \subset \cup_{i \in \sigma} I_i$ and $\sum_{i \in \sigma} \ell(I_i) \leq m^*(I \cup J) + \epsilon$. Assume, without loss of generality, that I lies to the left of J and let

$$\alpha = \frac{\sup I + \inf J}{2}$$

be the point midway between the two intervals.

Suppose $\alpha \in I_i = (a_i, b_i)$ for some $i \in \sigma$. Let $I_i^- = (a_i, \alpha)$ and $I_i^+ = (\alpha, b_i)$. Then, since $\alpha \notin I \cup J$, $(I \cup J) \cap I_i = (I \cup J) \cap (I_i^- \cup I_i^+)$, so that $I_i^- \cup I_i^+$ covers the same part of $I \cup J$ as I_i does, and, since

$$\ell(I_i) = b_i - a_i = (b_i - \alpha) + (\alpha - a_i) = \ell(I_i^-) + \ell(I_i^+),$$

replacing I_i by I_i^- and I_i^+ does not change the sum of the lengths of the intervals. Assume that every interval I_i that contains α is replaced by the two intervals I_i^- and I_i^+.

Let $\sigma(I) = \{i \in \sigma : I_i \cap J = \emptyset\}$ and $\sigma(J) = \{i \in \sigma : I_i \cap I = \emptyset\}$. It follows that $I \subset \cup_{i \in \sigma(I)} I_i$ and $J \subset \cup_{i \in \sigma(J)} I_i$. Thus,

$$m^*(I) + m^*(J) \leq \sum_{i \in \sigma(I)} \ell(I_i) + \sum_{i \in \sigma(J)} \ell(I_i) \leq \sum_{i \in \sigma} \ell(I_i) \leq m^*(I \cup J) + \epsilon.$$

Since this inequality is true for any $\epsilon > 0$, the proof is complete. □

Remark 3.13 *This result is true for intervals whose interiors are disjoint. If the intervals are open, this proof works whether the intervals touch or not. If any of the intervals are closed, we can replace them by their interiors, which does not change the measure of I, J or $I \cup J$, since the edge of an interval is a set of outer measure 0.*

The next result shows that condition (2) of Theorem 3.11 holds when E is an interval.

Lemma 3.14 *If $I \subset \mathbb{R}$ is a bounded interval and $J \subset I$ is an interval, then*

$$m^*(A) = m^*(A \cap J) + m^*(A \setminus J)$$

for all $A \subset I$.

Proof. Note first the conclusion holds if A is an interval in I. In this case, A and $A \cap J$ are both intervals, so their outer measures equal their lengths. Further, $A \setminus J$ is either an interval or a union of two disjoint intervals which are at a positive distance δ. In the first case, the equality is merely the fact that the length function is additive over disjoint intervals. In the second case, we write $A \setminus J = A_1 \cup A_2$, with A_1 and A_2 intervals at positive distance and use the previous lemma.

Let $A \subset I$ and $\epsilon > 0$. Choose a countable collection of open intervals $\{I_i\}_{i \in \sigma}$ such that $A \subset \cup_{i \in \sigma} I_i$ and $\sum_{i \in \sigma} \ell(I_i) \le m^*(A) + \epsilon$. As before, $m^*(I_i) = m^*(I_i \cap J) + m^*(I_i \setminus J)$ for all $i \in \sigma$. Therefore,

$$m^*(A \cap J) + m^*(A \setminus J) \le m^*((\cup_{i \in \sigma} I_i) \cap J) + m^*((\cup_{i \in \sigma} I_i) \setminus J)$$
$$\le \sum_{i \in \sigma} [m^*(I_i \cap J) + m^*(I_i \setminus J)]$$
$$= \sum_{i \in \sigma} m^*(I_i)$$
$$\le m^*(A) + \epsilon.$$

Since this is true for all $\epsilon > 0$, the result follows by countable subadditivity. $\qquad\qquad\qquad\qquad\qquad\qquad\qquad\qquad\qquad\qquad\qquad\square$

In the following lemma, we show that condition (1) of Theorem 3.11 implies condition (2) when A is an interval.

Lemma 3.15 *If $E \subset I$ satisfies condition (1) of Theorem 3.11, then*

$$m^*(J) = m^*(J \cap E) + m^*(J \setminus E)$$

for all intervals $J \subset I$.

Proof. By the previous lemma, for any interval $J \subset I$,

$$m^*(E) = m^*(E \cap J) + m^*(E \setminus J)$$

and

$$m^*(I \setminus E) = m^*((I \setminus E) \cap J) + m^*((I \setminus E) \setminus J).$$

By condition (1) and subadditivity, we see

$$m^*(I) = m^*(E) + m^*(I \setminus E)$$
$$= m^*(E \cap J) + m^*(E \setminus J) + m^*((I \setminus E) \cap J) + m^*((I \setminus E) \setminus J)$$
$$= \{m^*(E \cap J) + m^*((I \setminus E) \cap J)\}$$
$$+ \{m^*(E \setminus J) + m^*((I \setminus E) \setminus J)\}$$

Since $(I \setminus E) \cap J = J \setminus E$, $E \setminus J = (I \setminus J) \cap E$, and $(I \setminus E) \setminus J = (I \setminus J) \setminus E$, it follows from subadditivity that

$$
\begin{aligned}
m^*(I) &= \{m^*(J \cap E) + m^*(J \setminus E)\} \\
&\quad + \{m^*((I \setminus J) \cap E) + m^*((I \setminus J) \setminus E)\} \\
&\geq m^*(J) + m^*(I \setminus J) \\
&\geq m^*(I).
\end{aligned}
$$

Thus,

$$
\begin{aligned}
m^*(J) + m^*(I \setminus J) &= m^*(E \cap J) + m^*(E \setminus J) \\
&\quad + m^*((I \setminus E) \cap J) + m^*((I \setminus E) \setminus J).
\end{aligned}
$$

By subadditivity, $m^*(I \setminus J) \leq m^*(E \setminus J) + m^*((I \setminus E) \setminus J)$, which implies

$$
m^*(J) \geq m^*(E \cap J) + m^*((I \setminus E) \cap J) = m^*(E \cap J) + m^*(J \setminus E),
$$

and the proof now follows by subadditivity. $\qquad \Box$

We can now prove Theorem 3.11.

Proof. Setting $A = I$, we see that (2) implies (1). So, assume that (1) holds. Let $A \subset I$ and note that by subadditivity, it is enough to prove that

$$
m^*(A \cap E) + m^*(A \setminus E) \leq m^*(A).
$$

Fix $\epsilon > 0$ and choose a countable collection of open intervals $\{I_i\}_{i \in \sigma}$ such that $A \subset \cup_{i \in \sigma} I_i$ and $\sum_{i \in \sigma} \ell(I_i) \leq m^*(A) + \epsilon$. Then, by the previous lemma,

$$
\begin{aligned}
m^*(A \cap E) + m^*(A \setminus E) &\leq m^*((\cup_{i \in \sigma} I_i) \cap E) + m^*((\cup_{i \in \sigma} I_i) \setminus E) \\
&\leq \sum_{i \in \sigma} [m^*(I_i \cap E) + m^*(I_i \setminus E)] \\
&= \sum_{i \in \sigma} m^*(I_i) \\
&\leq m^*(A) + \epsilon.
\end{aligned}
$$

Thus, $m^*(A \cap E) + m^*(A \setminus E) \leq m^*(A)$ and the proof is complete. $\quad \Box$

Thus, for subsets of bounded intervals, measurability according to Lebesgue's definition is equivalent to measurability according to Carathéodory's definition. In order to include unbounded sets, we adapt Carathéodory's condition for our definition of measurable sets.

Definition 3.16 A subset $E \subset \mathbb{R}$ is *Lebesgue measurable* if for every set $A \subset \mathbb{R}$,

$$m^* (A) = m^* (A \cap E) + m^* (A \setminus E). \tag{3.2}$$

The set A is referred to as a *test set* for measurability. By subadditivity, we need only show that

$$m^* (A \cap E) + m^* (A \setminus E) \leq m^* (A)$$

in order to prove that E is measurable. We observe that we need only consider test sets with finite measure in (3.2) since if $m^* (A) = \infty$, then (3.2) follows from subadditivity. Set $E^c = \mathbb{R} \setminus E$. Note that condition (3.2) is the same as

$$m^* (A) = m^* (A \cap E) + m^* (A \cap E^c).$$

Definition 3.17 Let $\mathcal{M}$ be the collection of all Lebesgue measurable sets. The restriction of m^* to $\mathcal{M}$ is referred to as *Lebesgue measure* and denoted by $m = m^*|_{\mathcal{M}}$.

Thus, if $E \in \mathcal{M}$, then $m(E) = m^*(E)$.

We next study properties of m and $\mathcal{M}$. An immediate consequence of the definition is the following proposition.

Proposition 3.18 *The sets $\emptyset$ and $\mathbb{R}$ are measurable.*

Further, sets of outer measure 0 are measurable.

Proposition 3.19 *If $m^* (E) = 0$ then E is measurable.*

Proof. Let $A \subset \mathbb{R}$. By monotonicity, $0 \leq m^* (A \cap E) \leq m^* (E) = 0$. Thus,

$$m^* (A) \leq m^* (A \cap E) + m^* (A \setminus E) = m^* (A \setminus E) \leq m^* (A)$$

and E is measurable. $\qquad\square$

We say that a set E is a *null set* if $m(E) = 0$. Note that singleton sets are null, subsets of null sets are null, and countable unions of null sets are null. See Exercise 3.4.

Lemma 3.20 *Let $E_1, \ldots, E_n$ be pairwise disjoint and measurable sets. If $A \subset \mathbb{R}$, then*

$$m^* \left(A \cap \cup_{j=1}^n E_j \right) = \sum_{j=1}^n m^* (A \cap E_j).$$

Proof. Let $A \subset \mathbb{R}$. Proceeding by induction, we note that this statement is true for $n = 1$. Assume it is true for $n-1$ sets. Since the E_i's are pairwise disjoint,

$$A \cap \left(\cup_{j=1}^n E_i\right) \cap E_n = A \cap E_n \text{ and } A \cap \left(\cup_{j=1}^n E_i\right) \setminus E_n = A \cap \left(\cup_{j=1}^{n-1} E_i\right).$$

Since E_n is measurable, by the induction hypothesis,

$$m^* \left(A \cap \left(\cup_{j=1}^n E_i\right)\right) = m^* \left(A \cap E_n\right) + m^* \left(A \cap \left(\cup_{j=1}^{n-1} E_i\right)\right)$$

$$= m^* \left(A \cap E_n\right) + \sum_{i=1}^{n-1} m^* \left(A \cap E_i\right)$$

$$= \sum_{i=1}^n m^* \left(A \cap E_i\right).$$

$\square$

Let X be a nonempty set and $\mathcal{A} \subset \wp(X)$ a collection of subsets of X. We call $\mathcal{A}$ an *algebra* if $A, B \in \mathcal{A}$ implies that $A \cup B, A^c = X \setminus A \in \mathcal{A}$. Note that $\wp(X)$ is an algebra and so is the set $\{\emptyset, X\}$. In fact, every algebra contains $\emptyset$ and X since $A \in \mathcal{A}$ implies that $X = A \cup A^c \in \mathcal{A}$ and $\emptyset = X^c \in \mathcal{A}$. As a consequence of the definition and De Morgan's Laws, $\mathcal{A}$ is closed under finite unions and intersections. An algebra $\mathcal{A}$ is called a σ-*algebra* if it is closed under countable unions.

Example 3.21 Let

$$\mathcal{A} = \{F \subset (0,1) : F \text{ or } (0,1) \setminus F \text{ is a finite or empty set}\}.$$

Then, $\mathcal{A}$ is an algebra (see Exercise 3.5) which is not a σ-algebra. To see that $\mathcal{A}$ is not a σ-algebra, note that $\mathbb{Q} \cap (0,1)$ is a countable union of singleton sets, each of which is in $\mathcal{A}$, but neither $\mathbb{Q} \cap (0,1)$ nor its complement $(0,1) \setminus \mathbb{Q}$ is finite.

We want to show that $\mathcal{M}$ is a σ-algebra. We first prove that $\mathcal{M}$ is an algebra.

Theorem 3.22 *The set $\mathcal{M}$ of Lebesgue measurable sets is an algebra.*

Proof. We need to prove two things: $\mathcal{M}$ is closed under complementation; and, $\mathcal{M}$ is closed under finite unions. Since $A \cap E^c = A \cap (\mathbb{R} \setminus E) = A \setminus E$ and $A \setminus E^c = A \setminus (\mathbb{R} \setminus E) = A \cap E$, we see that the Carathéodory condition is symmetric in E and E^c, so if E is measurable then so is E^c.

Suppose that E and F are measurable. For $A \subset \mathbb{R}$, write $A \cap (E \cup F) = (A \cap E) \cup (A \cap E^c \cap F)$. Then, first using the measurability of F then the measurability of E,

$$
\begin{aligned}
m^*\left(A \cap (E \cup F)\right) + m^*\left(A \cap (E \cup F)^c\right) &= m^*\left(A \cap (E \cup F)\right) \\
&\quad + m^*\left(A \cap E^c \cap F^c\right) \\
&\leq m^*\left(A \cap E\right) + m^*\left(A \cap E^c \cap F\right) \\
&\quad + m^*\left(A \cap E^c \cap F^c\right) \\
&\leq m^*\left(A \cap E\right) + m^*\left(A \cap E^c\right) \\
&= m^*\left(A\right).
\end{aligned}
$$

Therefore, $E \cup F$ is measurable and $\mathcal{M}$ is an algebra. $\qquad \square$

Since $\mathcal{M}$ is an algebra, it satisfies the following proposition.

Proposition 3.23 *Let $\mathcal{A}$ be an algebra of sets and $\{A_i\}_{i=1}^\infty \subset \mathcal{A}$. Then, there is a collection $\{B_i\}_{i=1}^\infty \subset \mathcal{A}$ of pairwise disjoint sets so that $\cup_{i=1}^\infty A_i = \cup_{i=1}^\infty B_i$.*

Proof. Set $B_1 = A_1$ and for $j > 1$ set $B_j = A_j \setminus \cup_{i=1}^{j-1} A_i$. Since $\mathcal{A}$ is an algebra, $B_i \in \mathcal{A}$. Clearly, $B_i \subset A_i$ for all i, so for any index set $\sigma \subset \mathbb{N}$, $\cup_{i \in \sigma} B_i \subset \cup_{i \in \sigma} A_i$. Let $x \in \cup_{i=1}^\infty A_i$. Choose the smallest j so that $x \in A_j$. Then, $x \notin A_i$ for $i = 1, 2, \ldots, j-1$, which implies that $x \in B_j \subset \cup_{i=1}^\infty B_i$. Thus $\cup_{i=1}^\infty A_i \subset \cup_{i=1}^\infty B_i$ so that the two unions are equal. Finally, fix indices i and j and suppose that $j < i$. If $x \in B_j$, then $x \in A_j$, so that $x \notin B_i \subset A_i \setminus A_j$. Thus, $B_i \cap B_j = \emptyset$. $\qquad \square$

Note that the proof actually produces a collection of sets $\{B_i\}_{i=1}^\infty$ satisfying $\cup_{i=1}^N A_i = \cup_{i=1}^N B_i$ for every N. We can now prove that $\mathcal{M}$ is a σ-algebra.

Theorem 3.24 *The set $\mathcal{M}$ of Lebesgue measurable sets is a σ-algebra.*

Proof. We need to show that $\mathcal{M}$ is closed under countable unions. Let $\{E_i\}_{i=1}^\infty \subset \mathcal{M}$ and set $E = \cup_{i=1}^\infty E_i$. We want to show that $E \in \mathcal{M}$. By the previous proposition, there is a sequence $\{B_i\}_{i=1}^\infty \subset \mathcal{M}$ such that $E = \cup_{i=1}^\infty B_i$ and the B_i's are pairwise disjoint. Set $F_n = \cup_{i=1}^n B_i$. Then, $F_n \in \mathcal{M}$ and $F_n^c \supset E^c$.

Let $A \subset \mathbb{R}$. By Lemma 3.20,

$$
m^*\left(A \cap (\cup_{i=1}^n B_i)\right) = \sum_{i=1}^n m^*\left(A \cap B_i\right).
$$

Thus, for any $n \in \mathbb{N}$,

$$
\begin{aligned}
m^* (A) &= m^* (A \cap F_n) + m^* (A \cap F_n^c) \\
&\geq m^* (A \cap F_n) + m^* (A \cap E^c) \\
&= \sum_{i=1}^{n} m^* (A \cap B_i) + m^* (A \cap E^c).
\end{aligned}
$$

Since this is true for any n, by subadditivity we see that

$$
m^* (A) \geq \sum_{i=1}^{\infty} m^* (A \cap B_i) + m^* (A \cap E^c) \geq m^* (A \cap E) + m^* (A \cap E^c).
$$

Thus, E is measurable. Therefore, $\mathcal{M}$ is a σ-algebra. $\qquad \square$

A consequence of the translation invariance of m^* is that $\mathcal{M}$ is translation invariant; that is, if $E \in \mathcal{M}$ and $h \in \mathbb{R}$, then $E + h \in \mathcal{M}$. To see this, let $E \in \mathcal{M}$ and $A \subset \mathbb{R}$. Then,

$$
\begin{aligned}
m^* (A) &= m^* (A - h) = m^* ((A - h) \cap E) + m^* ((A - h) \setminus E) \\
&= m^* (((A - h) \cap E) + h) + m^* (((A - h) \setminus E) + h) \\
&= m^* (A \cap (E + h)) + m^* (A \setminus (E + h))
\end{aligned}
$$

which shows that $E + h \in \mathcal{M}$.

We saw above the $m^* (I) = \ell (I)$ for every interval $I \subset \mathbb{R}$. We now show that every interval is measurable.

Proposition 3.25 *Every interval $I \subset \mathbb{R}$ is a measurable set.*

Proof. Assume first that $I = (a, b)$. Fix a set $A \subset \mathbb{R}$ and set $A_1 = A \cap (-\infty, a]$, $A_2 = A \cap I$ and $A_3 = A \cap [b, \infty)$. Since

$$
m^* (A) \leq m^* (A \cap I) + m^* (A \setminus I) \leq m^* (A_1) + m^* (A_2) + m^* (A_3)
$$

it is enough to show

$$
m^* (A_1) + m^* (A_2) + m^* (A_3) \leq m^* (A).
$$

Without loss of generality, we may assume that $m^* (A) < \infty$.

Fix $\epsilon > 0$ and let $\{I_j\}_{j=1}^{\infty}$ be a collection of intervals such that $A \subset \cup_{j=1}^{\infty} I_j$ and $\sum_{j=1}^{\infty} \ell (I_j) \leq m^* (A) + \epsilon$. Set $I_j^1 = I_j \cap (-\infty, a]$, $I_j^2 = I_j \cap I$ and $I_j^3 = I_j \cap [b, \infty)$. Each I_j^n is either an interval or is empty, and $\ell (I_j) =$

$\ell\left(I_j^1\right) + \ell\left(I_j^2\right) + \ell\left(I_j^3\right) = m^*\left(I_j^1\right) + m^*\left(I_j^2\right) + m^*\left(I_j^3\right)$. For each n, we have $A_n \subset \cup_{j=1}^{\infty} I_j^n$, which implies that $m^*\left(A_n\right) \leq \sum_{j=1}^{\infty} m^*\left(I_j^n\right)$. Thus, we get

$$m^*\left(A_1\right) + m^*\left(A_2\right) + m^*\left(A_3\right) \leq \sum_{j=1}^{\infty} \left\{ m^*\left(I_j^1\right) + m^*\left(I_j^2\right) + m^*\left(I_j^3\right) \right\}$$

$$= \sum_{j=1}^{\infty} \ell\left(I_j\right)$$

$$\leq m^*\left(A\right) + \epsilon.$$

Since ϵ is arbitrary,

$$m^*\left(A_1\right) + m^*\left(A_2\right) + m^*\left(A_3\right) \leq m^*\left(A\right)$$

so that I is a measurable set.

Since $\mathcal{M}$ is a σ-algebra, $(a, \infty) = \cup_{n=1}^{\infty}(a, a+n)$ and $(-\infty, b) = \cup_{n=1}^{\infty}(b-n, b)$ are measurable, and so are their complements $(-\infty, a]$ and $[b, \infty)$. Since every interval is either the intersection or union of two such infinite intervals, all intervals are measurable. □

Thus, we see that Lebesgue measure extends the length function to the class of Lebesgue measurable sets.

We next study the open sets in $\mathbb{R}$ and the smallest σ-algebra that contains these sets. See Exercise 3.6.

Definition 3.26 Let $X \subset \mathbb{R}$. The collection of *Borel sets* in X is the smallest σ-algebra that contains all open subsets of X and is denoted $\mathcal{B}(X)$.

Since $\mathcal{B}(X)$ is a σ-algebra that contains the open subsets of X, by taking complements, $\mathcal{B}(X)$ contains all the closed subsets of X.

Let O be an open subset of $\mathbb{R}$. The next result shows that we can realize O as a countable union of open intervals.

Theorem 3.27 *Every open set in $\mathbb{R}$ is equal to the union of a countable collection of disjoint open intervals.*

Proof. Let $O \subset \mathbb{R}$ be an open set. For each $x \in O$, let I_x be the largest open interval contained in O that contains x. Clearly, $O \subset \cup_{x \in O} I_x$. Since $I_x \subset O$ for every $x \in O$, $\cup_{x \in O} I_x \subset O$, so that $O = \cup_{x \in O} I_x$. If $x, y \in O$, then either $I_x = I_y$ or $I_x \cap I_y = \emptyset$. To see this, note that if $I_x \cap I_y \neq \emptyset$, then $I_x \cup I_y$ is an open interval contained in O and containing both I_x and I_y. By the definition of the intervals I_x, we see that $I_x = I_x \cup I_y = I_y$. Thus, O is a union of disjoint open intervals. Since each of the intervals contains a distinct rational number, there are countably many distinct maximal intervals.   □

Thus, we can view $\mathcal{B}(\mathbb{R})$ as the smallest σ-algebra that contains the open intervals in $\mathbb{R}$. Since $\mathcal{M}$ is also a σ-algebra that contains the open intervals, we get

Corollary 3.28   *Every Borel set is measurable; that is,* $\mathcal{B}(\mathbb{R}) \subset \mathcal{M}$.

Remark 3.29   *The two sets* $\mathcal{B}(\mathbb{R})$ *and* $\mathcal{M}$ *are not equal; there are Lebesgue measurable sets that are not Borel sets. See, for example [Ha, Exercise 6, page 67], [Mu, pages 148-149], [Ru, page 53], and [Sw1, page 54]. Also, note that* $\mathcal{M}$ *is a proper subset of* $\wp(\mathbb{R})$ *as we show in Example 3.31 below.*

Let $\mathcal{F}_\sigma$ be the collection of all countable unions of closed sets. Then, $\mathcal{F}_\sigma \subset \mathcal{B}(\mathbb{R})$. Clearly, $\mathcal{F}_\sigma$ contains all the closed sets. It also contains all the open sets, since, for example, $(a,b) = \cup_{k=1}^\infty \left[a + \frac{1}{k}, b - \frac{1}{k} \right]$. Similarly, the collection of all countable intersections of open sets, $\mathcal{G}_\delta$, is contained in the Borel sets and contains all the open and closed sets.

So far, we have defined a nonnegative function m^* that is defined on all subsets of $\mathbb{R}$ and satisfies properties (1) and (3) of the problem of measure. This function does not satisfy property (2), as we saw in Example 3.9. Next, we defined a collection of sets, $\mathcal{M}$, and called m the restriction of m^* to $\mathcal{M}$. Consequently, m is translation invariant, and satisfies properties (1) and (3). We now show that m satisfies property (2), that is, m is *countably additive*.

Proposition 3.30   *Let* $\{E_i\}_{i=1}^\infty \subset \mathcal{M}$. *Then,* $m\left(\cup_{i=1}^\infty E_i\right) \leq \sum_{i=1}^\infty m\left(E_i\right)$. *If the sets* E_i *are pairwise disjoint, then* $m\left(\cup_{i=1}^\infty E_i\right) = \sum_{i=1}^\infty m\left(E_i\right)$.

Proof.   Since $\mathcal{M}$ is a σ-algebra, $\cup_{i=1}^\infty E_i \in \mathcal{M}$ and the inequality follows since it is true for outer measure. Assume the sets are pairwise disjoint. We need to show that

$$m\left(\cup_{i=1}^\infty E_i\right) \geq \sum_{i=1}^\infty m\left(E_i\right).$$

By Lemma 3.20, $m^*\left(A \cap \left(\cup_{i=1}^n E_i\right)\right) = \sum_{i=1}^n m^*\left(A \cap E_i\right)$. Let $A = \mathbb{R}$. Then, for all n,

$$m\left(\cup_{i=1}^\infty E_i\right) \geq m\left(\cup_{i=1}^n E_i\right) = \sum_{i=1}^n m\left(E_i\right).$$

Thus, $m\left(\cup_{i=1}^{\infty} E_i\right) = \sum_{i=1}^{\infty} m\left(E_i\right)$ as we wished to prove.   □

Thus, m solves *le problème de la mesure des ensembles* for the collection of measurable sets. But, not all sets are measurable, as the next example shows.

Example 3.31   The set P defined in Example 3.9 is not measurable. Suppose P were measurable. Then $P + r$ would be measurable for all $r \in \mathbb{R}$ and $m\left(P + r\right) = m\left(P\right)$. Thus,

$$m\left(\cup_{r \in \mathbb{Q}_0}\left(P + r\right)\right) = \sum_{r \in \mathbb{Q}_0} m\left(P + r\right) = \sum_{r \in \mathbb{Q}_0} m\left(P\right).$$

We saw in Example 3.9 that $1 \leq m\left(\cup_{r \in \mathbb{Q}_0}\left(P + r\right)\right) \leq 3$. If $m\left(P\right) = 0$, then the right hand side equals 0; if $m\left(P\right) > 0$, then the right hand side is infinite. In either case, the equality fails. Thus, P is not measurable.

Definition 3.32   Let $\mathcal{B}$ be a σ-algebra of sets. A nonnegative set function μ defined for all $A \in \mathcal{B}$ is called a *measure* if:

(1) $\mu\left(\emptyset\right) = 0$;
(2) μ is *countably additive*; that is,

$$\mu\left(\cup_{i=1}^{\infty} E_i\right) = \sum_{i=1}^{\infty} \mu\left(E_i\right)$$

for all sequences of pairwise disjoint sets $\{E_i\}_{i=1}^{\infty} \subset \mathcal{B}$.

Note that both $\sum_{i=1}^{\infty} \mu\left(E_i\right) = \infty$ and $\mu\left(E_i\right) = \infty$ for some i are allowed. Examples of measures include m defined on $\mathcal{M}$ and, also, m defined on $\mathcal{B}\left(\mathbb{R}\right)$.

Example 3.33   Define the *counting measure*, $\#$, by setting $\#\left(A\right)$ equal to the number of elements of A if A is a finite set and equal to ∞ if A is an infinite set. Then, $\#$ is a measure on the σ-algebra $\wp\left(X\right)$ of X, for any set X.

Suppose μ is a measure on $\mathcal{B}$ and $A, B \in \mathcal{B}$ with $A \subset B$. Then, by countable additivity, $\mu\left(B\right) = \mu\left(A\right) + \mu\left(B \setminus A\right)$, which also shows that μ is monotone. We use this identity in the following proof.

Proposition 3.34   *Let μ be a measure on a σ-algebra of sets $\mathcal{B}$. Suppose that $\{E_i\}_{i=1}^{\infty} \subset \mathcal{B}$.*

(1) If $E_i \subset E_{i+1}$, then,

$$\mu\left(\cup_{i=1}^{\infty} E_i\right) = \lim_{i \to \infty} \mu\left(E_i\right).$$

(2) If $E_i \supset E_{i+1}$, and there is a K so that $\mu\left(E_K\right) < \infty$, then,

$$\mu\left(\cap_{i=1}^{\infty} E_i\right) = \lim_{i \to \infty} \mu\left(E_i\right).$$

Proof. Suppose first that $E_i \subset E_{i+1}$. If $\mu\left(E_i\right) = \infty$ for some i, then the equality in (1) follows since both sides are infinite. So, assume $\mu\left(E_i\right) < \infty$ for all i. Set $E = \cup_{i=1}^{\infty} E_i$. Let $E_0 = \emptyset$. Since the sets are increasing, $E = (E_1 \setminus E_0) \cup (E_2 \setminus E_1) \cup (E_3 \setminus E_2) \cup \cdots$, which is a union of pairwise disjoint sets. Thus,

$$\mu\left(E\right) = \lim_{n \to \infty} \sum_{i=1}^{n} \mu\left(E_i \setminus E_{i-1}\right)$$

$$= \lim_{n \to \infty} \sum_{i=1}^{n} \left(\mu\left(E_i\right) - \mu\left(E_{i-1}\right)\right) = \lim_{n \to \infty} \mu\left(E_n\right).$$

Now, assume that $E_i \supset E_{i+1}$ and there is a K so that $\mu\left(E_K\right) < \infty$. Set $E = \cap_{i=1}^{\infty} E_i$. Since the sets are decreasing, $E_K \setminus E = (E_K \setminus E_{K+1}) \cup (E_{K+1} \setminus E_{K+2}) \cup \cdots$, where the sets on the right hand side are pairwise disjoint. It follows that

$$\mu\left(E_K\right) - \mu\left(E\right) = \lim_{n \to \infty} \sum_{i=K}^{n} \left[\mu\left(E_i\right) - \mu\left(E_{i+1}\right)\right] = \mu\left(E_K\right) - \lim_{n \to \infty} \mu\left(E_n\right).$$

Thus, $\mu\left(E\right) = \lim_{n \to \infty} \mu\left(E_n\right)$, proving the proposition. $\square$

Notice that we cannot drop the assumption in (2) that one of the sets E_k has finite measure.

Example 3.35 Let $E_i = [i, \infty)$. Then, $m\left(E_i\right) = \infty$ for all i while $E = \cap_{i=1}^{\infty} E_i = \emptyset$ has measure zero.

There are many ways to define Lebesgue measurability. The one we have chosen is useful for generalizing measurability to abstract settings. A common definition of measurability in Euclidean spaces is in terms of open sets. The following theorem gives four alternate characterizations of measurability. The second characterization is the classical definition in terms of open sets.

Theorem 3.36 *Let $E \subset \mathbb{R}$. The following are equivalent:*

(1) $E \in \mathcal{M}$;
(2) for all $\epsilon > 0$, there is an open set $G \supset E$ such that $m^* (G \setminus E) < \epsilon$;
(3) for all $\epsilon > 0$, there is a closed set $F \subset E$ such that $m^* (E \setminus F) < \epsilon$;
(4) there is a $\mathcal{G}_\delta$ set $G \supset E$ such that $m^* (G \setminus E) = 0$;
(5) there is an $\mathcal{F}_\sigma$ set $F \subset E$ such that $m^* (E \setminus F) = 0$.

Proof. We first show that (1) implies (2). Assume that E is a measurable set of finite measure. Fix $\epsilon > 0$ and choose a countable collection of open intervals $\{I_i\}_{i \in \sigma}$ such that $E \subset \cup_{i \in \sigma} I_i$ and $\sum_{i \in \sigma} \ell (I_i) < m (E) + \epsilon$. Set $G = \cup_{i \in \sigma} I_i$. Then, G is an open set containing E such that

$$m (G) \leq \sum_{i \in \sigma} \ell (I_i) \leq m (E) + \epsilon.$$

Therefore,

$$m^* (G \setminus E) = m (G \setminus E) = m (G) - m (E) < \epsilon.$$

If $m (E) = \infty$, set $E_k = E \cap (-k, k)$ and choose open sets $G_k \supset E_k$ so that $m^* (G_k \setminus E_k) < \epsilon 2^{-k}$. Then, the open set $G = \cup_{k=1}^\infty G_k \supset E$ and since

$$G \setminus E = \cup_{k=1}^\infty (G_k \setminus E) \subset \cup_{k=1}^\infty (G_k \setminus E_k)$$

we have that

$$m^* (G \setminus E) \leq \sum_{k=1}^\infty m^* (G_k \setminus E_k) < \sum_{k=1}^\infty \epsilon 2^{-k} = \epsilon,$$

as we wished to show.

To show that (2) implies (4), observe that for each k, there is an open set $G_k \supset E$ such that $m^* (G_k \setminus E) < \frac{1}{k}$. Then, $G = \cap_{k=1}^\infty G_k$ is the desired set.

Finally, we show that (1) is a consequence of (4). Let $G \in \mathcal{G}_\delta$ be such that $E \subset G$ and $m^* (G \setminus E) = 0$. Then, $G, G \setminus E \in \mathcal{M}$ which implies that $(G \setminus E)^c \in \mathcal{M}$. This implies that $E = G \setminus (G \setminus E) = G \cap (G \setminus E)^c \in \mathcal{M}$, as we wished to show.

It remains to show that (1), (3) and (5) are equivalent. To show that (1) implies (3), note that $E \in \mathcal{M}$ implies $E^c \in \mathcal{M}$. Thus, there is an open set $G \supset E^c$ with $m^* (G \setminus E^c) < \epsilon$. The set $F = G^c$ is the desired set, since $E \setminus F = E \setminus G^c = G \setminus E^c$. The other implications are similar. $\qquad \square$

Suppose that E is a measurable set. For each $\epsilon > 0$, there is an open set $G \supset E$ such that $m(G \setminus E) = m^*(G \setminus E) < \epsilon$, which implies

$$m(G) = m(E) + m(G \setminus E) \leq m(E) + \epsilon.$$

It follows that

Corollary 3.37 *Let $E \subset \mathbb{R}$ be a measurable set. Then,*

$$m(E) = \inf \{m(G) : E \subset G, G \ open\}.$$

If we wish to generalize the concept of length to general sets, we need a function that is defined on all of the Borel sets (and, in fact, many more sets). We call a measure μ defined on $\mathcal{B}(\mathbb{R})$ that is finite valued for all bounded intervals a *Borel measure*. We will show that every translation invariant Borel measure is a multiple of Lebesgue measure.

Definition 3.38 A measure μ defined for all elements of $\mathcal{B}(\mathbb{R})$ is called *outer regular* if

$$\mu(E) = \inf \{\mu(G) : E \subset G, G \ open\}$$

for all $E \in \mathcal{B}(\mathbb{R})$.

By the corollary, Lebesgue measure restricted to the Borel sets is an outer regular measure. In fact, every Borel measure on $\mathbb{R}$ is outer regular. (See [Sw1, Remark 7, page 64].) We show next that a translation invariant Borel measure is a constant multiple of Lebesgue measure.

Theorem 3.39 *If μ is a translation-invariant outer-regular Borel measure, then $\mu = cm$ for some constant c.*

In fact, we have already seen the proof of much of this theorem. It is a repetition of the argument proving $\int_a^b 1\,dx = b - a$ for all $a, b \in \mathbb{R}$ from Lebesgue's descriptive properties of the integral. The only properties used to show that equality were translation invariance (1), finite additivity (2), and $\int_0^1 1\,dx = 1$ (3), and our measures are translation invariant and finitely additive.

Proof. Set $c = \mu((0,1))$. We claim that $\mu(E) = cm(E)$ for all $E \in \mathcal{B}(\mathbb{R})$. Since μ is finite on bounded intervals, by translation invariance and countable additivity, $\mu(\{x\}) = 0$ for all $x \in \mathbb{R}$. Thus, $\mu([a,b]) = \mu((a,b)) = \mu([a,b)) = \mu((a,b])$. By this observation and translation invariance, if I is an interval with $\ell(I) = 1$, then

$$\mu(I) = \mu((0,1)) = c = cm(I).$$

Since μ is finitely additive, if $a_0 < a_1 < \cdots < a_n$, then

$$\mu\left((a_0, a_1)\right) + \mu\left((a_1, a_2)\right) + \cdots + \mu\left((a_{n-1}, a_n)\right)$$
$$= \mu\left((a_0, a_n)\right) - \sum_{i=1}^{n-1} \mu\left(\{a_i\}\right) = \mu\left((a_0, a_n)\right).$$

Setting $a_i = \dfrac{i}{n}$ shows that $\mu\left(\left(0, \dfrac{1}{n}\right)\right) = c\dfrac{1}{n}$, which in turn implies $\mu\left((0, q)\right) = cq$ for any rational number q. Finally, if $r \in \mathbb{R}$, let p and q be rational numbers such that $p < r < q$. Then, since $(0, p) \subset (0, r) \subset (0, q)$,

$$0 \le \mu\left((0, q)\right) - \mu\left((0, r)\right) = cq - \mu\left((0, r)\right) \le cq - \mu\left((0, p)\right) = c\left(q - p\right).$$

Letting p and q approach r, we conclude that for all real numbers r,

$$\mu\left((0, r)\right) = cr,$$

so that $\mu\left((a, b)\right) = cm\left((a, b)\right)$ for all $a, b \in \mathbb{R}$ and $\mu\left(I\right) = cm\left(I\right)$ for all open intervals $I \subset \mathbb{R}$.

Next, if G is an open set in $\mathbb{R}$, by Theorem 3.27, $G = \cup_{i \in \sigma} I_i$, a countable union of disjoint open intervals. By countable additivity,

$$\mu\left(G\right) = \sum_{i \in \sigma} \mu\left(I_i\right) = \sum_{i \in \sigma} cm\left(I_i\right) = cm\left(G\right).$$

Finally, since μ is outer regular, if $E \in \mathcal{B}\left(\mathbb{R}\right)$,

$$\mu\left(E\right) = \inf\left\{\mu\left(G\right) : E \subset G, G \text{ open}\right\}$$
$$= \inf\left\{cm\left(G\right) : E \subset G, G \text{ open}\right\} = cm\left(E\right),$$

since Lebesgue measure is regular. Thus, $\mu = cm$. $\qquad\square$

3.2.3   The Cantor set

The Cantor set is an important example for understanding some of the concepts related to Lebesgue measure. In particular, the Cantor set is an uncountable set with measure zero.

To create the Cantor set, we begin with the closed unit interval $[0, 1]$. Remove the open middle third of the interval, $\left(\frac{1}{3}, \frac{2}{3}\right)$, and call the remainder of the set $C_1 = \left[0, \frac{1}{3}\right] \cup \left[\frac{2}{3}, 1\right]$. Notice that C_1 consists of two intervals and has measure $\frac{2}{3}$. Next, remove the open middle third interval of each piece of C_1. Call the remainder $C_2 = \left[0, \frac{1}{9}\right] \cup \left[\frac{2}{9}, \frac{3}{9}\right] \cup \left[\frac{6}{9}, \frac{7}{9}\right] \cup \left[\frac{8}{9}, 1\right]$. Note that C_2 consists of $4 = 2^2$ intervals and has measure $4\left(\frac{1}{9}\right) = \left(\frac{2}{3}\right)^2$. Continuing this process, after the kth division, we are left with a closed set C_k which is the union of 2^k closed and disjoint subintervals, each of length 3^{-k}. Thus, $m^*(C_k) = \left(\frac{2}{3}\right)^k$. By construction, $C_k \supset C_{k+1}$ for all k. The set $C = \cap_{k=1}^{\infty} C_k$ is known as the *Cantor set*.

We now make some observations about C. It is a closed set since it is an intersection of closed sets. If x is an endpoint of an interval in C_k, then it is also an endpoint of an interval in C_{k+j} for all $j \in \mathbb{N}$. Thus, $x \in C$ and $C \neq \emptyset$. Finally, since

$$m^*(C) \leq m^*(C_k) = \left(\frac{2}{3}\right)^k$$

for all k, it follows that $m^*(C) = 0$. Hence, C is measurable and $m(C) = 0$.

We next show that the Cantor set is uncountable. For $x \in [0,1]$, let $0.a_1 a_2 a_3 \ldots$ be its ternary expansion. Thus, $a_i \in \{0,1,2\}$ for all i. Further, we write our expansions so that they do not end with '1000...' or '1222...'. To do this, we write '0222...' for '1000...' and '2' for '1222...'. Then, $x \in C$ if, and only if, $a_i \neq 1$ for all i. For example, if $a_1 = 1$, then $x \in \left(\frac{1}{3}, \frac{2}{3}\right)$, the first interval removed. Thus, we can think of the ternary decimal expansion of an element of C as a sequence of 0's and 2's. Dividing each term of this sequence by 2 defines a one-to-one, onto mapping from C to the set of all sequences of 0's and 1's. As proved in [DS, Prop. 8, page 12], this set of sequences is uncountable, so that the Cantor set is uncountable.

The Cantor set is an uncountable set of measure 0. One can also prove that its complement is dense in $[0,1]$. See Exercise 3.9. We define *generalized Cantor sets* as follows. Fix an $\alpha \in (0,1)$. At the kth step, remove 2^{k-1} open intervals of length $\alpha 3^{-k}$, instead of 3^{-k}. The rest of the construction is the same. The resulting set is a closed set of measure $1 - \alpha$ whose complement is dense in $[0,1]$.

3.3 Lebesgue measure in $\mathbb{R}^n$

In the previous section, we showed how the natural length function in the real line could be extended to a translation invariant measure on the measurable subsets of $\mathbb{R}$. In this section, we extend the result to Euclidean n-space. In particular, these results extend the natural area function in the plane and the natural volume function in Euclidean 3-space. Our procedure is very analogous to that employed in the one-dimensional case. We begin by defining Lebesgue outer measure for arbitrary subsets of $\mathbb{R}^n$, showing that Lebesgue outer measure extends the volume (area, when $n = 2$) function, and then restricting the outer measure to a class of subsets of $\mathbb{R}^n$ called the (Lebesgue) measurable sets to obtain Lebesgue measure on $\mathbb{R}^n$. Many of the statements and proofs of results for $\mathbb{R}^n$ are identical to those in $\mathbb{R}$ and will not be repeated.

The space $\mathbb{R}^n$ is the set of all real-valued n-tuples $x = (x_1, \ldots, x_n)$, where $x_i \in \mathbb{R}$. If $x, y \in \mathbb{R}^n$ and $t \in \mathbb{R}$, we define $x + y$ and tx to be

$$x + y = (x_1 + y_1, \ldots, x_n + y_n) \text{ and } tx = (tx_1, \ldots, tx_n).$$

We define the *norm*, $\|\cdot\|$, of x by $\|x\| = \left(\sum_{i=1}^{n} |x_i|^2 \right)^{1/2}$. The *distance*, d, between points $x, y \in \mathbb{R}^n$ is then the norm of their difference, $d(x, y) = \|x - y\| = \left(\sum_{i=1}^{n} |x_i - y_i|^2 \right)^{1/2}$. Let $B(x_0, r) = \{x \in \mathbb{R}^n : d(x, x_0) < r\}$ be the *ball centered at x_0 with radius r*. A set $G \subset \mathbb{R}^n$ is called *open* if for each $x \in G$, there is an $r > 0$ so that $B(x, r) \subset G$. Let $\{x_k\}_{k=1}^{\infty} \subset \mathbb{R}^n$ be a sequence in $\mathbb{R}^n$. We say that $\{x_k\}_{k=1}^{\infty}$ *converges* to $x_0 \in \mathbb{R}^n$ if $\lim_{k \to \infty} d(x_k, x_0) = 0$. A set $F \subset \mathbb{R}^n$ is called *closed* if every convergent sequence in F converges to a point in F; that is, if $\{x_k\}_{k=1}^{\infty} \subset F$ and $x_k \to x_0$, then $x_0 \in F$. Finally, a set H is called *bounded* if there is an $M > 0$ such that $\|x\| \leq M$ for all $x \in H$. We define the *symmetric difference* of sets $E_1, E_2 \subset \mathbb{R}^n$, denoted $E_1 \Delta E_2$, to be the set $E_1 \Delta E_2 = (E_1 \setminus E_2) \cup (E_2 \setminus E_1)$.

An important collection of subsets of $\mathbb{R}^n$ consists of the *compact* sets. By the Heine-Borel Theorem, a set $K \subset \mathbb{R}^n$ is compact if, and only if, K is closed and bounded. Below, we will use the following characterization of compact sets. A set $K \subset \mathbb{R}^n$ is compact if, and only if, given any collection of open sets $\{G_i\}_{i \in \Lambda}$ such that $K \subset \cup_{i \in \Lambda} G_i$, there is a finite subset $\{G_1, G_2, \ldots, G_m\} \subset \{G_i\}_{i \in \Lambda}$ such that $K \subset \cup_{i=1}^{m} G_i$. That is, every open cover of K contains a finite subcover. See [DS, pages 76-79].

An *interval* in $\mathbb{R}^n$ is a set of the form $I = I_1 \times \cdots \times I_n$, where each I_i, $i = 1, \ldots, n$, is an interval in $\mathbb{R}$. We say I is *open* (*closed*) if each I_i is open (closed). If each I_i is a half-closed interval of the form $[a, b)$, we call I a *brick*. If $I \subset \mathbb{R}^n$ is an interval, we define the *volume* of I to be

$$v(I) = \prod_{i=1}^{n} \ell(I_i),$$

with the convention that $0 \cdot \infty = 0$, so that if some interval I_i has infinite length and another interval $I_{i'}$, $i \neq i'$, is degenerate and has length 0, then $v(I) = 0$. In particular, the edge of an interval is a degenerate interval and, hence, has volume 0. Finally, note that if B is a brick which is a union of pairwise disjoint bricks $\{B_i : 1 \le i \le k\}$, then

$$v\left(B\right) = \sum_{i=1}^{k} v\left(B_i\right).$$

In the figure below, the brick B_1 is the union of bricks $b_1, \ldots, b_{11}$.

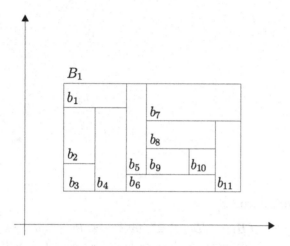

Figure 3.1

Analogous to the case of outer measure in the line, we define the outer measure of a subset of $\mathbb{R}^n$ by using covers of the subset by open intervals in $\mathbb{R}^n$.

Definition 3.40 Let $E \subset \mathbb{R}$. We define the *(Lebesgue) outer measure* of E, $m_n^*\left(E\right)$, by

$$m_n^*\left(E\right) = \inf \left\{ \sum_{j \in \sigma} v\left(J_j\right) \right\},$$

where the infimum is taken over all countable collections of open intervals $\{J_j\}_{j \in \sigma}$ such that $E \subset \cup_{j \in \sigma} J_j$.

It is straightforward to extend results (1), (2) and (4) of Theorem 3.6 to m_n^*. We show that the analogue of property (3) of Theorem 3.6 also holds. For this result, we need the observations that the intersection of two bricks is a brick and the difference of two bricks is a finite union of pairwise disjoint bricks. In the following figure, the difference of bricks B_1 and B_2 is the union of bricks $b_1, \ldots, b_4$.

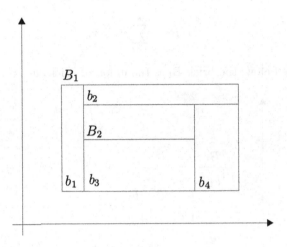

Figure 3.2

We begin with a lemma.

Lemma 3.41   *If $B_1, \ldots, B_m \subset \mathbb{R}^n$ are bricks, then there is a finite family $\mathcal{F} = \{F_1, \ldots, F_k\}$ of pairwise disjoint bricks such that each B_i is a union of members of $\mathcal{F}$.*

Proof.   Assume that $m = 2$. Then, $B_1 \cap B_2$, $B_1 \setminus (B_1 \cap B_2)$ and $B_2 \setminus (B_1 \cap B_2)$ are pairwise disjoint and since $B_1 \cap B_2$ is a brick and the difference of bricks is a union of pairwise disjoint bricks, the existence of the family $\mathcal{F}$ follows.

Note that this result implies that the union of two bricks is a finite union of pairwise disjoint bricks. Since

$$B_1 \cup B_2 = (B_1 \cap B_2) \cup (B_1 \setminus (B_1 \cap B_2)) \cup (B_2 \setminus (B_1 \cap B_2)),$$

we can decompose $B_1 \cup B_2$ into three pairwise disjoint sets, each of which is a finite union of pairwise disjoint bricks.

Proceeding by induction, assume we have proved the result for sets of m bricks. Suppose we have $m + 1$ bricks $B_1, \ldots, B_{m+1}$. By the induction hypothesis, there exist pairwise disjoint bricks $C_1, \ldots, C_l$ such that $B_1, \ldots, B_m$ are unions of members of $\{C_i : 1 \leq i \leq l\}$. Note that $B = B_{m+1} \setminus \cup_{i=1}^{m} B_i = \cap_{i=1}^{m} (B_{m+1} \setminus B_i)$ is an intersection of finite unions of disjoint bricks. Consequently, B is a finite union of disjoint bricks, $B = \cup_{j=1}^{k} C_j'$

where $\{C'_1, \ldots, C'_k\}$ is a collection of pairwise disjoint bricks. Therefore, we may replace the set $\{B_1, \ldots, B_{m+1}\}$ by $\{C_1, \ldots, C_l, C'_1, \ldots, C'_k\}$, and the members of this collection can be replaced by the pairwise disjoint sets $C_i \cap C'_j$, $C_i \setminus (C_i \cap C'_j)$ and $C'_j \setminus (C_i \cap C'_j)$, $i = 1, \ldots, l$ and $j = 1, \ldots, k$, each of which is a union of pairwise disjoint bricks. The result follows by induction. $\qquad\square$

We now prove that the outer measure of an interval equals its volume.

Theorem 3.42 *If $I \subset \mathbb{R}^n$ is an interval, then*

$$m_n^* (I) = v (I).$$

Proof. Suppose first that $I = I_1 \times \cdots \times I_n$ is a closed and bounded interval. To see that $m_n^* (I) \leq v (I)$, let I_i^* be an open interval with the same center as I_i such that $\ell (I_i^*) = (1 + \epsilon) \ell (I_i)$. Then, $I^* = I_1^* \times \cdots \times I_n^*$ is an open set containing I and $v (I^*) = (1 + \epsilon)^n v (I)$. It follows that $m_n^* (I) \leq v (I)$.

To complete the proof, we need to know that if $\{J_i : i \in \sigma\}$ is a countable cover of I by open intervals, then $v (I) \leq \sum_{j \in \sigma} v (J_i)$. Since I is compact, I is covered by a finite number of the intervals $\{J_1, \ldots, J_k\}$, say. Let K_i be the smallest brick containing J_i and K the largest brick contained in I. These bricks exist because I is a closed interval and each J_i is an open interval. It follows that $v (J_i) = v (K_i)$, $v (I) = v (K)$ and $K \subset \cup_{i=1}^k K_i$. By the lemma, there is a family $\mathcal{F} = \{F_1, \ldots, F_l\}$ of pairwise disjoint bricks such that K and each K_i is a union of members of $\mathcal{F}$. Suppose $K = \cup_{j=1}^p F_j$, $F_j \in \mathcal{F}$. Then,

$$v (I) = v (K) = \sum_{j=1}^p v (F_j) \leq \sum_{j=1}^l v (F_j) = \sum_{j=1}^k v (K_j) = \sum_{j=1}^k v (J_j)$$

as desired. The case of a general interval can be treated as in the proof of Theorem 3.6. $\qquad\square$

We note that since the edge of an interval is a degenerate interval, the outer measure of the surface of an interval is 0.

We define (Lebesgue) measurability for subsets of $\mathbb{R}^n$ as in Definition 3.16.

Definition 3.43 A subset $E \subset \mathbb{R}^n$ is *Lebesgue measurable* if for every set $A \subset \mathbb{R}^n$,

$$m_n^* (A) = m_n^* (A \cap E) + m_n^* (A \setminus E).$$

We denote the collection of measurable subsets of $\mathbb{R}^n$ by $\mathcal{M}_n$ and define *Lebesgue measure* m_n on $\mathbb{R}^n$ to be m_n^* restricted to $\mathcal{M}_n$. As in Theorem 3.24, Proposition 3.25 and Proposition 3.30, $\mathcal{M}_n$ is a σ-algebra containing all n-dimensional intervals and m_n is countably additive. The collection of Borel sets of $\mathbb{R}^n$, $\mathcal{B}(\mathbb{R}^n)$, comprises the smallest σ-algebra generated by the open subsets of $\mathbb{R}^n$. As in the one-dimensional case (Corollary 3.28), we see by using Lemma 3.44 below that, $\mathcal{B}(\mathbb{R}^n) \subset \mathcal{M}_n$, and the regularity conditions of Theorem 3.36 and its corollary hold.

Since the Lebesgue measure of an interval in $\mathbb{R}^n$ is equal to its n-dimensional volume, we use Lebesgue measure to define area of planar regions in two dimensions and volume of solid regions in three dimensions, extending these concepts from intervals to more general sets. We will discuss computing these quantities using Fubini's Theorem below.

Before leaving this section, we show that every open set $G \subset \mathbb{R}^n$ can be decomposed into an countable collection of disjoint bricks.

Lemma 3.44   *If $G \subset \mathbb{R}^n$ is an open set, then G is the union of a countable collection of pairwise disjoint cubic bricks.*

Proof.   Let B_k be the family of all cubic bricks with edge length 2^{-k} whose vertices are integral multiples of 2^{-k}. Note that B_k is a countable set. We need the following observations, which follow from the definitions of the sets B_k:

(1) if $x \in \mathbb{R}^n$, then there is a unique $B \in B_k$ such that $x \in B$;
(2) if $B \in B_j$ and $B' \in B_k$ with $j < k$, then either $B' \subset B$ or $B \cap B' = \emptyset$.

Since G is open, if $x \in G$ then x is contained in an open sphere contained in G. Thus, for large enough k, there is a brick $B \in B_k$ such that $B \subset G$ and $x \in B$. Set $B_k(G) = \{B \in B_k : B \subset G\}$. Thus, it follows that $G = \cup_{k=1}^{\infty} \cup_{B \in B_k(G)} B$. Choose all the bricks in $B_1(G)$. Next, choose all the bricks in $B_2(G)$ that are not contained in any brick in $B_1(G)$. Continuing, we keep all the bricks in $B_j(G)$ that are not contained in any of the bricks chosen in the previous steps. This construction produces a countable family of pairwise disjoint cubic bricks whose union is G. $\square$

Using Lemma 3.44, we can prove an extension of Theorem 3.39.

Theorem 3.45   *If μ is a translation invariant measure on $\mathcal{B}(\mathbb{R}^n)$ which is finite on compact sets, then $\mu = cm_n$ for some constant c.*

Proof.   Let $I = [0, 1) \times \cdots \times [0, 1)$ be the unit brick in $\mathbb{R}^n$ and set $c = \mu(I)$. For any $k \in \mathbb{N}$, I is the union of 2^{nk} pairwise disjoint bricks of side length

2^{-k}, so by translation invariance, each of these bricks has the same μ-measure. If B is any brick with side length 2^{-k}, we have

$$\mu(B) = \frac{1}{2^{nk}}\mu(I) = \frac{1}{2^{nk}}cm_n(I) = cm_n(B).$$

Hence, $\mu(B) = cm_n(B)$ for any such B. By Lemma 3.44, $\mu(G) = cm_n(G)$ for any open set $G \subset \mathbb{R}^n$. Since μ is outer regular ([Sw1, Remark 7, page 64]), we have that $\mu = cm_n$ by the analog of Corollary 3.37. $\square$

3.4 Measurable functions

Lebesgue's descriptive definition of the integral led us, in a very natural way, to consider the measure of sets, which in turn forced us to consider a proper subset of the set of all subsets of $\mathbb{R}$. We already know that if E is an interval, then

$$\int \chi_E dx = \ell(E) = m(E).$$

In fact, in order for χ_E to have an integral, E must be a measurable set. But, then, by linearity, if $E_1, \ldots, E_n \subset \mathbb{R}$ are pairwise disjoint, measurable sets, then $\varphi(x) = \sum_{i=1}^n a_i \chi_{E_i}(x)$ is also integrable. For property (6) of Lebesgue's definition to hold, monotonic limits of such simple functions must also be integrable. We now investigate such functions.

To begin, we extend the real numbers by adjoining two distinguished elements, $-\infty$ and ∞. We call the set $\mathbb{R}^* = \mathbb{R} \cup \{-\infty, \infty\}$ the *extended real numbers*. The extended real numbers satisfy the following properties for all $x \in \mathbb{R}$:

(1) $-\infty < x < \infty$;
(2) $\infty + x = x + \infty$;
(3) $-\infty + x = x + (-\infty)$;
(4) if $a > 0$ then $\infty \cdot a = a \cdot \infty = \infty$ and $(-\infty) \cdot a = a \cdot (-\infty) = -\infty$;
(5) if $a < 0$ then $\infty \cdot a = a \cdot \infty = -\infty$ and $(-\infty) \cdot a = a \cdot (-\infty) = \infty$.

While $\infty + \infty = \infty$ and $-\infty + (-\infty) = -\infty$, both $\infty + (-\infty)$ and $(-\infty) + \infty$ are undefined. Also, (4) and (5) remain valid if a equals ∞ or $-\infty$. Recall that we follow the convention $\infty \cdot 0 = 0 \cdot \infty = 0$.

Our study of measurable functions will involve simple functions. We recall their definition.

Definition 3.46 A *simple function* is a function which assumes a finite number of finite values.

Let φ be a simple function which takes on the distinct values $a_1, \ldots, a_m$ on the sets $E_i = \{x : \varphi(x) = a_i\}$, $i = 1, \ldots, m$. Then, the *canonical form* of φ is

$$\varphi(x) = \sum_{i=1}^{m} a_i \chi_{E_i}(x).$$

Let $E \subset \mathbb{R}^n$. We call f an *extended real-valued function* if $f : E \to \mathbb{R}^*$. Suppose that $\{\varphi_k\}_{k=1}^{\infty}$ is a monotonically increasing sequence of simple functions defined on some set E. Then, for each $x \in E$, $\lim_{k \to \infty} \varphi_k(x)$ exists in $\mathbb{R}^*$; the limit exists, but may not be finite. Thus, a monotonic limit of simple functions is an extended real-valued function.

Definition 3.47 Let E be a measurable subset of $\mathbb{R}^n$. We say that an extended real-valued function $f : E \to \mathbb{R}^*$ is *(Lebesgue) measurable* if $\{x \in E : f(x) > \alpha\} \in \mathcal{M}_n$ for all $\alpha \in \mathbb{R}$.

We first observe that a simple function φ is measurable if, and only if, each set E_i is measurable.

Example 3.48 Let $\varphi(x) = \sum_{i=1}^{m} a_i \chi_{E_i}(x)$ be a simple function in canonical form, with $E_1, \ldots, E_m$ pairwise disjoint. Then

$$\{x \in E : \varphi(x) > \alpha\} = \cup_{a_i > \alpha} E_i$$

and it follows that φ is measurable if, and only if, each E_i is measurable. To see this, suppose that $a_1 < a_2 < \cdots < a_m$. If $a_{m-1} \leq \alpha < a_m$, then $\{x \in E : \varphi(x) > \alpha\} = E_m$, so the measurability of φ requires that $E_m \in \mathcal{M}_n$. If $a_{m-2} \leq \alpha < a_{m-1}$, then $\{x \in E : \varphi(x) > \alpha\} = E_{m-1} \cup E_m$. Thus, if φ is measurable, then $E_{m-1} \cup E_m$ is measurable, and since $E_m \in \mathcal{M}_n$, $E_{m-1} = E_{m-1} \cup E_m \setminus E_m \in \mathcal{M}_n$. Continuing in this manner, we see that each $E_i \in \mathcal{M}_n$. On the other hand, if each $E_i \in \mathcal{M}_n$ then, $\{x \in E : \varphi(x) > \alpha\} \in \mathcal{M}_n$ for each $\alpha \in \mathbb{R}$ since it is a (finite) union of measurable sets, and consequently φ is a measurable function.

Below, we will always assume that a simple function is measurable, unless explicitly stated otherwise.

Example 3.49 Let $f : \mathbb{R} \to \mathbb{R}$ be continuous. Since

$$\{x \in \mathbb{R} : f(x) > \alpha\} = f^{-1}((\alpha, \infty))$$

is an open set, it is measurable. Thus, every continuous function defined on $\mathbb{R}$ is measurable.

The measurability of a function, like the measurability of a set, has several characterizations.

Proposition 3.50 *Let $E \in \mathcal{M}_n$ and $f : E \to \mathbb{R}^*$. The following are equivalent:*

(1) f is measurable; that is, $\{x \in E : f(x) > \alpha\} \in \mathcal{M}_n$ for all $\alpha \in \mathbb{R}$;
(2) $\{x \in E : f(x) \geq \alpha\} \in \mathcal{M}_n$ for all $\alpha \in \mathbb{R}$;
(3) $\{x \in E : f(x) < \alpha\} \in \mathcal{M}_n$ for all $\alpha \in \mathbb{R}$;
(4) $\{x \in E : f(x) \leq \alpha\} \in \mathcal{M}_n$ for all $\alpha \in \mathbb{R}$.

Proof. Since $\{x \in E : f(x) \leq \alpha\} = E \setminus \{x \in E : f(x) > \alpha\}$ and $\mathcal{M}_n$ is a σ-algebra, (1) and (4) are equivalent; similarly, (2) and (3) are equivalent. Since $\{x \in E : f(x) > \alpha\} = \cup_{k=1}^{\infty} \{x \in E : f(x) \geq \alpha + \frac{1}{k}\}$ and $\{x \in E : f(x) \geq \alpha\} = \cap_{k=1}^{\infty} \{x \in E : f(x) > \alpha - \frac{1}{k}\}$, (1) and (2) are equivalent, completing the proof. $\qquad\square$

Remark 3.51 *We can replace the condition "for all $\alpha \in \mathbb{R}$" by "for every α in a dense subset of $\mathbb{R}$" in Definition 3.47 and Proposition 3.50. See Exercise 3.20.*

Since

$$\{x \in E : f(x) = \alpha\} = \{x \in E : f(x) \leq \alpha\} \cap \{x \in E : f(x) \geq \alpha\}$$
$$\{x \in E : f(x) = \infty\} = \cap_{n=1}^{\infty} \{x \in E : f(x) > n\}$$

and

$$\{x \in E : f(x) = -\infty\} = \cap_{n=1}^{\infty} \{x \in E : f(x) < -n\},$$

we see that

Corollary 3.52 *Let $E \in \mathcal{M}_n$ and $f : E \to \mathbb{R}^*$ be measurable. Then, $\{x \in E : f(x) = \alpha\} \in \mathcal{M}_n$ for all $\alpha \in \mathbb{R}^*$.*

It is a bit surprising that the converse to this corollary is false.

Example 3.53 Let $P \subset (0,1)$ be a nonmeasurable set. Define $f : (0,1) \to \mathbb{R}$ by

$$f(x) = \begin{cases} t & \text{if } t \in P \\ -t & \text{if } t \notin P \end{cases}.$$

Then, f is one-to-one which implies that $\{x \in (0,1) : f(x) = \alpha\}$ is Lebesgue measurable for all $\alpha \in \mathbb{R}^*$, but since $\{x \in (0,1) : f(x) > 0\} = P$, f is not measurable.

The next result contains some of the algebraic properties of measurable functions.

Proposition 3.54   *Let $E \in \mathcal{M}_n$ and $f, g : E \to \mathbb{R}^*$ be measurable and assume that $f + g$ is defined for all $x \in E$. Let $c \in \mathbb{R}$. Then:*

(1) $\{x \in E : f(x) > g(x)\}$ is a measurable set;
(2) $f + c$, cf, $f + g$, fg, $f \vee g$ and $f \wedge g$ are measurable functions.

Proof.   To prove (1), notice that

$$\{x \in E : f(x) > g(x)\} = \cup_{r \in \mathbb{Q}} \{x \in E : f(x) > r > g(x)\}$$
$$= \cup_{r \in \mathbb{Q}} \{x \in E : f(x) > r\} \cap \{x \in E : r > g(x)\}.$$

Since each of these sets is measurable, $\{x \in E : f(x) > g(x)\}$ is a measurable set.

Consider (2). Fix $\alpha \in \mathbb{R}$. Since

$$\{x \in E : f(x) + c > \alpha\} = \{x \in E : f(x) > \alpha - c\},$$

the function $f + c$ is measurable. If $c \neq 0$, then

$$\{x \in E : cf(x) > \alpha\} = \begin{cases} \{x \in E : f(x) > \frac{\alpha}{c}\} \text{ if } c > 0 \\ \{x \in E : f(x) < \frac{\alpha}{c}\} \text{ if } c < 0 \end{cases}.$$

If $c = 0$, then

$$\{x \in E : cf(x) > \alpha\} = \begin{cases} E \text{ if } \alpha < 0 \\ \emptyset \text{ if } \alpha \geq 0 \end{cases}.$$

Thus, cf is measurable function.

Note that

$$\{x \in E : f(x) + g(x) > \alpha\} = \{x \in E : f(x) > \alpha - g(x)\}.$$

Since g is measurable, $\alpha - g$ is a measurable function by (2) and so, by (1), $f + g$ is measurable.

To see that fg is measurable, note that for $\alpha < 0$,

$$\{x \in E : f^2(x) > \alpha\} = E,$$

while for $\alpha \geq 0$,

$$\{x \in E : f^2(x) > \alpha\} = \{x \in E : f(x) > \sqrt{\alpha}\} \cup \{x \in E : f(x) < -\sqrt{\alpha}\}.$$

Since all of these sets are measurable, f^2 is measurable. Writing

$$fg = \frac{(f+g)^2 - (f-g)^2}{4},$$

we see that fg is a measurable function.

Finally, since

$$\{x \in E : (f \vee g)(x) > \alpha\} = \{x \in E : f(x) > \alpha\} \cup \{x \in E : (g)(x) > \alpha\}$$

and

$$\{x \in E : (f \wedge g)(x) > \alpha\} = \{x \in E : f(x) > \alpha\} \cap \{x \in E : (g)(x) > \alpha\},$$

it follows that $f \vee g$ and $f \wedge g$ are measurable. $\square$

Consequently, we get the following result.

Corollary 3.55 *Let $E \in \mathcal{M}_n$ and $f : E \to \mathbb{R}^*$. Then, f is measurable if, and only if, f^+ and f^- are measurable functions. If f is measurable, then $|f|$ is measurable.*

The converse to the last statement is false. See Exercise 3.21.

As in the case $n = 1$, a statement about the points of a measurable set E is said to hold *almost everywhere* in E if the set of points in E for which the statement fails to hold has Lebesgue measure 0. Additionally, we use phrases like "almost every x" or "almost all x" to mean that a property holds almost everywhere in the set being considered.

Proposition 3.56 *Let $E \in \mathcal{M}_n$ and $f, g : E \to \mathbb{R}^*$. Suppose that f is measurable and $f = g$ a.e.. Then, g is measurable.*

Proof. Let $Z = \{x \in E : f(x) \neq g(x)\}$. Then, Z is measurable and $m_n(Z) = 0$. Fix $\alpha \in \mathbb{R}$. Then,

$$\{x \in E : g(x) > \alpha\} = \{x \in E \setminus Z : g(x) > \alpha\} \cup \{x \in Z : g(x) > \alpha\}$$
$$= \{x \in E \setminus Z : f(x) > \alpha\} \cup \{x \in Z : g(x) > \alpha\}.$$

Since Z has measure 0, all of its subsets are measurable. Thus, the measurability of f and the equality $\{x \in E \setminus Z : f(x) > \alpha\} = \{x \in E : f(x) > \alpha\} \setminus \{x \in Z : f(x) > \alpha\}$ imply that $\{x \in E : g(x) > \alpha\} \in \mathcal{M}_n$. $\square$

We next investigate limits of measurable functions. To do this, we first define some special limits. Given a sequence $\{x_i\}_{i=1}^{\infty} \subset \mathbb{R}$, we define the *limit superior* and the *limit inferior* of $\{x_i\}_{i=1}^{\infty}$ by:

$$\limsup_i x_i = \inf_i \left\{ \sup_{k \geq i} x_k \right\} = \lim_{i \to \infty} \left\{ \sup_{k \geq i} x_k \right\}$$

and

$$\liminf_i x_i = \sup_i \left\{ \inf_{k \geq i} x_k \right\} = \lim_{i \to \infty} \left\{ \inf_{k \geq i} x_k \right\}.$$

We always have that $-\infty \leq \liminf_i x_i \leq \limsup_i x_i \leq +\infty$. When they are finite, $\liminf_i x_i$ is the smallest accumulation point of $\{x_i\}_{i=1}^{\infty}$ and $\limsup_i x_i$ is the largest accumulation point. Further, by Exercise 3.22, $\lim_{i \to \infty} x_i$ exists if, and only if, $\limsup_i x_i$ and $\liminf_i x_i$ are equal, in which case $\lim_{i \to \infty} x_i$ equals their common value.

Example 3.57   The sequence $\left\{ (-1)^i \right\}_{i=1}^{\infty}$ satisfies $\limsup_i (-1)^i = 1$ and $\liminf_i (-1)^i = -1$. Thus, the sequence does not have a limit.

We now consider limits of sequences of measurable functions.

Theorem 3.58   *Let $E \in \mathcal{M}_n$ and suppose $f_k : E \to \mathbb{R}^*$ is a measurable function for all $k \in \mathbb{N}$. Then $\sup_k f_k$, $\inf_k f_k$, $\limsup_k f_k$ and $\liminf_k f_k$ are measurable functions. If $\lim f_k$ exists a.e., then it is measurable.*

Proof.   Fix $\alpha \in \mathbb{R}$. Note that

$$\left\{ x \in E : \sup_k f_k(x) > \alpha \right\} = \cup_{k=1}^{\infty} \left\{ x \in E : f_k(x) > \alpha \right\},$$

which implies that $\sup_k f_k$ is measurable. Next, the equality $\inf_k f_k = -\sup(-f_k)$ proves that $\inf_k f_k$ is measurable. By definition, $\limsup_k f_k = \inf_k \sup_{j \geq k} f_j$ and $\liminf_k f_k = \sup_k \inf_{j \geq k} f_j$, which shows that $\limsup_k f_k$ and $\liminf_k f_k$ are measurable. Finally, if $\lim_k f_k$ exists a.e., then it equals the $\limsup f_k$ a.e. and, consequently, is measurable. $\square$

The following result, which is due to D. F. Egoroff (1869-1931), shows that when a sequence of measurable functions converges, it almost converges uniformly; that is, the sequence converges uniformly except on a set of small measure.

Theorem 3.59   *(Egoroff's Theorem) Let $m_n(E) < \infty$. Suppose that $f_k : E \to \mathbb{R}^*$ is a measurable function for each k, $\lim_{k \to \infty} f_k(x) = f(x)$*

a.e. on E and f is finite valued a.e. on E. Then, given any $\epsilon > 0$, there is a measurable set $F \subset E$ such that $m_n (E \setminus F) < \epsilon$ and $\{f_k\}_{k=1}^{\infty}$ converges uniformly to f on F.

Proof. The function f is measurable since it is the pointwise limit a.e. of a sequence of measurable functions. For all $m, i \in \mathbb{N}$, set

$$E_{mi} = \cap_{k=i}^{\infty} \left\{ x \in E : |f_k (x) - f (x)| < \frac{1}{m} \right\}$$

and

$$H = \left\{ x \in E : \lim_{k \to \infty} f_k (x) = f (x) \right\}.$$

Then, E_{mi} and H are measurable sets and, for all m, $H \subset \cup_{i=1}^{\infty} E_{mi}$. Fix m. Since $E_{mi} \subset E_{m(i+1)}$, by Proposition 3.34

$$\lim_{i \to \infty} m_n (E_{mi}) = m_n (\cup_{i=1}^{\infty} E_{mi}) \geq m_n (H) = m_n (E),$$

and, since E has finite measure, $\lim_{i \to \infty} m_n (E \setminus E_{mi}) = 0$.

Therefore, given $\epsilon > 0$, for each m there is an i_m such that $m_n (E \setminus E_{mi_m}) < \epsilon 2^{-m}$. Set $F = \cap_{m=1}^{\infty} E_{mi_m}$, so that F is measurable and

$$m_n (E \setminus F) \leq \sum_{m=1}^{\infty} m_n (E \setminus E_{mi_m}) < \sum_{m=1}^{\infty} \epsilon 2^{-m} = \epsilon.$$

Finally, given $\eta > 0$, choose m so that $\frac{1}{m} < \eta$. If $k \geq i_m$ and $x \in F \subset E_{mi_m}$, then by the definition of E_{mi_m}, $|f_k (x) - f (x)| < \eta$. Therefore, $\{f_k\}_{k=1}^{\infty}$ converges uniformly to f on F. $\square$

The next two examples show that we cannot relax the conditions that E have finite measure or f be finite valued.

Example 3.60 Let $E = \mathbb{R}^n$ and let f_k be the characteristic function of the ball centered at the origin and having radius k. Then, $f_k (x) \to 1$ for all x but the convergence is not uniform on sets whose complements have finite measure. Thus, we need E to have finite measure.

Example 3.61 Let $E = [0, 1]$ and $f_k (x) \equiv k$. Then, f_k converges to the function which is identically ∞ on $[0, 1]$, so the convergence cannot be uniform. We need f to be finite valued a.e..

The British mathematician J. E. Littlewood (1885-1977) summed up how nice Lebesgue measurable functions and sets are with his "three principles":

(1) Every measurable set is nearly a finite union of intervals.
(2) Every measurable function is nearly continuous.
(3) Every convergent sequence of measurable functions is nearly uniformly convergent.

The third principle is Egoroff's Theorem. The first principle follows from condition (2) of Theorem 3.36. Given $E \in \mathcal{M}_n$ and $\epsilon > 0$, there is an open set G containing E such that $m_n (G \setminus E) < \epsilon$. By Lemma 3.44, $G = \cup_{i \in \sigma} B_i$, a countable union of disjoint bricks. If σ is finite, we can approximate E by the union of all the bricks. If σ is infinite, since $m_n (G) = \lim_{k \to \infty} m_n \left(\cup_{i=1}^k B_i \right)$, we can approximate E by a finite set of the B_i's (at least when the measure of E is finite). Finally, since the surface of a brick has measure 0, replacing B_i by the largest open interval contained inside of B_i, which has the same measure as B_i, we can approximate E by a finite union of open intervals. We now turn our attention the second condition.

Let f be a nonnegative and measurable function on $E \subset \mathbb{R}^n$. We can define a sequence of simple functions that converges pointwise to f. To see this, for $k \in \mathbb{N}$, define measurable sets A_i^k and A_i^∞ by

$$A_i^k = \left\{ x \in E : \frac{i-1}{2^k} \le f(x) < \frac{i}{2^k} \right\} \text{ and } A_\infty^k = \{ x \in E : k \le f(x) \}.$$

Then, the function

$$f_k(x) = \sum_{i=1}^{k2^k} \frac{i-1}{2^k} \chi_{A_i^k}(x) + k \chi_{A_\infty^k}(x)$$

is nonnegative and simple and $\{f_k\}_{k=1}^\infty$ increases monotonically to f for all $x \in E$. Further, if f is bounded then, once k is greater than the bound on $|f|$, $|f_k(x) - f(x)| < \frac{1}{2^k}$ for all $x \in E$. Thus, we have proved

Theorem 3.62 *Let $E \in \mathcal{M}_n$ and suppose $f : E \to \mathbb{R}^*$ is nonnegative and measurable. There is a sequence of nonnegative, simple functions $\{f_k\}_{k=1}^\infty$ which increases to f pointwise on E. If f is bounded, then the convergence is uniform on E.*

If f is a measurable function on E, then $f = f^+ - f^-$ and both f^+ and f^- are nonnegative and measurable. Applying the theorem to each function separately, we get the following corollary.

Corollary 3.63 *Let $E \in \mathcal{M}_n$ and suppose $f : E \to \mathbb{R}^*$ is measurable. There is a sequence of simple functions $\{f_k\}_{k=1}^{\infty}$ which converges to f point-wise on E. If f is bounded, then the convergence is uniform on E.*

Littlewood's second principle is contained in the following theorem of N. N. Lusin (1883-1950).

Theorem 3.64 *(Lusin's Theorem) Let $E \in \mathcal{M}_n$ and suppose $f : E \to \mathbb{R}^*$ is measurable and finite valued almost everywhere. Given $\epsilon > 0$, there is a closed set $F \subset E$ such that $m_n (E \setminus F) < \epsilon$ and $f|_F$, the restriction of f to F, is continuous.*

Proof. Assume that f is a simple function with canonical form $f(x) = \sum_{i=1}^{m} a_i \chi_{E_i}(x)$, where the a_i's are distinct, the E_i's are measurable and pairwise disjoint, and $E = \cup_{i=1}^{m} E_i$. (If $f(x) = 0$ for some x, then $a_j = 0$ for some j.) Fix $\epsilon > 0$. By Theorem 3.36, there are closed sets $F_i \subset E_i$ such that $m_n (E_i \setminus F_i) < \dfrac{\epsilon}{m}$. Set $F = \cup_{i=1}^{m} F_i$. Then, F is a closed set, and since the sets F_i are pairwise disjoint and f is constant on each of these sets, $f|_F$ is continuous. Since $E = \cup_{i=1}^{m} E_i$, we have $E \setminus F \subset \cup_{i=1}^{m} (E_i \setminus F) \subset \cup_{i=1}^{m} (E_i \setminus F_i)$ which implies that

$$m_n (E \setminus F) \leq m_n (\cup_{i=1}^{m} (E_i \setminus F_i)) = \sum_{i=1}^{m} m_n (E_i \setminus F_i) < \epsilon.$$

Next, suppose that f is measurable and $m_n (E) < \infty$. Choose a sequence of simple functions $\{f_k\}_{k=1}^{\infty}$ that converges pointwise to f. Choose closed sets $F_k \subset E$ such that $m_n (E \setminus F_k) < \epsilon 2^{-(k+1)}$ and $f_k|_{F_k}$ is continuous. By Egoroff's Theorem and Theorem 3.36, there is a closed set $F_0 \subset E$ such that $m_n (E \setminus F_0) < \epsilon 2^{-1}$ and $\{f_k\}_{k=1}^{\infty}$ converges uniformly to f on F_0. Set $F = \cap_{k=0}^{\infty} F_k$. Then $f_k|_F$ is continuous and $\{f_k\}_{k=1}^{\infty}$ converges uniformly to f on F. Since a uniform limit of continuous functions is continuous, $f|_F$ is continuous. Further,

$$m_n (E \setminus F) < \sum_{k=0}^{\infty} \epsilon 2^{-(k+1)} = \epsilon.$$

Finally, suppose $m_n (E) = \infty$. Let $A_j = \{x \in \mathbb{R}^n : j - 1 \leq \|x\| < j\}$ and write $E = \cup_{j=1}^{\infty} E \cap A_j$. Since $m_n (A_j) < \infty$, there is a closed set

$F_j \subset E \cap A_j$ such that $f|_{F_j}$ is continuous and $m_n (E \cap A_j \setminus F_j) < \epsilon 2^{-j}$. Set $F = \cup_{j=1}^{\infty} F_j$. Note that by construction F_j and F_l are at a positive distance for $j \neq l$. Thus, F is closed, $m_n (E \setminus F) < \epsilon$ and $f|_F$ is continuous. $\square$

Remark 3.65 *The conclusion is not that f is continuous on F but that the restriction of f to F is continuous. See the next example.*

Example 3.66 Let f be the Dirichlet function defined on all of $\mathbb{R}$. Let G be an open set containing $\mathbb{Q}$ with $m(G) < \epsilon$. Set $F = G^c$. Then, $m(\mathbb{R} \setminus F) = m(G) < \epsilon$ and since $f|_F \equiv 0$, $f|_F$ is continuous. However, when considered as a function on $\mathbb{R}$, f is not continuous on F.

In the one-dimensional case, a step function is a finite-valued function which is constant on a finite number of open intervals of finite length. We can define a step function on the entire real line by setting it equal to 0 on the complement of the union of these open intervals. We extend this idea to higher dimensions by calling φ a *step function* if there are finite sets of pairwise disjoint bricks, $\{B_i\}_{i=1}^{m}$, and scalars $\{a_i\}_{i=1}^{m}$ such that $\varphi(x) = a_i$ for $x \in B_i$ and $\varphi(x) = 0$ for $x \notin \cup_{i=1}^{m} B_i$. We now show that a measurable function defined on a set of finite measure can be approximated by a sequence of step functions.

Theorem 3.67 *Let $E \in \mathcal{M}_n$ and suppose $f : E \to \mathbb{R}^*$ is measurable. Then, there is a sequence of step functions $\{\varphi_k\}_{k=1}^{\infty}$ that converges to f a.e. in E. Moreover, if $|f(x)| \leq M$ for all $x \in E$, then $|\varphi_k(x)| \leq M$ for all $x \in E$ and $k \in \mathbb{N}$.*

Proof. Suppose, first, that $m(E) < \infty$ and f is bounded. Let M be the bound on f and suppose $k \geq M$. Let f_k be the simple function

$$f_k(x) = \sum_{i=1}^{k2^k} \frac{i-1}{2^k} \chi_{A_i^k}(x)$$

where $A_i^k = \left\{ x \in E : \frac{i-1}{2^k} \leq f(x) < \frac{i}{2^k} \right\}$. By construction, we have $|f_k(x) - f(x)| < 2^{-k}$ for all x. Since each A_i^k is a measurable set, there is an open set $H_i^k \supset A_i^k$ such that $m_n (H_i^k \setminus A_i^k) < \frac{1}{k2^k} 2^{-k}$. By Lemma 3.44, $H_i^k = \cup_{j \in \sigma_{k,i}} B_j^{k,i}$, where $\left\{ B_j^{k,i} \right\}_{j \in \sigma_{k,i}}$ is a countable union of disjoint bricks. If $\sigma_{k,i}$ is finite, we set $G_i^k = \cup_{j=1}^{l_{k,i}} B_j^{k,i}$, where $l_{k,i}$ equals the number of bricks in $\sigma_{k,i}$. If $\sigma_{k,i}$ is infinite, since $m_n (H_i^k) < m_n (A_i^k) + \frac{1}{k2^k} 2^{-k} \leq m_n (E) + \frac{1}{k2^k} 2^{-k} < \infty$, we can choose $l_{k,i}$ such that

$m_n \left(\cup_{j=l_{k,i}+1}^\infty B_j^{k,i} \right) < \frac{1}{k2^k} 2^{-k}$. Set $G_i^k = \cup_{j=1}^{l_{k,i}} B_j^{k,i}$. Then,

$$m_n \left(G_i^k \Delta A_i^k \right) \leq m_n \left(H_i^k \setminus A_i^k \right) + m_n \left(\cup_{j=l_{k,i}+1}^\infty B_j^{k,i} \right) < \frac{2}{k2^k} 2^{-k}.$$

Set

$$\varphi_k(x) = \sum_{i=1}^{k2^k} \frac{i-1}{2^k} \chi_{G_i^k}(x) = \sum_{i=1}^{k2^k} \sum_{j=1}^{l_{k,i}} \frac{i-1}{2^k} \chi_{B_j^{k,i}}(x),$$

so that φ_k is a step function and $\varphi_k(x) = f_k(x)$ for all $x \notin \cup_{i=1}^{k2^k} \left(G_i^k \Delta A_i^k \right)$. Further,

$$m_n \left(\cup_{i=1}^{k2^k} \left(G_i^k \Delta A_i^k \right) \right) \leq \sum_{i=1}^{k2^k} m_n \left(G_i^k \Delta A_i^k \right) < \sum_{i=1}^{k2^k} \frac{2}{k2^k} 2^{-k} = 2^{-k+1}.$$

We now show that $\varphi_k \to f$ a.e.. Let $F_k = \left\{ x : |\varphi_k(x) - f(x)| \geq 2^{-k} \right\}$. Then, $F_k \subset \cup_{i=1}^{k2^k} \left(G_i^k \Delta A_i^k \right)$ so that $m_n(F_k) \leq 2^{-k+1}$. If $x \notin \cup_{k=m}^\infty F_k$, then $|\varphi_k(x) - f(x)| < 2^{-k}$ for all $k > m$ so that $\varphi_k(x) \to f(x)$. Consequently, if $x \notin \cap_{m=1}^\infty \cup_{k=m}^\infty F_k$, then $\varphi_k(x) \to f(x)$. Finally, since

$$m_n \left(\cap_{m=1}^\infty \cup_{k=m}^\infty F_k \right) \leq m_n \left(\cup_{k=m}^\infty F_k \right)$$
$$\leq \sum_{k=m}^\infty m_n(F_k) < \sum_{k=m}^\infty 2^{-k+1} = 2^{-m+2}$$

for all m, $m_n \left(\cap_{m=1}^\infty \cup_{k=m}^\infty F_k \right) = 0$ and $\{\varphi_k\}_{k=1}^\infty$ converges to f a.e.. By construction, $|\varphi_k(x)| \leq M$ for all $x \in E$ and $k \in \mathbb{N}$.

Now, suppose that f is a measurable function defined on a measurable set E. Let I_N be the interval in $\mathbb{R}^n$ that is the n-fold product of the interval $[-N, N]$. Set $E_N = E \cap I_N$ and define f_N by

$$f_N(x) = \begin{cases} N & \text{if } x \in E_N \text{ and } f(x) \geq N \\ f(x) & \text{if } x \in E_N \text{ and } |f(x)| < N \\ -N & \text{if } x \in E_N \text{ and } f(x) \leq -N \\ 0 & \text{if } \qquad x \notin E_N \end{cases}.$$

Note that $E = \cup_{N=1}^\infty E_N$, $E_N \subset E_{N+1}$, $m(E_N) < \infty$, f_N is bounded on E_N and $\{f_N\}_{N=1}^\infty$ converges to f for all $x \in E$. By the previous part of the proof, there is a step function φ_N, supported in E_N, and a set $F_N \subset E_N$ such that $|\varphi_N(x) - f_N(x)| < 2^{-N}$ for all $x \in E_N \setminus F_N$ and

$m_n(F_N) \leq 2^{-(N+1)}$. We claim that for each fixed K, $\{\varphi_N\}_{N=1}^{\infty}$ converges to f on E_K except for a set of measure 0. For, if that were true, then

$$m_n(\{x \in E : \varphi_N(x) \nrightarrow f(x)\}) \leq \sum_{K=1}^{\infty} m_n(\{x \in E_K : \varphi_N(x) \nrightarrow f(x)\}) = 0$$

and $\{\varphi_N\}_{N=1}^{\infty}$ converges to f a.e. in E.

So, fix K and argue as in the previous part. Set $Z_K = \cap_{M=K}^{\infty} \cup_{N=M}^{\infty} F_N$. Since

$$m_n(Z_K) \leq m_n(\cup_{N=M}^{\infty} F_N) \leq \sum_{N=M}^{\infty} m_n(F_N) < \sum_{N=M}^{\infty} 2^{-N+1} = 2^{-M+2},$$

$m_n(Z_K) = 0$. It remains to show that $\varphi_N(x) \to f(x)$ for all $x \notin Z_K$.

If $M \geq K$ and $x \notin \cup_{N=M}^{\infty} F_N$, then $|\varphi_N(x) - f_N(x)| < 2^{-N}$ for all $N > M$ so that $\varphi_N(x) - f_N(x) \to 0$. Consequently, if $x \notin \cap_{M=K}^{\infty} \cup_{N=M}^{\infty} F_N$, then $\varphi_N(x) - f_N(x) \to 0$. If $|f(x)| < \infty$, then $f_N(x) = f(x)$ for all sufficiently large N and $\varphi_N(x) \to f(x)$; if $|f(x)| = \infty$, then $\{\varphi_N(x)\}_{N=1}^{\infty}$ tends to ∞ (or $-\infty$) so that $\varphi_N(x) \to f(x)$. This completes the proof of the proposition. $\qquad\square$

3.5   Lebesgue integral

Lebesgue's descriptive definition of the integral led, in a very natural way, to the development of the Lebesgue measure of sets in $\mathbb{R}^n$ and, via limits of simple functions, to a study of measurable functions. If f is a step function (on $\mathbb{R}$), $f(x) = \sum_{i=1}^{k} a_i \chi_{I_i}(x)$ where the I_i's are pairwise disjoint intervals, then by using properties (1), (2), (3) and (5) of Lebesgue's descriptive definition, we see

$$\int_{\mathbb{R}} f(x)\, dx = \sum_{i=1}^{k} a_i \ell(I_i) = \sum_{i=1}^{k} a_i m(I_i),$$

as long as $\ell(I_i) < \infty$ for all i. This equality will guide our definition of the Lebesgue integral.

Recall that a simple function φ takes on a finite number of distinct nonzero values $a_1, a_2, \ldots, a_k$. If $A_i = \{x : \varphi(x) = a_i\}$, then φ has the canonical form

$$\varphi(x) = \sum_{i=1}^{k} a_i \chi_{A_i}(x).$$

Definition 3.68   Let $\varphi : \mathbb{R}^n \to \mathbb{R}$ be a nonnegative, simple function with canonical form $\sum_{i=1}^{k} a_i \chi_{A_i} (x)$. We define the *Lebesgue integral* of φ to be

$$\int \varphi = \int \varphi (x) \, dx = \int_{\mathbb{R}^n} \varphi (x) \, dx = \sum_{i=1}^{k} a_i m_n (A_i).$$

If $E \in \mathcal{M}_n$, set

$$\int_E \varphi = \int_E \varphi (x) \, dx = \int \chi_E (x) \varphi (x) \, dx.$$

Since $\chi_E \chi_{A_i} = \chi_{E \cap A_i}$, we see that $\int_E \varphi = \sum_{i=1}^{k} a_i m_n (A_i \cap E)$.

Remark 3.69   *For the remainder of this chapter, we will use $\int$ to denote the Lebesgue integral.*

The definition of the Lebesgue integral of a simple function is independent of its representation.

Proposition 3.70   *Let $\varphi (x) = \sum_{j=1}^{m} b_j \chi_{F_j} (x)$ where the sets F_j are pairwise disjoint measurable sets. Then,*

$$\int \varphi = \sum_{j=1}^{m} b_j m_n (F_j).$$

Proof.   Let $\sum_{i=1}^{k} a_i \chi_{A_i} (x)$ be the canonical form of φ. Then, $A_i = \cup_{b_j = a_i} F_j$, and $\sum_{b_j = a_i} m_n (F_j) = m_n (A_i)$. Thus,

$$\int \varphi = \sum_{i=1}^{k} a_i m_n (A_i) = \sum_{i=1}^{k} a_i \sum_{b_j = a_i} m_n (F_j) = \sum_{j=1}^{m} b_j m_n (F_j),$$

as we wished to show.   □

The next result collects some of the basic properties of the Lebesgue integral of nonnegative, simple functions.

Proposition 3.71   *Let φ and ψ be nonnegative, simple functions and $\alpha \geq 0$. Then,*

(1) $\int \alpha \varphi = \alpha \int \varphi$;
(2) $\int \varphi \geq 0$;
(3) $\int (\varphi + \psi) = \int \varphi + \int \psi$;
(4) The mapping $\Phi : \mathcal{M}_n \to [0, \infty]$ defined by $\Phi (E) = \int_E \varphi$ is countably additive.

Proof.   Let $\varphi(x) = \sum_{i=1}^{k} a_i \chi_{A_i}(x)$ and $\psi(x) = \sum_{j=1}^{m} b_j \chi_{B_j}(x)$ be the canonical forms of φ and ψ. To prove (1), since $\alpha\varphi(x) = \sum_{i=1}^{k} \alpha a_i \chi_{A_i}(x)$,

$$\int \alpha\varphi = \sum_{i=1}^{k} \alpha a_i m_n(A_i) = \alpha \sum_{i=1}^{k} a_i m_n(A_i) = \alpha \int \varphi.$$

To prove (2), we need only note that since φ is nonnegative, $a_i \geq 0$ for all i, so that

$$\int \varphi = \sum_{i=1}^{k} a_i m_n(A_i) \geq 0.$$

For (3), we set $E_{ij} = A_i \cap B_j$ for $1 \leq i \leq k$ and $1 \leq j \leq m$. Let $S = \{(i,j) : 1 \leq i \leq k, 1 \leq j \leq m\}$. Then, $\varphi(x) = \sum_{(i,j) \in S} a_i \chi_{E_{ij}}(x)$ and $\psi(x) = \sum_{(i,j) \in S} b_j \chi_{E_{ij}}(x)$. By the proposition above,

$$\int (\varphi + \psi) = \sum_{(i,j) \in S} (a_i + b_j) m_n(E_{ij})$$

$$= \sum_{(i,j) \in S} a_i m_n(E_{ij}) + \sum_{(i,j) \in S} b_j m_n(E_{ij}) = \int \varphi + \int \psi.$$

Finally, to prove (4), let $\{E_j\}_{j \in \sigma} \subset \mathcal{M}_n$ be a countable collection of pairwise disjoint sets. Thus,

$$\Phi(\cup_{j \in \sigma} E_j) = \int_{\cup_{j \in \sigma} E_j} \varphi = \sum_{i=1}^{k} a_i m_n(A_i \cap \cup_{j \in \sigma} E_j)$$

$$= \sum_{i=1}^{k} a_i \sum_{j \in \sigma} m_n(A_i \cap E_j)$$

$$= \sum_{j \in \sigma} \sum_{i=1}^{k} a_i m_n(A_i \cap E_j)$$

$$= \sum_{j \in \sigma} \int_{E_i} \varphi = \sum_{j \in \sigma} \Phi(E_j).$$

$\square$

Applying part (2) to the function $\psi - \varphi$, we get the following corollary.

Corollary 3.72   *Let φ and ψ be nonnegative, simple functions. If $\varphi \leq \psi$, then $\int \varphi \leq \int \psi$.*

In fact, it is only necessary that $\varphi \leq \psi$ except on a null set, as we will discuss below.

Suppose φ is a nonnegative, simple function on $\mathbb{R}^n$. Then

$$\Phi(\emptyset) = \sum_{i=1}^{k} a_i m_n (A_i \cap \emptyset) = 0.$$

Since Φ is countably additive, we see

Corollary 3.73 *If φ is a nonnegative, simple function on $\mathbb{R}^n$, then Φ : $\mathcal{M}_n \to [0, \infty]$ defined by $\Phi(E) = \int_E \varphi$ is a measure on $\mathcal{M}_n$.*

Using simple functions, we extend the definition of the Lebesgue integral to nonnegative, measurable functions.

Definition 3.74 Let $E \in \mathcal{M}_n$ and $f : E \to \mathbb{R}$ be nonnegative and measurable. Define the *Lebesgue integral of f over E* by

$$\int_E f = \int_E f(x)\, dx = \sup \left\{ \int_E \varphi : 0 \leq \varphi \leq f \text{ and } \varphi \text{ is simple} \right\}. \quad (3.3)$$

If A is a measurable subset of E, we define

$$\int_A f = \int_A f(x)\, dx = \int_E \chi_A(x) f(x)\, dx.$$

Remark 3.75 *Equation (3.3) is analogous to a "lower integral". Since we are considering functions which may be unbounded, there may be no simple functions that dominate f, so it would then be impossible to define an "upper integral". However, even for bounded functions, it is not necessary to compare upper and lower integrals. This is pointed out in Proposition 3.102 after we have developed some of the basic properties of the Lebesgue integral.*

The next result shows that the Lebesgue integral is a positive operator on nonnegative measurable functions.

Proposition 3.76 *Let $E \in \mathcal{M}_n$ and $f, h : E \to \mathbb{R}$ be nonnegative and measurable and $\alpha \geq 0$. Then,*

(1) If $h \leq f$, then $\int_E h \leq \int_E f$;
(2) If $0 \leq f$, then $0 \leq \int_E f$;
(3) $\int_E \alpha f = \alpha \int_E f$.

Proof.   To prove (1), note that if $\varphi \leq h$ then $\varphi \leq f$, so the Lebesgue integral of f is the supremum over a bigger set. Setting $h \equiv 0$ in (1) proves (2). For (3), note first that if $\alpha = 0$, then $\alpha f = 0$ and by our convention that $0 \cdot \infty = 0$,

$$\int_E \alpha f = \int_E 0 dx = 0 = \alpha \int_E f.$$

If $\alpha > 0$, we see that if φ is a simple function and $0 \leq \varphi \leq f$, then $\alpha \varphi$ is a simple function and $0 \leq \alpha \varphi \leq \alpha f$. Further, if ψ is a simple function and $0 \leq \psi \leq \alpha f$, then $\frac{1}{\alpha}\psi$ simple function and $0 \leq \frac{1}{\alpha}\psi \leq f$. Thus

$$\int_E \alpha f = \sup\left\{\int_E \psi : 0 \leq \psi \leq \alpha f \text{ and } \psi \text{ is simple}\right\}$$

$$= \sup\left\{\int_E \alpha\left(\frac{1}{\alpha}\psi\right) : 0 \leq \frac{1}{\alpha}\psi \leq f \text{ and } \frac{1}{\alpha}\psi \text{ is simple}\right\}$$

$$= \sup\left\{\int_E \alpha\varphi : 0 \leq \varphi \leq f \text{ and } \varphi \text{ is simple}\right\}$$

$$= \alpha\sup\left\{\int_E \varphi : 0 \leq \varphi \leq f \text{ and } \varphi \text{ is simple}\right\} = \alpha\int_E f. \qquad \square$$

Note that (2) is the statement that the Lebesgue integral is a positive operator on nonnegative measurable functions

We now come to our first convergence theorem for the Lebesgue integral, the *Monotone Convergence Theorem*.

Theorem 3.77   *(Monotone Convergence Theorem) Let $E \in \mathcal{M}_n$ and $\{f_k\}_{k=1}^{\infty}$ be an increasing sequence of nonnegative, measurable functions defined on E. Set $f(x) = \lim_{k\to\infty} f_k(x)$. Then,*

$$\int_E f = \lim_{k\to\infty} \int_E f_k.$$

Proof.   Note first that f is nonnegative and measurable since it is a limit of measurable functions. Since $0 \leq f_k \leq f_{k+1} \leq f$, by the previous proposition, $\left\{\int_E f_k\right\}_{k=1}^{\infty}$ is a monotonic sequence and $\lim_k \int_E f_k \leq \int_E f$.

To prove the reverse inequality, fix $0 < a < 1$ and let φ be a simple function with $0 \leq \varphi \leq f$. Set $E_k = \{x \in E : f_k(x) \geq a\varphi(x)\}$. Since $f_k(x)$ increases to $f(x)$ pointwise, it follows that $E_k \subset E_{k+1}$ for all k and $E = \cup_{k=1}^{\infty} E_k$. Thus,

$$\int_E f_k \geq \int_{E_k} f_k \geq a\int_{E_k} \varphi.$$

By part (4) of Proposition 3.71, $\Phi(E) = \int_E \varphi$ defines a measure, so by Proposition 3.34,

$$a \int_E \varphi = a \lim_{k \to \infty} \int_{E_k} \varphi \leq \lim_{k \to \infty} \int_E f_k.$$

If we let $a \to 1$, we see that $\lim_k \int_E f_k \geq \int_E \varphi$. Since this is true for all simple functions $\varphi \leq f$, we get

$$\lim_{k \to \infty} \int_E f_k \geq \int_E f,$$

which completes the proof. $\square$

Suppose that f and g are nonnegative and measurable. By Theorem 3.62, there are sequences of nonnegative, simple functions $\{\varphi_i\}_{i=1}^{\infty}$ and $\{\psi_i\}_{i=1}^{\infty}$ which increase to f and g, respectively. Thus, $0 \leq \varphi_i + \psi_i$ and $\{\varphi_i + \psi_i\}_{i=1}^{\infty}$ increases to $f + g$. By Proposition 3.71,

$$\int_E (\varphi_i + \psi_i) = \int_E \varphi_i + \int_E \psi_i,$$

so by the Monotone Convergence Theorem,

$$\int_E (f + g) = \int_E f + \int_E g.$$

Thus, we see that the Lebesgue integral is linear when restricted to nonnegative, measurable functions.

Using this result, we can easily show that the Lebesgue integral is countably additive.

Corollary 3.78 *Let $E \in \mathcal{M}_n$ and $\{f_k\}_{k=1}^{\infty}$ be a sequence of nonnegative, measurable functions defined on E. Then,*

$$\int_E \sum_{k=1}^{\infty} f_k = \sum_{k=1}^{\infty} \int_E f_k.$$

Proof. The proof is almost done. We use linearity and induction to show that

$$\int_E \sum_{k=1}^{N} f_k = \sum_{k=1}^{N} \int_E f_k$$

for all $N \in \mathbb{N}$. Since all the functions are nonnegative, we can apply the Monotone Convergence Theorem to complete the proof. $\square$

In fact, this corollary is equivalent to the Monotone Convergence Theorem. See Exercise 3.42.

We saw above that the Lebesgue integral of a nonnegative, simple function defines a measure. The same is true for all nonnegative, measurable functions. This will follow from the next two results.

Proposition 3.79   *Let f be a nonnegative, measurable function on $\mathbb{R}^n$. Then, the mapping $\Phi : \mathcal{M}_n \to \mathbb{R}^*$ defined by $\Phi(E) = \int_E f$ is countably additive.*

Proof.   Pick a sequence of nonnegative simple functions $\{\varphi_k\}_{k=1}^{\infty}$ that increases to f. By the Monotone Convergence Theorem, $\int_E \varphi_k \to \int_E f$. Suppose $\{E_j\}_{j \in \sigma}$ is a countable collection of pairwise disjoint measurable sets and $E = \cup_{j \in \sigma} E_j$. By part (4) of Proposition 3.71 and Exercise 3.3,

$$\int_E f = \lim_{k \to \infty} \int_E \varphi_k = \lim_{k \to \infty} \sum_{j \in \sigma} \int_{E_j} \varphi_k = \sum_{j \in \sigma} \lim_{k \to \infty} \int_{E_j} \varphi_k = \sum_{j \in \sigma} \int_{E_j} f. \qquad \square$$

Proposition 3.80   *Let $E \in \mathcal{M}_n$ and f be a nonnegative, measurable function on $\mathbb{R}^n$. Then, $\int_E f = 0$ if, and only if, $f = 0$ a.e. in E.*

Proof.   Suppose $f(x) = \sum_{i=1}^{k} a_i \chi_{A_i}(x)$ is simple function. If $f = 0$ a.e. in E and $a_i > 0$, then $m_n(A_i \cap E) = 0$. Thus, $\int_E f = 0$. For general, nonnegative functions f, the result follows by approximating f by simple functions. Thus, if $f = 0$ a.e. in E, then $\int_E f = 0$.

Now, suppose that $\int_E f = 0$. Set $A_k = \{x \in E : f(x) \geq \frac{1}{k}\}$, so that $A = \{x \in E : f(x) > 0\} = \cup_{k=1}^{\infty} A_k$. If $m_n(A) > 0$, then $m_n(A_k) > 0$ for some k which implies

$$\int_E f \geq \int_{A_k} f \geq \frac{1}{k} m_n(A_k) > 0.$$

This contradiction shows that $m_n(A) = 0$ and $f = 0$ a.e. in E. $\square$

Consequently, $\Phi(\emptyset) = \int_{\emptyset} f = 0$ and the Lebesgue integral of a nonnegative, measurable function defines a measure.

Remark 3.81   *The previous proof uses a very important inequality in analysis, know as Tchebyshev's inequality after P. L. Tchebyshev (1821-1894). Suppose that f is a nonnegative, measurable function on a measurable set E. Let $\lambda > 0$. Then, $\lambda \chi_{\{x \in E : f(x) > \lambda\}}(x) \leq f(x)$ for all $x \in E$. Thus,*

$$\lambda m_n(\{x \in E : f(x) > \lambda\}) = \int_E \lambda \chi_{\{x \in E : f(x) > \lambda\}}(x)\, dx \leq \int_E f,$$

from which we get Tchebyshev's inequality,

$$m_n \left(\{x \in E : f(x) > \lambda\} \right) \leq \frac{1}{\lambda} \int_E f.$$

Example 3.82 Let $A \in \mathcal{M}_n$ and set $f(x) = \chi_A(x)$. Then, the measure Φ defined by $\Phi(E) = \int_E \chi_A = m_n(A \cap E)$ is the *restriction* of Lebesgue measure m_n to A.

For a general measurable function, we can use the Lebesgue integrals of f^+ and f^- to define the Lebesgue integral of f, whenever we can make sense of their difference.

Definition 3.83 Let $E \in \mathcal{M}_n$ and f be a measurable function on $\mathbb{E}$. We say that f has a *Lebesgue integral* over E if at least one of $\int_E f^+$ and $\int_E f^-$ is finite and in this case we define the Lebesgue integral of f over E to be

$$\int_E f = \int_E f^+ - \int_E f^-.$$

We say that f is *Lebesgue integrable* over E if the Lebesgue integral of f over E exists and is finite.

Remark 3.84 *If f has a Lebesgue integral over E, then $\int_E f$ may equal $\pm\infty$. In order for f to be Lebesgue integrable, the integral must exist and be finite.*

Note that if φ is a simple function, then φ has a Lebesgue integral over E if, and only if, (at least) one of the sets $\{t \in E : \varphi(t) > 0\}$ and $\{t \in E : \varphi(t) < 0\}$ has finite measure. When this is the case and $\varphi(x) = \sum_{i=1}^k a_i \chi_{A_i}(x)$,

$$\int_E \varphi = \sum_{i=1}^k a_i m_n(E \cap A_i).$$

Further, φ is Lebesgue integrable over E if, and only if, $m_n(E \cap A_i) < \infty$ for all $i = 1, \ldots, k$.

An important consequence of Tchebyshev's inequality is that a Lebesgue integrable function is finite almost everywhere.

Proposition 3.85 *Let $E \in \mathcal{M}_n$ and f be a Lebesgue integrable function on E. Then,*

(1) for all $\alpha > 0$, the set $E_\alpha = \{t \in E : |f(t)| > \alpha\}$ has finite measure;
(2) f is finite valued a.e. in E.

Proof. By hypothesis, both f^+ and f^- are Lebesgue integrable. Fix $\alpha > 0$. We see that

$$E_\alpha = \{t \in E : f^+(t) > \alpha\} \cup \{t \in E : f^-(t) > \alpha\},$$

so it is enough to prove (1) for nonnegative functions f. Then, by Tchebyshev's inequality,

$$m_n(\{x \in E : f(x) > \alpha\}) \leq \frac{1}{\alpha} \int_E f < \infty.$$

To show the second part, it is, again, enough to show that f^+ and f^- are finite valued a.e. in E, so we assume that f is nonnegative. Since

$$\{t \in E : f(t) = \infty\} \subset \{t \in E : f(t) > \alpha\}$$

for all $\alpha > 0$,

$$m_n(\{t \in E : f(t) = \infty\}) \leq m_n(\{t \in E : f(t) > \alpha\}) \leq \frac{1}{\alpha} \int_E f,$$

so letting α tend to ∞ shows that $m_n(\{t \in E : f(t) = \infty\}) = 0$ and f is finite a.e. in E. $\square$

For Lebesgue integrable functions, we can get an improvement of the Monotone Convergence Theorem. See Exercise 3.43.

Corollary 3.86 *Let $E \in \mathcal{M}_n$ and $\{f_k\}_{k=1}^\infty$ be an increasing sequence of nonnegative, Lebesgue integrable functions defined on E. Set $f(x) = \lim_{k \to \infty} f_k(x)$. Then, f is Lebesgue integrable if, and only if, $\sup_k \int_E f_k < \infty$. In this case, f is finite a.e..*

Example 3.87 In Example 3.3, we defined a sequence of simple functions that increase pointwise to the Dirichlet function on $[0, 1]$. Since each f_k is Lebesgue integrable, the Monotone Convergence Theorem implies that the Dirichlet function is Lebesgue integrable with integral 0. Note that each f_k is also Riemann integrable while the limit function is not, which shows that the Riemann integral does not satisfy the Monotone Convergence Theorem.

Using the relationships between f, $|f|$, f^+ and f^-, we get the following result which shows that Lebesgue integrable functions are *absolutely integrable*.

Proposition 3.88 *Let $E \in \mathcal{M}_n$ and $f : E \to \mathbb{R}^*$ be measurable. Then, f is Lebesgue integrable over E if, and only if $|f|$ is Lebesgue integrable over E. In this case, $\left|\int_E f\right| \leq \int_E |f|$.*

Proof. If f is Lebesgue integrable, then both $\int_E f^+$ and $\int_E f^-$ are finite, so that $\int_E |f| = \int_E f^+ + \int_E f^- < \infty$ and $|f|$ is Lebesgue integrable. On the other hand, if $|f|$ is Lebesgue integrable, by the positivity of the integral and the pointwise inequalities $f^+ \le |f|$ and $f^- \le |f|$, the integrals $\int_E f^+$ and $\int_E f^-$ are finite, and so is $\int_E f$. Finally,

$$\left| \int_E f \right| = \left| \int_E f^+ - \int_E f^- \right| \le \int_E f^+ + \int_E f^- = \int_E (f^+ + f^-) = \int_E |f|$$

and the proof is complete. $\qquad\square$

The null sets, that is, sets of measure 0, play an important role in integration theory. The next few results examine some of the properties of null sets.

Proposition 3.89 *Suppose that $E \in \mathcal{M}_n$ and $f : E \to \mathbb{R}^*$ is measurable. If $m_n(E) = 0$, then f is Lebesgue integrable over E and $\int_E f = 0$.*

Proof. Since $0 = f = f^+ = f^-$ a.e. in E, $\int_E f^+ = \int_E f^- = 0$ and the result follows. $\qquad\square$

On the other hand, for general measurable functions, it is not enough to assume that $\int_E f = 0$ to derive $m_n(E) = 0$, or that $f = 0$ a.e. in E.

Example 3.90 Let $f : [-1, 1] \to \mathbb{R}$ be defined by $f(x) = \frac{x}{|x|}$ for $x \ne 0$. Then, $\int_{[-1,1]} f = 0$ while $f \ne 0$ a.e. in $[-1, 1]$ and $m([-1, 1]) \ne 0$.

However, if the Lebesgue integral of f is 0 over enough subsets of E, then it follows that $f = 0$ a.e. in E.

Proposition 3.91 *Suppose that f has a Lebesgue integral over E. If $\int_A f = 0$ for all measurable sets $A \subset E$, then $f = 0$ a.e. in E.*

Proof. Since $\int_A f^+ = \int_{A \cap \{x \in E : f(x) > 0\}} f$ and $\int_A f^- = \int_{A \cap \{x \in E : f(x) < 0\}} f$, we see that $\int_A f^+ = 0$ and $\int_A f^- = 0$ for all measurable subsets $A \subset E$. Thus, we may assume that f is nonnegative. As above, set $A_k = \{x \in E : f(x) \ge \frac{1}{k}\}$. Then, by Tchebyshev's inequality,

$$\frac{1}{k} m_n(A_k) \le \int_{A_k} f = 0,$$

so that $m_n(A_k) = 0$. Thus, $m_n(\{x \in E : f(x) > 0\}) \le \sum_{k=1}^{\infty} m_n(A_k) = 0$ which implies $f = 0$ a.e. in E. $\qquad\square$

If f is measurable, then $\int_E f = 0$ for every set E of measure 0. When f is Lebesgue integrable, this happens in a continuous way.

Theorem 3.92 *Let $E \in \mathcal{M}_n$ and $f : E \to \mathbb{R}^*$ be Lebesgue integrable. Then,*

$$\lim_{m_n(A) \to 0} \int_A f = 0,$$

where the limit is taken over measurable sets $A \subset E$.

Proof. Since $\left| \int_E f \right| \le \int_E |f|$ for Lebesgue integrable functions f, it is enough to prove the result for $f \ge 0$. Set $f_k(x) = \min \{ f(x), k \}$ so that $f_k \ge 0$ and $\{ f_k \}_{k=1}^{\infty}$ increases to f. By the Monotone Convergence Theorem,

$$\lim_{k \to \infty} \int_E f_k = \int_E f.$$

Fix $\epsilon > 0$ and choose k such that $\int_E (f - f_k) < \frac{\epsilon}{2}$. Fix δ, $0 < \delta < \frac{\epsilon}{2k}$. If $m_n(A) < \delta$, then

$$\int_A f \le \int_E (f - f_k) + \int_A f_k < \frac{\epsilon}{2} + k \frac{\epsilon}{2k} = \epsilon,$$

as we wished to show. □

Remark 3.93 *If $f \ge 0$ is Lebesgue integrable, then $\Phi(E) = \int_E f$ defines a measure. This theorem says that given any $\epsilon > 0$, there is a $\delta > 0$ so that if $m_n(E) < \delta$ then $\Phi(E) < \epsilon$. When this condition is satisfied, we say that the measure Φ is absolutely continuous with respect to m_n.*

If f and g are equal a.e. in E, then $f - g = 0$ a.e. in E so $\int_E (f - g) = 0$. When one of the functions is Lebesgue integrable, so is the other and their integrals are equal.

Proposition 3.94 *Suppose that $E \in \mathcal{M}_n$ and $f, g : E \to \mathbb{R}^*$ are measurable.*

(1) *If $|f| \le g$ a.e. in E and g is Lebesgue integrable over E, then f is Lebesgue integrable over E.*

(2) *If f is Lebesgue integrable over E and $f = g$ a.e. in E, then g is Lebesgue integrable over E and $\int_E f = \int_E g$.*

Proof. To prove (1), set $Z = \{x \in E : f(x) > g(x)\}$ and note that $m_n(Z) = 0$ so that $\int_Z |f| = \int_Z g = 0$. Since $|f|$ and g are nonnegative and measurable functions,

$$\int_E |f| = \int_{E\setminus Z} |f| + \int_Z |f| = \int_{E\setminus Z} |f| \leq \int_{E\setminus Z} g = \int_{E\setminus Z} g + \int_Z g = \int_E g.$$

Thus, $|f|$ is Lebesgue integrable over E and, consequently, f is Lebesgue integrable over E.

For the second part, note that by hypothesis, $f^+ = g^+$ a.e. in E and $f^- = g^-$ a.e. in E and both f^+ and f^- are Lebesgue integrable. By part (1), since $g^+ \leq f^+$ and $g^- \leq f^-$ a.e. in E, g^+ and g^- are Lebesgue integrable over E. Thus, g is Lebesgue integrable over E. Moreover, $\int_E g^+ \leq \int_E f^+$ and $\int_E g^- \leq \int_E f^-$. Reversing the roles of f and g, we conclude that $\int_E g^+ = \int_E f^+$ and $\int_E g^- = \int_E f^-$. It follows that

$$\int_E g = \int_E g^+ - \int_E g^- = \int_E f^+ - \int_E f^- = \int_E f.$$

$\square$

Suppose that $h \leq f \leq g$ a.e. in E. Then, $|f| \leq |g| + |h|$. If g and h are Lebesgue integrable over E, then so is $|g| + |h|$. Thus, we have the following corollary.

Corollary 3.95 *Suppose that $E \in \mathcal{M}_n$ and $f, g, h : E \to \mathbb{R}^*$ are measurable. If $h \leq f \leq g$ a.e. in E and g and h are Lebesgue integrable over E, then f is integrable over E.*

The sum of measurable functions is measurable if the sum is defined but, since that is not always the case, in general we cannot integrate the sum of measurable functions. However, if a function is Lebesgue integrable, then we have seen that it is finite almost everywhere. Thus, the sum of Lebesgue integrable functions is defined almost everywhere and, since sets of measure 0 do not effect the value of the Lebesgue integral, we may assume that the Lebesgue integral of the sum is well defined. The next result shows that the Lebesgue integral is linear for Lebesgue integrable functions.

Theorem 3.96 *(Linearity) Suppose f and g are Lebesgue integrable over a measurable set E. Then, for all $\alpha, \beta \in \mathbb{R}$, $\alpha f + \beta g$ is Lebesgue integrable and*

$$\int_E (\alpha f + \beta g) = \alpha \int_E f + \beta \int_E g.$$

Proof. We have already proved this result when f and g are nonnegative and $\alpha, \beta \geq 0$. If $\alpha < 0$, then $(\alpha f)^+ = -\alpha f^-$ and $(\alpha f)^- = -\alpha f^+$, so that

$$\int_E \alpha f = \int_E (\alpha f)^+ - \int_E (\alpha f)^-$$

$$= -\alpha \int_E f^- + \alpha \int_E f^+ = \alpha \left(\int_E f^+ - \int_E f^- \right) = \alpha \int_E f.$$

Thus, we only need consider the sum of Lebesgue integrable functions.

If f and g are nonnegative and $h = f - g$, then $h^+ = f$ and $h^- = g$ so that

$$\int_E (f - g) = \int_E h = \int_E h^+ - \int_E h^- = \int_E f - \int_E g,$$

since h is defined and finite almost everywhere. Consequently, for Lebesgue integrable functions f and g,

$$\int_E (f + g) = \int_E \left(f^+ - f^- + g^+ - g^- \right)$$

$$= \int_E \left(f^+ + g^+ - (f^- + g^-) \right) = \int_E (f^+ + g^+) - \int_E (f^- + g^-)$$

$$= \int_E f^+ + \int_E g^+ - \int_E f^- - \int_E g^- = \int_E f + \int_E g$$

since all the integrals are finite. $\qquad\square$

Suppose f and g are Lebesgue integrable functions over a measurable set E. It follows that $f - g$ and, hence, $|f - g|$ are Lebesgue integrable. Consequently, $f \vee g = \frac{1}{2} (f + g + |f - g|)$ and $f \wedge g = \frac{1}{2} (f + g - |f - g|)$ are Lebesgue integrable over E. Thus, analogous to the set of Riemann integrable functions on an interval $[a, b]$, the set of Lebesgue integrable functions on a measurable set E is a vector lattice. See Theorem 2.23 and the following paragraph.

As we have seen, the Monotone Convergence Theorem is a very useful tool in analysis. However, in many situations, the monotonicity condition is not satisfied by a convergent sequence and other conditions which guarantee the exchange of the limit and the integral are desirable. We next consider Lebesgue's Dominated Convergence Theorem. This result replaces the monotonicity condition of the Monotone Convergence Theorem by the requirement that the convergent sequence of functions be bounded by a Lebesgue integrable function. As a corollary of the Dominated Convergence Theorem, we will get the Bounded Convergence Theorem. We begin with a result due to P. Fatou (1878-1929), known a *Fatou's Lemma*.

Lemma 3.97 *(Fatou's Lemma) Suppose that $E \in \mathcal{M}_n$ and $f_k : E \to \mathbb{R}^*$ is nonnegative and measurable for all k. Then,*

$$\int_E \liminf_{k \to \infty} f_k \leq \liminf_{k \to \infty} \int_E f_k.$$

Proof. Set $h_k(x) = \inf_{j \geq k} f_j(x)$, so that h_k is nonnegative and measurable, and $\{h_k\}_{k=1}^{\infty}$ increases to $\liminf_{k \to \infty} f_k$. By the Monotone Convergence Theorem,

$$\int_E \liminf_{k \to \infty} f_k = \lim_{k \to \infty} \int_E h_k.$$

Since $h_k \leq f_k$ for all $x \in E$,

$$\lim_{k \to \infty} \int_E h_k \leq \liminf_{k \to \infty} \int_E f_k,$$

and the proof is complete. $\qquad \square$

Suppose that g is a Lebesgue integrable function and each f_k is a measurable function such that $f_k \geq g$ for a.e. $x \in E$ and all $k \in \mathbb{N}$. Then $f_k - g$ is a nonnegative function and we can apply Fatou's Lemma to get that

$$\int_E \liminf_{k \to \infty} f_k - \int_E g = \int_E \liminf_{k \to \infty} (f_k - g)$$

$$\leq \liminf_{k \to \infty} \int_E (f_k - g) = \liminf_{k \to \infty} \int_E f_k - \int_E g.$$

Since g is Lebesgue integrable, we have

Corollary 3.98 *Suppose that $E \in \mathcal{M}_n$ and $f_k, g : E \to \mathbb{R}^*$ are measurable and $g \leq f_k$ for all k. If g is Lebesgue integrable over E, then,*

$$\int_E \liminf_{k \to \infty} f_k \leq \liminf_{k \to \infty} \int_E f_k.$$

There is also a result dual to Corollary 3.98. See Exercise 3.44.

Corollary 3.99 *Suppose that $E \in \mathcal{M}_n$ and $f_k, g : E \to \mathbb{R}^*$ are measurable and $f_k \leq g$ for all k. If g is Lebesgue integrable over E, then,*

$$\int_E \limsup_{k \to \infty} f_k \geq \limsup_{k \to \infty} \int_E f_k.$$

We can now prove Lebesgue's *Dominated Convergence Theorem.*

Theorem 3.100   *(Dominated Convergence Theorem) Let $\{f_k\}_{k=1}^{\infty}$ be a sequence of measurable functions defined on a measurable set E. Suppose that $\{f_k\}_{k=1}^{\infty}$ converges to f pointwise almost everywhere and there is a Lebesgue integrable function g such that $|f_k(x)| \leq g(x)$ for all k and almost every $x \in E$. Then, f is Lebesgue integrable and*

$$\int_E f = \lim \int_E f_k.$$

Moreover,

$$\lim \int_E |f - f_k| = 0.$$

Proof.   By hypothesis, $-g \leq f_k \leq g$ a.e., so Corollaries 3.98 and 3.99 apply. Since $\{f_k\}_{k=1}^{\infty}$ converges to f pointwise almost everywhere,

$$\limsup_{k\to\infty} \int_E f_k \leq \int_E \limsup_{k\to\infty} f_k = \int_E f = \int_E \liminf_{k\to\infty} f_k \leq \liminf_{k\to\infty} \int_E f_k.$$

Thus, $\int_E f = \lim \int_E f_k$.

To complete the proof, note that $|f - f_k|$ converges to 0 pointwise a.e. and $|f(x) - f_k(x)| \leq 2g(x)$ for all k and almost every x. Thus, by the first part of the theorem, $\lim \int_E |f - f_k| = 0$ and the proof is complete.   □

For a more traditional proof of the Dominated Convergence Theorem, see [Ro, pages 91-92].

If the measure of E is finite, then constant functions are Lebesgue integrable over E. From the Dominated Convergence Theorem we get the *Bounded Convergence Theorem.*

Corollary 3.101   *(Bounded Convergence Theorem) Let $\{f_k\}_{k=1}^{\infty}$ be a sequence of measurable functions on a set E of finite measure. Suppose there is a number M so that $|f_k(x)| \leq M$ for all k and for almost all $x \in E$. If $f(x) = \lim_{k\to\infty} f_k(x)$ almost everywhere, then*

$$\int_E f = \lim_{k\to\infty} \int_E f_k.$$

We defined the Lebesgue integral by approximating nonnegative, measurable functions from below by simple functions. At the time, we mentioned that for bounded functions we could also consider approximation from above by simple functions. We now show that the two constructions lead to the same value for the integral. Thus, we do not need to show that an upper integral equals a lower integral to conclude that a function is Lebesgue integrable.

Proposition 3.102 *Let E be a measurable set with finite measure and $f : E \to \mathbb{R}^*$ be bounded. Then, f is measurable on E if, and only if,*

$$\sup\left\{\int_E \varphi : \varphi \le f, \ \varphi \ \text{is simple}\right\} = \inf\left\{\int_E \psi : f \le \psi, \ \psi \ \text{is simple}\right\}.$$
(3.4)

Proof. Suppose that f is measurable. Choose l and L such that $l \le f(x) < L$ for all $x \in E$. Let $\epsilon > 0$ and $\mathcal{P} = \{y_0, y_1, \ldots, y_m\}$ be a partition of $[l, L]$ with mesh $\mu(\mathcal{P}) < \epsilon$. Set $E_i = \{x \in E : y_{i-1} \le f(x) < l_i\}$, for $i = 1, \ldots, m$, and define simple functions φ and ψ by

$$\varphi(x) = \sum_{i=1}^m l_{i-1}\chi_{E_i}(x) \ \text{and} \ \psi(x) = \sum_{i=1}^m l_i\chi_{E_i}(x).$$

Then, $\varphi \le f \le \psi$ and

$$\int_E (\psi - \varphi) = \sum_{i=1}^m (l_i - l_{i-1}) m_n(E_i) < \epsilon m_n(E),$$

which implies (3.4).

On the other hand, suppose (3.4) holds. Then, there exist simple functions φ_k and ψ_k such that $\varphi_k \le f \le \psi_k$ on E and $\int_E (\psi_k - \varphi_k) < \frac{1}{k}$. Define φ and ψ by $\varphi(x) = \sup_k \varphi_k(x)$ and $\psi(x) = \inf_k \psi_k(x)$. Then, φ and ψ are measurable and $\varphi \le f \le \psi$ on E. Further,

$$\int_E (\psi - \varphi) \le \int_E (\psi_k - \varphi_k) < \frac{1}{k}$$

for all k, so that $\int_E (\psi - \varphi) = 0$. Thus, $\psi - \varphi = 0$ a.e. in E. Therefore, $\psi = f = \varphi$ a.e. in E and it follows that f is measurable. $\square$

3.5.1 *Integrals depending on a parameter*

As an application of the convergence theorems, we consider integrals depending on a parameter. Let $S \subset \mathbb{R}^n$, $T \subset \mathbb{R}^m$ and $f : S \times T \to \mathbb{R}$. If $f(s, \cdot) : T \to \mathbb{R}$ is integrable for every $s \in S$ and $F : S \to \mathbb{R}$ is defined by $F(s) = \int_T f(s, t)dt$, we say that the integral $F(s)$ depends on the parameter s. We consider how the properties of f are inherited by F.

We first consider the continuity of integrals depending on parameters.

Theorem 3.103   *Assume $f(s, \cdot)$ is integrable for every $s \in S$ and $f(\cdot, t)$ is continuous at $s_0 \in S$ for every $t \in T$. If there exists an integrable function $g : T \to \mathbb{R}$ such that $|f(s,t)| \leq g(t)$ for all $s \in S, t \in T$, then $F(s) = \int_T f(s,t) \, dt$ is continuous at s_0.*

Proof.   *Let $\{s_k\}_{k=1}^{\infty}$ be a sequence from S converging to s_0. By the continuity assumption, $f(s_k, t) \to f(s_0, t)$. Since $|f(s_k, \cdot)| \leq g(\cdot)$, the Dominated Convergence Theorem implies that $F(s_k) = \int_T f(s_k, t) \, dt \to F(s_0) = \int_T f(s_0, t) \, dt$ and F is continuous at s_0, as we wished to show.* $\square$

See Exercise 3.49.

Our second application applies to "differentiating under the integral sign".

Theorem 3.104   *(Leibniz' Rule) Let $S = [a, b]$ and let I be a closed interval. Suppose that*

(1) $\dfrac{\partial f}{\partial s} = D_1 f$ *exist for all $s \in S$ and $t \in I$;*
(2)   *each function $f(s, \cdot)$ is integrable over I and there is an integrable function $g : I \to \mathbb{R}$ such that*

$$\left| \frac{\partial f}{\partial s}(s, t) \right| = |D_1 f(s,t)| \leq g(t)$$

for all $s \in S$ and $t \in I$.

Then, $F(s) = \int_I f(s,t) \, dt$ is differentiable on S and

$$F'(s) = \int_I \frac{\partial f}{\partial s}(s, t) \, dt = \int_I D_1 f(s, t) \, dt.$$

Proof.   *Fix $s_0 \in S$ and let $\{s_k\}$ be a sequence in S converging to s_0 with $s_k \neq s_0$. For $t \in I$, define $\{h_k\}$ by*

$$h_k(t) = \frac{f(x_k, t) - f(x_0, t)}{x_k - x_0}$$

and note that h_k is integrable over I. By the Mean Value Theorem, for each pair (k, t) there is a $z_{(k,t)}$ between s_k and s_0 so that

$$\frac{f(s_k, t) - f(s_0, t)}{s_k - s_0} = D_1 f\left(z_{(k,t)}, t\right)$$

which implies that

$$|h_k(t)| = \left| \frac{f(s_k, t) - f(s_0, t)}{s_k - s_0} \right| \le g(t).$$

The Dominated Convergence Theorem implies that

$$\begin{aligned} F'(s_0) &= \lim_{k \to \infty} \frac{F(s_k) - F(s_0)}{s_k - s_0} \\ &= \lim_{k \to \infty} \int_I \frac{f(s_k, t) - f(s_0, t)}{s_k - s_0} dt \\ &= \int_I D_1 f(s_0, t)\, dt \end{aligned}$$

as we wished to show. □

3.6 Riemann and Lebesgue integrals

The Dirichlet function, which is 0 on the irrationals, provides an example of a function that is Lebesgue integrable but is not Riemann integrable on any interval. Thus, Lebesgue integrability does not imply Riemann integrability. The next result shows that the Lebesgue integral is a proper extension of the Riemann integral. In the proof below, we use $\mathcal{R} \int$ for the Riemann integral.

Theorem 3.105 *Let $f : [a, b] \to \mathbb{R}$ be Riemann integrable. Then, f is Lebesgue integrable and the two integrals are equal.*

Proof. Let $\{\mathcal{Q}_k\}_{k=1}^{\infty}$ be a sequence of partitions of $[a, b]$ such that:

(1) $\lim_{k \to \infty} \mu(\mathcal{Q}_k) = 0$;
(2) $\lim_{k \to \infty} L(f, \mathcal{Q}_k) = \underline{\int_a^b} f$;
(3) $\lim_{k \to \infty} U(f, \mathcal{Q}_k) = \overline{\int_a^b} f$.

If we then set $\mathcal{P}_k = \cup_{j=1}^{k} \mathcal{Q}_j$, then $\{\mathcal{P}_k\}_{k=1}^{\infty}$ is a sequence of nested partitions, $\mathcal{P}_k \subset \mathcal{P}_{k+1}$, that satisfy conditions (1), (2), and (3).

Fix k and suppose $\mathcal{P}_k = \{x_0, x_1, \ldots, x_j\}$. Set $m_i = \inf\{f(t) : x_{i-1} \le t \le x_i\}$ and $M_i = \sup\{f(t) : x_{i-1} \le t \le x_i\}$, and define simple functions l_k and u_k by

$$l_k(x) = \sum_{i=1}^{j-1} m_i \chi_{[x_{i-1}, x_i)}(x) + m_j \chi_{[x_{j-1}, x_j]}(x)$$

and

$$u_k(x) = \sum_{i=1}^{j-1} M_i \chi_{[x_{i-1}, x_i)}(x) + M_j \chi_{[x_{j-1}, x_j]}(x),$$

so that $\int_{[a,b]} l_k = L(f, \mathcal{P}_k)$ and $\int_{[a,b]} u_k = U(f, \mathcal{P}_k)$. Since the partitions are nested, it follows that $l_k \leq f \leq u_k$ and the sequence $\{l_k\}_{k=1}^{\infty}$ increases and $\{u_k\}_{k=1}^{\infty}$ decreases. Define l and u by $l(x) = \lim_k l_k(x)$ and $u(x) = \lim_k u_k(x)$. By the Monotone Convergence Theorem,

$$\int_{[a,b]} l = \lim_{k \to \infty} \int_{[a,b]} l_k = \lim_{k \to \infty} L(f, \mathcal{P}_k) = \int_{\underline{a}}^b f$$

and

$$\int_{[a,b]} u = \lim_{k \to \infty} \int_{[a,b]} u_k = \lim_{k \to \infty} U(f, \mathcal{P}_k) = \overline{\int_a}^b f.$$

Since f is Riemann integrable, $\int_{\underline{a}}^b f = \overline{\int_a}^b f$, so that

$$\int_{[a,b]} l = \int_{[a,b]} u.$$

Thus, because $l \leq f \leq u$, $l = f = u$ a.e. in $[a, b]$. Hence, f is Lebesgue integrable over $[a, b]$ and

$$\int_{[a,b]} f = \mathcal{R} \int_a^b f. \qquad \square$$

There is no direct comparison of the Lebesgue and Cauchy-Riemann integrals. Again, the Dirichlet function is an example of a Lebesgue integrable function that is not Cauchy-Riemann integrable. The function $f(x) = \frac{\sin x}{x}$ of Example 2.49 is Cauchy-Riemann integrable but, as shown in that example, is not absolutely integrable. Thus, it is not Lebesgue integrable.

3.7 Mikusinski's characterization of the Lebesgue integral

We next give a characterization of the Lebesgue integral due to J. Mikusinski (see [Mi1]; see also [MacN]). The characterization is of interest because it involves no mention of Lebesgue measure or the measurability of functions. The characterization will be utilized in the next section where we discuss Fubini's Theorem on the equality of integrals on $\mathbb{R}^n$ for $n \geq 2$ and iterated integrals.

We saw in Theorem 3.67 that a measurable function can be approximated a.e. by step functions on a set of finite measure. When the function is Lebesgue integrable, we can say more, that the Lebesgue integrals of the step functions converge in a very strong sense. In the following proof, we refer to the notation used in the proof of Theorem 3.67. In this proof, we use the *support* of a function f, which we define as the closure of the set $\{x \in \mathbb{R}^n : f(x) \neq 0\}$ and denote by $supp(f)$.

Theorem 3.106 *Let $f : \mathbb{R}^n \to \mathbb{R}^*$ be Lebesgue integrable. Then, there is a sequence of step functions $\{\varphi_k\}_{k=1}^{\infty}$ that converges to f a.e. such that*

$$\lim_{k \to \infty} \int |\varphi_k - f| = 0.$$

Proof. By considering f^+ and f^- separately, we may assume that $f \geq 0$. Also, f is finite valued a.e. since f is Lebesgue integrable. Without loss of generality, we may assume that f is finite valued on all of $\mathbb{R}^n$. Let

$$B_k = \{x \in \mathbb{R}^n : f(x) < k \text{ and } x_i \in [-k, k], i = 1, \dots, n\}.$$

Then, the sets B_k are measurable with finite measure and increase to $\mathbb{R}^n$.

Using the notation of Theorem 3.67, we define the function f_k and the sets A_i^k, H_i^k and G_i^k relative to the function $f\chi_{B_k}$. Then, the support of f_k is contained in B_k and $\{f_k\}_{k=1}^{\infty}$ increases to f a.e. in $\mathbb{R}^n$. Define step functions φ_k by

$$\varphi_k(x) = \sum_{i=1}^{k2^k} \frac{i-1}{2^k} \chi_{G_i^k}(x)$$

and set $F_k = \{x \in B_k : |\varphi_k(x) - f(x)| \geq 2^{-k}\}$. Since $F_k \subset \cup_{i=1}^{k2^k}(G_i^k \triangle A_i^k)$, we see that $m_n(F_k) \leq 2^{-k+1}$. Since

$$f_k(x) - \varphi_k(x) = \sum_{i=1}^{k2^k} \frac{i-1}{2^k} \left(\chi_{A_i^k}(x) - \chi_{G_i^k}(x)\right),$$

it follows that

$$\int |f_k - \varphi_k| \le \sum_{i=1}^{k2^k} \frac{i-1}{2^k} \int |\chi_{A_i^k} - \chi_{G_i^k}| \le \sum_{i=1}^{k2^k} \frac{i-1}{2^k} \int \chi_{G_i^k \Delta A_i^k}$$

$$\le \sum_{i=1}^{k2^k} \frac{i-1}{2^k} m_n \left(G_i^k \Delta A_i^k\right) \le \sum_{i=1}^{k2^k} \frac{i-1}{2^k} \frac{2}{k2^k} 2^{-k}$$

$$\le k^2 2^k \frac{2}{k2^k} 2^{-k} = \frac{k}{2^{k-1}},$$

so that $\lim_{k\to\infty} \int |f_k - \varphi_k| = 0$.

Set $Z = \cap_{m=1}^{\infty} \cup_{k=m}^{\infty} F_k$. Since $Z \subset \cup_{k=m}^{\infty} F_k$ for all m, we see that

$$m_n(Z) \le \sum_{k=m}^{\infty} m_n(F_k) \le \sum_{k=m}^{\infty} \frac{1}{2^{k-1}} = \frac{4}{2^m},$$

so that $m_n(Z) = 0$. We claim that $\{\varphi_k(x)\}_{k=1}^{\infty}$ converges to $f(x)$ for almost every $x \notin Z$. For, suppose that $x \notin Z$ and $f(x)$ is finite. Then, there is an m such that $x \notin \cup_{k=m} F_k$ and a j such that $x \in B_j$. Set $N = \max\{m, j\}$. Then, for $k \ge N$, $x \notin F_k$ and $x \in B_k$, which implies that $|\varphi_k(x) - f(x)| < \frac{1}{2^k}$. Thus, $\{\varphi_k(x)\}_{k=1}^{\infty}$ converges to $f(x)$ for almost every $x \notin Z$. By the Monotone Convergence Theorem, $\lim_{k\to\infty} \int |f_k - f| = 0$, so that

$$\lim_{k\to\infty} \int |\varphi_k - f| \le \lim_{k\to\infty} \int |\varphi_k - f_k| + \lim_{k\to\infty} \int |f_k - f| = 0,$$

as we wished to show. $\qquad\square$

To prove the Mikusinski characterization, we will use the following two lemmas.

Lemma 3.107 *Let E be a null set and $\epsilon > 0$. Then, there is a sequence of bricks $\{B_k\}_{k=1}^{\infty}$ such that $\sum_{k=1}^{\infty} m_n(B_k) < \infty$ and $\sum_{k=1}^{\infty} \chi_{B_k}(t) = \infty$ for all $t \in E$.*

Thus, the sum of the measures of the bricks is finite but each $t \in E$ belongs to infinitely many of the bricks.

Proof. Since $m_n(E) = 0$, for each $i \in \mathbb{N}$, there is a countable collection of open intervals $\{I_{ij} : j \in \sigma_i\}$ covering E such that $\sum_{j\in\sigma_i} m_n(I_{ij}) < \epsilon/2^i$. Let K_{ij} be the smallest brick containing I_{ij}, so that $E \subset \cup_{j\in\sigma_i} K_{ij}$ and $\sum_{j\in\sigma_i} m_n(K_{ij}) < \epsilon/2^i$ for each i. Arrange the doubly-indexed sequence

$\{K_{ij}\}_{i\in\mathbb{N}, j\in\sigma_i}$ into a sequence $\{B_k\}_{k=1}^{\infty}$. Since $t \in E \subset \cup_{j\in\sigma_i} K_{ij}$ for all i, t belongs to infinitely many bricks B_k so that $\sum_{k=1}^{\infty} \chi_{B_k}(t) = \infty$. Finally,

$$\sum_{k=1}^{\infty} m_n(B_k) = \sum_{i=1}^{\infty} \sum_{j\in\sigma_i} m_n(I_{ij}) < \sum_{i=1}^{\infty} \frac{\epsilon}{2^i} = \epsilon.$$

□

Suppose that a series of functions $\sum_{k=1}^{\infty} \psi_k$ converges to a function f pointwise (almost everywhere). If the series converges absolutely, that is, if $\sum_{k=1}^{\infty} |\psi_k(x)|$ is finite for almost every x, we say that the series is *absolutely convergent* to f a.e..

Lemma 3.108 *Suppose $f : \mathbb{R}^n \to \mathbb{R}^*$ is Lebesgue integrable. Then, there exists a sequence of step functions $\{\psi_k\}_{k=1}^{\infty}$ such that the series $\sum_{k=1}^{\infty} \psi_k$ converges to f absolutely a.e. and*

$$\sum_{k=1}^{\infty} \int_{\mathbb{R}^n} |\psi_k| < \infty.$$

Proof. By Theorem 3.106, there is a sequence of step functions $\{\varphi_k\}_{k=1}^{\infty}$ which converges to f a.e. and $\int |\varphi_k - f| \to 0$. Thus, there is a subsequence $\left\{\varphi_{k_j}\right\}_{j=1}^{\infty}$ of $\{\varphi_k\}_{k=1}^{\infty}$ such that $\int \left|\varphi_{k_{j+1}} - \varphi_{k_j}\right| < \frac{1}{2^j}$. Set $\psi_j = \varphi_{k_j} - \varphi_{k_{j-1}}$, for $j \geq 1$, where we define $\varphi_{k_0} = 0$. Then, $\sum_{j=1}^{K} \psi_j = \varphi_{k_K} \to f$ a.e., or $\sum_{k=1}^{\infty} \psi_k = f$ a.e.. Since

$$\sum_{j=1}^{\infty} \int |\psi_j| < \int |\varphi_{k_1}| + \sum_{j=2}^{\infty} \frac{1}{2^j} < \infty,$$

by Corollary 3.78, $\sum_{j=1}^{\infty} |\psi_j|$ is Lebesgue integrable and hence $\sum_{j=1}^{\infty} |\psi_j(x)|$ converges in $\mathbb{R}$ for almost all $x \in \mathbb{R}^n$. Thus, the series $\sum_{j=1}^{\infty} \psi_j$ is absolutely convergent to f a.e.. □

Mikusinski characterized Lebesgue integrable functions as absolutely convergent series of step functions.

Theorem 3.109 *Let $f : \mathbb{R}^n \to \mathbb{R}^*$. Then, f is Lebesgue integrable if, and only if, there is a sequence $\{\varphi_k\}_{k=1}^{\infty}$ of step functions satisfying:*

(1) $\sum_{k=1}^{\infty} \int |\varphi_k| < \infty$;

(2) if $\sum_{k=1}^{\infty} |\varphi_k(x)| < \infty$ then $f(x) = \sum_{k=1}^{\infty} \varphi_k(x)$.

In either case,

$$\int f = \sum_{k=1}^{\infty} \int \varphi_k.$$

Proof. Suppose first that such a sequence of functions exists. By Corollary 3.78, $\sum_{j=1}^{\infty} |\varphi_j|$ is Lebesgue integrable and the series $\sum_{j=1}^{\infty} |\varphi_j(x)|$ converges in $\mathbb{R}$ for almost all $x \in \mathbb{R}^n$. By (2), $f(x) = \sum_{k=1}^{\infty} \varphi_k(x)$ at such points and f, the almost everywhere limit of a sequence of measurable functions, is measurable. Since $|f| \leq \sum_{j=1}^{\infty} |\varphi_j|$ a.e., the Dominated Convergence Theorem implies that f is Lebesgue integrable and $\int f = \sum_{k=1}^{\infty} \int \varphi_k$.

Now, suppose that f is Lebesgue integrable. Choose $\{\psi_k\}_{k=1}^{\infty}$ by Lemma 3.108 and let E be the null set of points at which $\sum_{k=1}^{\infty} |\psi_k(x)|$ diverges. Let $\{B_k\}_{k=1}^{\infty}$ be the bricks corresponding to E in Lemma 3.107. Define a sequence of step functions $\{\varphi_k\}_{k=1}^{\infty}$ by

$$\varphi_k = \begin{cases} \psi_l & \text{if } k = 3l - 2 \\ \chi_{B_l} & \text{if } k = 3l - 1 \\ -\chi_{B_l} & \text{if } k = 3l \end{cases}.$$

If $x \in E$, then the series $\sum_{k=1}^{\infty} |\varphi_k(x)| = \infty$ by construction. If $\sum_{k=1}^{\infty} |\varphi_k(x)| < \infty$, then $x \notin E$ and $\sum_{k=1}^{\infty} \chi_{B_l}$ is finite and, hence, equal to 0, so that $\sum_{k=1}^{\infty} \varphi_k(x) = \sum_{k=1}^{\infty} \psi_k(x) = f(x)$ a.e.. Moreover,

$$\sum_{k=1}^{\infty} \int |\varphi_k| \leq \sum_{k=1}^{\infty} \int_{\mathbb{R}^n} |\psi_k| + 2 \sum_{k=1}^{\infty} \int_{\mathbb{R}^n} \chi_{B_k}$$

$$= \sum_{k=1}^{\infty} \int_{\mathbb{R}^n} |\psi_k| + 2 \sum_{k=1}^{\infty} m_n(B_k) < \infty$$

by Lemmas 3.107 and 3.108. $\qquad \square$

Remark 3.110 *Note that if* (1) *and* (2) *hold, then* $f = \sum_{k=1}^{\infty} \varphi_k$ *a.e. and f is measurable. Note also that the conditions* (1) *and* (2) *contain no statements involving Lebesgue measure. These conditions can be utilized to give a development of the Lebesgue integral in $\mathbb{R}^n$ which depends only on properties of step functions and not on a development of Lebesgue measure. For such an exposition, see* [DM], [Mi2], *or* [MM].

3.8 Fubini's Theorem

The most efficient way to evaluate integrals in $\mathbb{R}^n$ for $n \geq 2$ is to calculate iterated integrals. Theorems which assert the equality of integrals in $\mathbb{R}^n$ with iterated integrals are often referred to as "Fubini Theorems" after G. Fubini (1879-1943). In this section, we will use Mikusinski's characterization of the Lebesgue integral in $\mathbb{R}^n$ to establish a very general form of Fubini's Theorem.

For convenience, we will treat the case $n = 2$; the results remain valid in $\mathbb{R}^{n+m} = \mathbb{R}^n \times \mathbb{R}^m$. Suppose $f : \mathbb{R}^2 = \mathbb{R} \times \mathbb{R} \to \mathbb{R}^*$. We are interested in equalities of the form

$$\int_{\mathbb{R}^2} f = \int_{\mathbb{R}} \left(\int_{\mathbb{R}} f(x, y) \, dy \right) dx,$$

in which the integral on the left is a Lebesgue integral in $\mathbb{R}^2$ and the expression on the right is an iterated integral. If $f = \chi_I$ is the characteristic function of an interval in $\mathbb{R}^2$, then $I = I_1 \times I_2$, where I_i is an interval in $\mathbb{R}$, $i = 1, 2$. Since

$$\int_{\mathbb{R}^2} \chi_I = m_2(I) = m(I_1) m(I_2) = \left(\int_{\mathbb{R}} \chi_{I_1} \right) \left(\int_{\mathbb{R}} \chi_{I_2} \right)$$

$$= \int_{\mathbb{R}} \int_{\mathbb{R}} \chi_{I_1}(x) \chi_{I_2}(y) \, dy dx = \int_{\mathbb{R}} \int_{\mathbb{R}} \chi_I(x, y) \, dy dx,$$

Fubini's Theorem holds for characteristic functions of intervals and, by linearity, it holds for Lebesgue integrable step functions.

If f is a function on $\mathbb{R}^2$, we can view f as a function of two real variables, $f(x, y)$, where $x, y \in \mathbb{R}$. For each $x \in \mathbb{R}$, define a function $f_x : \mathbb{R} \to \mathbb{R}^*$ by $f_x(y) = f(x, y)$. Similarly, for each $y \in \mathbb{R}$, we define $f^y : \mathbb{R} \to \mathbb{R}^*$ by $f^y(x) = f(x, y)$. For the remainder of this section, we make the agreement that if a function g is defined almost everywhere, then g is defined to be equal to 0 on the null set where g fails to be defined. Thus, if $\{g_k\}_{k=1}^{\infty}$ is a sequence of measurable functions which converges a.e., we may assume that there exists a measurable function $g : \mathbb{R}^n \to \mathbb{R}^*$ such that $\{g_k\}_{k=1}^{\infty}$ converges to g a.e. in $\mathbb{R}^n$. This situation is encountered several times in the proof of *Fubini's Theorem*, which we now prove.

Theorem 3.111 *(Fubini's Theorem) Let $f : \mathbb{R} \times \mathbb{R} \to \mathbb{R}^*$ be Lebesgue integrable. Then:*

(1) f_x is Lebesgue integrable in $\mathbb{R}$ for almost every $x \in \mathbb{R}$;

(2) the function $x \longmapsto \int_{\mathbb{R}} f_x = \int_{\mathbb{R}} f(x,y)\, dy$ is Lebesgue integrable over $\mathbb{R}$;

(3) the following equality holds:

$$\int_{\mathbb{R} \times \mathbb{R}} f = \int_{\mathbb{R}} \left(\int_{\mathbb{R}} f_x \right) dx = \int_{\mathbb{R}} \int_{\mathbb{R}} f(x,y)\, dy dx.$$

Proof. By Mikusinski's Theorem, there is a sequence of step functions $\{\varphi_k\}_{k=1}^{\infty}$ on $\mathbb{R} \times \mathbb{R}$ such that:

   i. $\sum_{k=1}^{\infty} \int_{\mathbb{R} \times \mathbb{R}} |\varphi_k| < \infty$;

   ii. if $\sum_{k=1}^{\infty} |\varphi_k(x,y)| < \infty$ then $f(x,y) = \sum_{k=1}^{\infty} \varphi_k(x,y)$;

   iii. $\int_{\mathbb{R} \times \mathbb{R}} f = \sum_{k=1}^{\infty} \int_{\mathbb{R} \times \mathbb{R}} \varphi_k$.

By Corollary 3.78, the fact that Fubini's Theorem holds for step functions, and (i),

$$\int_{\mathbb{R}} \sum_{k=1}^{\infty} \int_{\mathbb{R}} |\varphi_k(x,y)|\, dy dx = \sum_{k=1}^{\infty} \int_{\mathbb{R}} \int_{\mathbb{R}} |\varphi_k(x,y)|\, dy dx$$

$$= \sum_{k=1}^{\infty} \int_{\mathbb{R} \times \mathbb{R}} |\varphi_k| < \infty, \qquad (3.5)$$

which implies that there is a null set $E \subset \mathbb{R}$ such that $\sum_{k=1}^{\infty} \int_{\mathbb{R}} |\varphi_k(x,y)|\, dy < \infty$ for all $x \notin E$. Now, for $x \notin E$, Corollary 3.78 implies

$$\int_{\mathbb{R}} \sum_{k=1}^{\infty} |\varphi_k(x,y)|\, dy = \sum_{k=1}^{\infty} \int_{\mathbb{R}} |\varphi_k(x,y)|\, dy < \infty$$

so that $\sum_{k=1}^{\infty} |\varphi_k(x,y)| < \infty$ for almost all $y \in \mathbb{R}$, where the null set may depend on $x \notin E$. For such a pair (x,y), $f(x,y) = \sum_{k=1}^{\infty} \varphi_k(x,y)$ by (ii) and, in particular, for $x \notin E$, $f_x = \sum_{k=1}^{\infty} (\varphi_k)_x$ a.e.. Since $\left| \sum_{k=1}^{N} (\varphi_k)_x \right| \leq \sum_{k=1}^{\infty} |(\varphi_k)_x|$, the Dominated Convergence Theorem implies that f_x is Lebesgue integrable over $\mathbb{R}$, proving (1), and $\int_{\mathbb{R}} f_x = \sum_{k=1}^{\infty} \int_{\mathbb{R}} \varphi(x,y)\, dy$.

If $x \notin E$, then

$$\left| \sum_{k=1}^{N} \int_{\mathbb{R}} \varphi_k(x,y)\, dy \right| \leq \sum_{k=1}^{\infty} \int_{\mathbb{R}} |\varphi_k(x,y)|\, dy,$$

and the function on the right hand side of the inequality is Lebesgue integrable over $\mathbb{R}$ by (3.5). By the Dominated Convergence Theorem, (2) holds, and by (iii), we have

$$\int_{\mathbb{R}\times\mathbb{R}} f = \sum_{k=1}^{\infty} \int_{\mathbb{R}\times\mathbb{R}} \varphi_k = \sum_{k=1}^{\infty} \int_{\mathbb{R}} \int_{\mathbb{R}} \varphi_k(x,y)\, dy dx$$

$$= \int_{\mathbb{R}} \int_{\mathbb{R}} \sum_{k=1}^{\infty} \varphi_k(x,y)\, dy dx = \int_{\mathbb{R}} \int_{\mathbb{R}} f(x,y)\, dy dx,$$

so that (3) holds. □

Fubini's Theorem could also be stated in terms of f^y. Thus, if f is Lebesgue integrable on $\mathbb{R}^2$, then f^y is Lebesgue integrable on $\mathbb{R}$ for almost every $y \in \mathbb{R}$, the function $y \longmapsto \int_{\mathbb{R}} f(x,y)\, dx$ in Lebesgue integrable over $\mathbb{R}$, and

$$\int_{\mathbb{R}\times\mathbb{R}} f = \int_{\mathbb{R}} \left(\int_{\mathbb{R}} f^y \right) dy = \int_{\mathbb{R}} \int_{\mathbb{R}} f(x,y)\, dx dy.$$

The main difficulty in applying Fubini's Theorem is establishing the integrability of the function f on $\mathbb{R}^2$. However, when f is nonnegative, we get the equality of the double integral with the iterated integral. Thus, in this case, f is Lebesgue integrable if either the integral of f or the iterated integral is finite. *Tonelli's Theorem*, named after L. Tonelli (1885-1946), guarantees the equality of multiple integrals and iterated integrals for nonnegative functions.

Theorem 3.112 *(Tonelli's Theorem) Let $f : \mathbb{R} \times \mathbb{R} \to \mathbb{R}^*$ be nonnegative and measurable. Then:*

(1) f_x is measurable on $\mathbb{R}$ for almost every $x \in \mathbb{R}$;

(2) the function $x \longmapsto \int_{\mathbb{R}} f_x = \int_{\mathbb{R}} f(x,y)\, dy$ is measurable on $\mathbb{R}$;

(3) the following equality holds:

$$\int_{\mathbb{R}\times\mathbb{R}} f = \int_{\mathbb{R}} \left(\int_{\mathbb{R}} f_x \right) dx = \int_{\mathbb{R}} \int_{\mathbb{R}} f(x,y)\, dy dx.$$

Proof. Let $I_k = [-k,k] \times [-k,k]$ so that $\cup_{k=1}^{\infty} I_k = \mathbb{R}^2$. For each k, set $f_k(x,y) = (\max\{f(x,y),k\}) \chi_{I_k}(x,y)$, so that f_k is Lebesgue integrable over $\mathbb{R}^2$. By Fubini's Theorem, $(f_k)_x$ is Lebesgue integrable for almost all x and since $\{(f_k)_x\}_{k=1}^{\infty}$ increases to f_x on $\mathbb{R}$, f_x is measurable for almost every x. By the Monotone Convergence Theorem

$$\int_{\mathbb{R}} (f_k)_x = \int_{\mathbb{R}} f_k(x,y)\, dy \uparrow \int_{\mathbb{R}} f_x = \int_{\mathbb{R}} f(x,y)\, dy. \qquad (3.6)$$

By Theorem 3.58, the function $x \longmapsto \int_{\mathbb{R}} f_x = \int_{\mathbb{R}} f(x,y) \, dy$ is measurable and the Monotone Convergence Theorem applied to (3.6) yields

$$\lim_{k \to \infty} \int_{\mathbb{R}} \int_{\mathbb{R}} f_k(x,y) \, dy dx = \int_{\mathbb{R}} \int_{\mathbb{R}} f(x,y) \, dy dx. \qquad (3.7)$$

By Fubini's Theorem, $\int_{\mathbb{R}^2} f_k = \int_{\mathbb{R}} \int_{\mathbb{R}} f_k(x,y) \, dy dx$ and since $\{f_k\}_{k=1}^{\infty}$ increases to f pointwise, by the Monotone Convergence Theorem, $\int_{\mathbb{R}^2} f = \lim_k \int_{\mathbb{R}^2} f_k$. Combining this with (3.7) implies

$$\int_{\mathbb{R}^2} f = \int_{\mathbb{R}} \int_{\mathbb{R}} f(x,y) \, dy dx.$$

$\square$

Note that we cannot drop the nonnegativity condition in Tonelli's Theorem. See Exercise 3.55. For alternate proofs of the Fubini and Tonelli Theorems, see [Ro, pages 303-309].

Tonelli's Theorem can be used to check the integrability of a measurable function $f : \mathbb{R}^2 \to \mathbb{R}^*$. If the iterated integral $\int_{\mathbb{R}} \int_{\mathbb{R}} |f(x,y)| \, dy dx$ is finite, then $|f|$ and, consequently, f are Lebesgue integrable by Tonelli's Theorem and then, by Fubini's Theorem, $\int_{\mathbb{R}^2} f = \int_{\mathbb{R}} \int_{\mathbb{R}} f(x,y) \, dy dx$.

As an application of Fubini's Theorem, we show how the area of a bounded subset of $\mathbb{R}^2$ can be calculated as a one-dimensional integral. If $E \subset \mathbb{R}^2$ and $x \in \mathbb{R}$, the *x-section* of E at x is defined to be $E_x = \{y : (x,y) \in E\}$. Similarly, for $y \in \mathbb{R}$, the *y-section* of E at y is defined to be $E^y = \{x : (x,y) \in E\}$. We have the following elementary observations.

Proposition 3.113   *Let $E, E_\alpha \subset \mathbb{R}^2$, $\alpha \in A$, and $x \in \mathbb{R}$. Then,*

(1) $\chi_E(x,y) = \chi_{E_x}(y)$;

(2) $(E^c)_x = (E_x)^c$;

(3) $(\cup_{\alpha \in A} E_\alpha)_x = \cup_{\alpha \in A} (E_\alpha)_x$;

(4) $(\cap_{\alpha \in A} E_\alpha)_x = \cap_{\alpha \in A} (E_\alpha)_x$.

For example, $\chi_E(x,y) = (\chi_E)_x(y) = \chi_{E_x}(y)$ since all three equal 1 if, and only if, $(x,y) \in E$.

From this proposition and Tonelli's Theorem, we have

Theorem 3.114   *Let $E \subset \mathbb{R}^2$ be measurable. Then,*

(1) for almost every $x \in \mathbb{R}$, the sections E_x are measurable;

(2) the function $x \longmapsto m(E_x)$ is Lebesgue integrable over $\mathbb{R}$;
(3) $m_2(E) = \int_{\mathbb{R}} m(E_x)\,dx$.

When f is a continuous function on an interval $[a, b]$, we can use this result to compute the area under the graph of f.

Example 3.115 Let $f : [a, b] \to \mathbb{R}$ be nonnegative and continuous. Then, the region under the graph of f is the set $E = \{(x, y) : x \in [a, b] \text{ and } 0 \leq y \leq f(x)\}$. By considering the points where $x < a$, $x > b$, $y < 0$ and $y > f(x)$ separately, one sees that the complement of E is an open set, so that E is closed and hence measurable. Thus, by the previous theorem, $m_2(E) = \int_{\mathbb{R}} m(E_x)\,dx$. Notice that

$$E_x = \{y : (x, y) \in E\} = \begin{cases} [0, f(x)] & \text{if } x \in [a, b] \\ \emptyset & \text{if } x \notin [a, b] \end{cases},$$

which implies that

$$m_2(E) = \int_a^b f(x)\,dx.$$

This result can be used to compute the area and volume of familiar regions. See Exercises 3.56 and 3.57.

3.8.1 *Convolution*

Let f and g be integrable functions on $\mathbb{R}^n$. The (formal) definition of the convolution product of f and g is given by the integral

$$f * g(x) = \int_{\mathbb{R}^n} f(x - y)g(y)\,dy.$$

Although the product of integrable functions does not exist in general, it is interesting that the convolution of two integrable functions does always exist (a.e.). We use Tonelli's Theorem 3.112 to show that this integral exists for almost all $x \in \mathbb{R}^n$ and defines a function in $L^1(\mathbb{R}^n)$ with

$$\int_{\mathbb{R}^n} |f * g| \leq \int_{\mathbb{R}^n} \int_{\mathbb{R}^n} |f(x - y)g(y)|\,dy\,dx \tag{3.8}$$

$$= \int_{\mathbb{R}^n} |g(y)|\,dy \int_{\mathbb{R}^n} |f(x)|\,dx < \infty.$$

First assume that the function $(x, y) \to f(x - y)g(y)$ is measurable as a function from $\mathbb{R}^{2n} \to \mathbb{R}$. Then we have

$$\int_{\mathbb{R}^n} \int_{\mathbb{R}^n} |f(x - y)g(y)| \, dy dx = \int_{\mathbb{R}^n} \int_{\mathbb{R}^n} |f(x - y)g(y)| \, dx dy$$

$$= \int_{\mathbb{R}^n} |g(y)| \, dy \int_{\mathbb{R}^n} |f(x)| \, dx < \infty.$$

It follows from Tonelli's Theorem that $f * g$ exists for almost all $x \in \mathbb{R}^n$ and defines a function belonging to $L^1(\mathbb{R}^n)$ (after it has been extended to be 0 where the integral fails to exist). The computation above shows (3.8) holds. The measurability of the function $(x, y) \to f(x - y)g(y)$ is established below by a sequence of results.

More specifically, we need to prove two things:

(1) the mapping $(x, y) \to g(y)$ is measurable on $\mathbb{R}^{2n}$ whenever g is a measurable function on $\mathbb{R}^n$;
(2) the mapping $(x, y) \to f(x - y)$ is measurable on $\mathbb{R}^{2n}$ whenever f is a measurable function on $\mathbb{R}^n$.

The first result is simple to prove.

Lemma 3.116   *Let $f : \mathbb{R}^n \to \mathbb{R}$ be measurable on $\mathbb{R}^n$ and set $F(x, t) = f(x)$. Then, $F : \mathbb{R}^{2n} \to \mathbb{R}$ is measurable on $\mathbb{R}^{2n}$.*

Proof.   The set

$$\{(x, t) : F(x, t) > \alpha\} = \{(x, t) : f(x) > \alpha\} = \{x : f(x) > \alpha\} \times \mathbb{R}^n.$$

Since this is a cross product of measurable sets, F is measurable on $\mathbb{R}^{2n}$. $\qquad\square$

To deduce the second result, we will prove the following more general result.

Theorem 3.117   *Let $f : \mathbb{R}^n \to \mathbb{R}$ be measurable and $T : \mathbb{R}^n \to \mathbb{R}^n$ be a nonsingular linear transformation. Then, $f \circ T = f(T)$ is a measurable function on $\mathbb{R}^n$.*

The proof of this theorem is based on the facts that nonsingular linear transformations on $\mathbb{R}^n$ are homeomorphisms and map null sets to null sets. The result about null sets is proved in the following two lemmas.

Lemma 3.118   *Let $T : \mathbb{R}^n \to \mathbb{R}^n$ be a linear transformation. There is a constant $C = C(n)$ so that given any cube $Q \subset \mathbb{R}^n$ there is a cube $Q' \subset \mathbb{R}^n$ such that $T(Q) \subset Q'$ and $m_n(Q') \leq Cm_n(Q)$.*

Proof.   Let Q be the cube centered at x_0 with edge length d. If $x \in Q$, then $\|x - x_0\| \leq \frac{\sqrt{n}}{2}d$. Since T is linear, it is bounded; let $B = \|T\|_{\mathbb{R}^n \to \mathbb{R}^n}$. So if $x, y \in \mathbb{R}^n$, then $\|Tx - Ty\| \leq B\|x - y\|$. If $x \in Q$ we have

$$\|Tx - Tx_0\| \leq B\|x - x_0\| \leq \frac{\sqrt{n}B}{2}d.$$

Let Q' be the cube centered at Tx_0 with edge length $\sqrt{n}Bd$. Then, $x \in Q$ implies that $Tx \in Q'$, so that $T(Q) \subset Q'$. Finally, $m_n(Q') = (\sqrt{n}B)^n m_n(Q)$. $\qquad \square$

We can now show that the linear image of a null set is a null set.

Lemma 3.119   *Let $T : \mathbb{R}^n \to \mathbb{R}^n$ be a linear transformation. If $E \subset \mathbb{R}^n$ is a null set, then $T(E)$ is measurable and null.*

Proof.   Let E be a null set. It is enough to show that $T(E)$ is null. Let $\epsilon > 0$. Let G be an open set containing E such that $m_n(G) < \epsilon$. By Lemma 3.44, we can write G as a countable union of disjoint cubic bricks, $\{Q_j\}$. Thus, $E \subset G = \cup_j Q_j$ and $\sum_j m_n(Q_j) = m_n(G) < \epsilon$. Let Q'_j be the cube corresponding to Q_j in the previous lemma. Then,

$$T(E) \subset T(\cup_j Q_j) \subset \cup_j T(Q_j) \subset \cup_j Q'_j,$$

which implies that $m_n^*(T(E)) \leq \sum_j m_n(Q'_j) \leq C\sum_j m_n(Q_j) < C\epsilon$. Since this is true for all $\epsilon > 0$, it follows that $T(E)$ has outer measure 0, so that $T(E)$ is measurable and null. $\qquad \square$

We are ready to show that $f \circ T$ is a measurable function.

Proof.   [of Theorem 3.117] Let $f : \mathbb{R}^n \to \mathbb{R}$ be measurable and $T : \mathbb{R}^n \to \mathbb{R}^n$ be a nonsingular linear transformation. Fix $\lambda \in \mathbb{R}$ and consider the set $E = \{x \in \mathbb{R}^n : f(Tx) > \lambda\}$. Set $E' = \{x \in \mathbb{R}^n : f(x) > \lambda\}$. Then, $E = T^{-1}E'$. To see this, let $x \in E$. Then, $f(Tx) > \lambda$, so that $Tx \in E'$. Therefore, $x \in T^{-1}E'$. Also, if $x \in T^{-1}E'$, then $Tx \in E'$ so $f(Tx) > \lambda$ and $x \in E$. Therefore, to see that $f \circ T$ is measurable, it is enough to show that T^{-1} maps measurable sets to measurable sets.

Let $H \subset \mathbb{R}^n$ be a measurable set. By the n-dimensional analog of Theorem 3.36, we can write $H = G \cup N$ with $G \in \mathcal{F}_\sigma$ and N a null set. Since T is continuous, $T^{-1}(G)$ is an $\mathcal{F}_\sigma$ set and therefore measurable. By Lemma 3.119, $T^{-1}(N)$ is a null set, so $T^{-1}(H) = T^{-1}(G) \cup T^{-1}(N)$ is measurable. Thus, T^{-1} maps measurable sets to measurable sets, which implies that $f \circ T$ is measurable. $\qquad \square$

As a corollary of this theorem, we can prove (2) above. Since the product of measurable functions is measurable, this will complete the proof to show that the function $(x, y) \to f(x - y)g(y)$ is measurable as a function from $\mathbb{R}^{2n} \to \mathbb{R}$.

Corollary 3.120   *Let* $f : \mathbb{R}^n \to \mathbb{R}$ *be measurable on* $\mathbb{R}^n$ *and set* $F(x, t) = f(x - t)$. *Then,* $F : \mathbb{R}^{2n} \to \mathbb{R}$ *is measurable on* $\mathbb{R}^{2n}$.

Proof.   The function $F_1(x, t) = f(x)$ is measurable on $\mathbb{R}^{2n}$ by Lemma 3.116. Since the transformation $T : \mathbb{R}^{2n} \to \mathbb{R}^{2n}$ defined by $T(x, t) = (x - t, x + t)$ in linear and nonsingular, it follows from the previous theorem that $f(x - t) = F_1(x - t, x + t) = F_1(T(x, t))$ is measurable. $\square$

As an application of the convolution product, we prove several approximation results. These results employ what is called an *approximate identity*.

Definition 3.121   A sequence $\{\varphi_k\} \subset L^1(\mathbb{R}^n)$ is called an *approximate identity* (or δ-*sequence*) if:

(1) $\varphi_k(x) \geq 0$ for all x;
(2) $\int_{\mathbb{R}^n} \varphi_k = 1$ for all k;
(3) for every neighborhood of 0, $U \subset \mathbb{R}^n$, $\lim_k \int_{U^c} \varphi_k = 0$.

In practice, many examples of approximate identities are formed by dilations of a specific functions. Suppose φ is an integrable, nonnegative function on $\mathbb{R}$ with $\int_{-\infty}^{\infty} \varphi = 1$ and set $\varphi_k(t) = k\varphi(kt)$, so that by a change of variables,

$$\int_{\mathbb{R}} \varphi_k = \int_{\mathbb{R}} k\varphi(kt)\, dt = \int_{\mathbb{R}} \varphi(s)\, ds = 1.$$

If φ has enough decay at ∞ for (3) to hold, then $\{\varphi_k\}$ forms an approximate identity. We give examples of approximate identities.

Example 3.122   Below are examples of approximate identities in $\mathbb{R}$, all given by dilations of a single generating function:

(1) Let $\varphi(t) = \frac{1}{2}\chi_{[-1,1]}(t)$. Then, $\varphi_k(t) = \frac{k}{2}\chi_{[-1/k,1/k]}$.
(2) Let $\varphi(t) = \sqrt{\frac{2}{\pi}}\frac{1}{1 + t^2}$. Then, $\varphi_k(t) = \sqrt{\frac{2}{\pi}}\frac{k}{1 + (kt)^2}$.
(3) Let $\varphi(t) = Ce^{-t^2}$, with $C = \left(\int_{\mathbb{R}} e^{-t^2} dt\right)^{-1}$, so that $\int_{\mathbb{R}} \varphi = 1$. Then,
$$\varphi_k(t) = \frac{k}{C}e^{-k^2 t^2}.$$

(4) Let $\varphi : \mathbb{R} \to \mathbb{R}$ be continuous, nonnegative, $\varphi(t) = 0$ for $|t| \geq 1$ and $\int_{-1}^{1} \varphi = 1$. Set $\varphi_k(t) = k\varphi(kt)$. Note that φ_k will satisfy (3) due to the compact support or φ.

(5) Let $\varphi(t) = Ce^{-1/(1-t^2)}$ for $|t| < 1$ and $\varphi(t) = 0$ otherwise, with C chosen to satisfy the integrability condition (2). Set $\varphi_k(t) = k\varphi(kt)$ as in the example above. Note that in this case the φ_k are infinitely differentiable and have compact support.

The next example discusses the approximate identity known as the *Landau kernels*, which will be used in the proof of the Weierstrass Approximation Theorem.

Example 3.123 The Landau kernels $\varphi_k : \mathbb{R} \to \mathbb{R}$. Let $c_k = \int_{-1}^{1}(1-t^2)^k dt$ and set $\varphi_k(t) = (1 - t^2)^k/c_k$ for $|t| < 1$ and $\varphi_k(t) = 0$ if $|t| \geq 1$. We claim that $\{\varphi_k\}$ is an approximate identity. By the choice of c_k, (1) and (2) are clear. For (3) note that by symmetry,

$$c_k = 2\int_0^1 (1 - t^2)^k dt = 2\int_0^1 (1 + t)^k (1 - t)^k dt \geq 2\int_0^1 (1 - t)^k dt = \frac{2}{k+1}.$$

If $0 < \delta < 1$,

$$\int_\delta^1 (1 - t^2)^k/c_k dt \leq \int_\delta^1 \frac{k+1}{2}(1 - \delta^2)^k dt \leq \frac{k+1}{2}(1 - \delta^2)^k(1 - \delta) \to 0$$

as $k \to \infty$. By symmetry, (3) follows.

We conclude with some approximate identities in $\mathbb{R}^n$.

Example 3.124 We can construct approximate identities in $\mathbb{R}^n$ using products of approximate identities in $\mathbb{R}$. Suppose that $\{\varphi_k\}$ is an approximate identity in $\mathbb{R}$ and set $\psi_k(x_1, ..., x_n) = \varphi_k(x_1) \cdots \varphi_k(x_n)$. Then $\{\psi_k\}$ is an approximate identity in $\mathbb{R}^n$.

As before, we can create an approximate identity in $\mathbb{R}^n$ by dilating a particular generating function. In this case, the dilation becomes $\varphi_k(t) = k^n \varphi(kt)$.

Example 3.125 Define φ on $\mathbb{R}^n$ by $\varphi(x) = Ce^{-1/(1-\|x\|^2)}$ if $\|x\| < 1$ and $\varphi(x) = 0$ otherwise, with C such that $\int_{\mathbb{R}^n} \varphi = 1$. Set $\varphi_k(x) = k^n \varphi(kx)$. Then $\{\varphi_k\}$ is an approximate identity in $\mathbb{R}^n$, where each φ_k is infinitely differentiable and has compact support. See Exercise 3.33 for a proof that $\int_{\mathbb{R}^n} \varphi_k = 1$.

We now establish an approximation result using approximate identities.

Theorem 3.126 *Let $\{\varphi_k\}_{k=1}^{\infty}$ be an approximate identity in $\mathbb{R}^n$. If $f :$ $\mathbb{R}^n \to \mathbb{R}$ is bounded and uniformly continuous, then $f * \varphi_k$ exists everywhere in $\mathbb{R}^n$, is bounded, uniformly continuous and $f * \varphi_k$ converges to f uniformly on compact subsets.*

Proof. By Exercise 3.61, we see that $f * \varphi_k$ exists everywhere in $\mathbb{R}^n$ and is bounded and uniformly continuous. Let M be a bound for $|f|$. Let $K \subset \mathbb{R}^n$ be compact and $\epsilon > 0$. There exists $\delta > 0$ such that $|f(x - y) - f(x)| < \epsilon/2$ when $x \in K$ and $\|y\| \leq \delta$. By Definition 3.121 (3), there exists N such that $\int_{\|y\| \geq \delta} \varphi_k < \epsilon/4M$ when $k \geq N$. Then for $x \in K$ and $k \geq N$, by (2) we have

$$|f * \varphi_k(x) - f(x)| = \left| \int_{\mathbb{R}^n} \{f(x - y) - f(x)\} \varphi_k(y) \, dy \right|$$

$$\leq \left(\int_{\{\|y\| \geq \delta\}} + \int_{\{\|y\| < \delta\}} \right) |f(x - y) - f(x)| \varphi_k(y) \, dy$$

$$< 2M \int_{\{\|y\| \geq \delta\}} \varphi_k + \frac{\epsilon}{2} \int_{\|y\| < \delta} \varphi_k < \frac{\epsilon}{2} + \frac{\epsilon}{2} = \epsilon.$$

Thus, $f * \varphi_k$ converges to f uniformly on K as we wished to show. $\square$

A similar result is true if f is continuous and measurable instead of uniformly continuous. See Exercise 3.65.

Suppose that f and g are integrable functions. Then, the convolution $f * g$ exists and it follows from the change of variables $t = x - y$ that

$$f * g(x) = \int_{\mathbb{R}} f(x - y)g(y)dy = \int_{\mathbb{R}} g(x - t) f(t) \, dt = g * f(x).$$

In other words, the convolution product is a commutative operation. Using this observation and the Landau kernels, we can use Theorem 3.126 to give a proof of the Weierstrass Approximation Theorem.

Theorem 3.127 *(Weierstrass Approximate Theorem) Let $f : [a, b] \to \mathbb{R}$ be continuous. Then there exists a sequence of polynomials $\{p_k\}_{k=1}^{\infty}$ which converges uniformly to f on $[a, b]$.*

Proof. Begin by making the change of variables $u = (x - a)/(b - a)$. Then $x = (b - a)u + a$ and $0 \leq u \leq 1$. Set $g(u) = f((b - a)u + a) = f(x)$. If we can find a polynomial p such that $|g(u) - p(u)| < \epsilon$ for $0 \leq u \leq 1$, then

$\left| p(\frac{x-a}{b-a}) - f(x) \right| < \epsilon$ for $a \le x \le b$ and the function defined by $p(\frac{x-a}{b-a})$ is a polynomial in x. Thus, we may assume $[a, b] = [0, 1]$.

Next, let $h(x) = f(x) - f(0) - x(f(1) - f(0))$. If we can approximate h by a polynomial on $[0, 1]$, then we can approximate f by a polynomial on $[0, 1]$. Since $h(0) = h(1) = 0$, we may assume $f(0) = f(1) = 0$. With this assumption, we extend f to $\mathbb{R}$ by setting $f = 0$ outside $[0, 1]$. Then the extension, which we still denote by f, is bounded and uniformly continuous.

Let $\{\varphi_k\}_{k=1}^\infty$ be the Landau kernels. By Theorem 3.126, $\{f * \varphi_k\}$ converges uniformly to f on $[0, 1]$. But, $\varphi_k(x - y) = \left(1 - (x - y)^2\right)^k / c_k$ is a polynomial in the variables x and y. By the commutativity of the convolution product, $f * \varphi_k(x) = \varphi_k * f(x) = \int_{\mathbb{R}} \varphi_k(x - t) f(t) \, dt$. Thus, Exercise 3.66 implies $f * \varphi_k(x)$ is a polynomial in x. Setting $p_k = \varphi_k * f$ completes the proof. $\qquad\square$

3.9 The space of Lebesgue integrable functions

The space of Lebesgue integrable functions possesses a natural distance function which we will study in this section and use to contrast the Lebesgue and Riemann integrals. If X is a nonempty set, a *semi-metric* on X is a function $d : X \times X \to [0, \infty)$ which satisfies for all $x, y, z \in X$:

(1) $d(x, y) = d(y, x)$ [symmetry];

(2) $d(x, z) \le d(x, y) + d(y, z)$ [triangle inequality].

A semi-metric d is a *metric* if

(3) $d(x, y) = 0$ if, and only if, $x = y$.

If d is a (semi-)metric on X, then the pair (X, d) is called a *(semi-)metric space*. Standard examples of metrics are the function $d(x, y) = |x - y|$ in $\mathbb{R}$ and $d(x, y) = \|x - y\|$ in $\mathbb{R}^n$. For a proof of the triangle inequality in $\mathbb{R}^n$, see Exercise 3.12.

Example 3.128 If S is any nonempty set, the function $d : S \times S \to [0, \infty)$ defined by

$$d(x, y) = \begin{cases} 0 \text{ if } x = y \\ 1 \text{ if } x \ne y \end{cases}$$

defines a metric on S. The metric d is called the *discrete metric* or the *distance-1 metric*.

It is common for a (semi-)metric in vector space to be induced by a function called a (semi-)norm. If X is a real vector, a *semi-norm* on X is a function $\| \ \| : X \to [0, \infty)$ which satisfies for all $x, y, z \in X$:

(1) $\|x\| \geq 0$;
(2) $\|tx\| = |t| \, \|x\|$ for all $t \in \mathbb{R}$;
(3) $\|x + y\| \leq \|x\| + \|y\|$.

Inequality (3) is known as the *triangle inequality*. From (2) it follows that $\|0\| = 0$. A semi-norm $\| \ \|$ is called a *norm* if, and only if:

(4) $\|x\| = 0$ if, and only if, $x = 0$.

If $\| \ \|$ is a (semi-)norm on X, then $\| \ \|$ induces a (semi-)metric d, often denoted $d_{\| \ \|}$, defined by $d(x, y) = \|x - y\|$ (see Exercise 3.68). For example, the standard distances in $\mathbb{R}$ and $\mathbb{R}^n$ are induced by the norms $\|t\| = |t|$ for $t \in \mathbb{R}$ and $\|x\| = \left(\sum_{i=1}^{n} |x_i|^2 \right)^{1/2}$ for $x \in \mathbb{R}^n$.

Let $E \subset \mathbb{R}^n$ be a measurable set and let $L^1(E)$ be the space of all real-valued Lebesgue integrable functions on E. We define an integral semi-norm $\| \ \|_1$ on $L^1(E)$, called the L^1-norm, by setting

$$\|f\|_1 = \int_E |f|.$$

The semi-metric d_1 induced by $\| \ \|_1$ is then $d_1(f, g) = \int_E |f - g|$, for all $f, g \in L^1(E)$. Since $\|f\|_1 = 0$ if, and only if, $f = 0$ a.e. in E, $\| \ \|_1$ is not a norm (and d_1 is not a metric). However, if we identify functions which are equal almost everywhere, then $\| \ \|_1$ is a norm and d_1 is a metric on $L^1(E)$.

Let d be a semi-metric on X. A sequence $\{x_k\}_{k=1}^{\infty} \subset X$ is said to *converge* to $x \in X$ if for every $\epsilon > 0$, there is an $N \in \mathbb{N}$ such that $d(x, x_k) < \epsilon$ whenever $k \geq N$. We call x the *limit* of the sequence $\{x_k\}_{k=1}^{\infty}$. A sequence $\{x_k\}_{k=1}^{\infty} \subset X$ is called a *Cauchy sequence* in X if for every $\epsilon > 0$, there is an $N \in \mathbb{N}$ such that $j, k \geq N$ implies that $d(x_j, x_k) < \epsilon$. By the triangle inequality, every convergent sequence is Cauchy. A semi-metric space is said to be *complete* if every Cauchy sequence in (X, d) converges to a point in X. For example, $\mathbb{R}$ is complete under its natural metric, while the subset $\mathbb{Q}$ of rational numbers is not complete under this metric. Similarly, $\mathbb{R}^n$ is complete under its natural metric. See Exercise 3.13.

Example 3.129 Let (S, d) be a discrete metric space. Then, every Cauchy sequence $\{x_k\}_{k=1}^{\infty} \subset S$ converges to an element of S since the sequence must eventually be constant. Thus, every discrete metric space is complete.

Completeness is an important property of a space since in a complete space, it suffices to show that a sequence is Cauchy in order to assert that the sequence converges.

We establish a theorem due to F. Riesz (1880-1956) and E. Fischer (1875-1954). The *Riesz-Fischer Theorem* asserts that $L^1(E)$ is complete under the semi-metric d_1.

Theorem 3.130 *(Riesz-Fischer Theorem) Let $E \in \mathcal{M}_n$ and let $\{f_k\}_{k=1}^{\infty}$ be a Cauchy sequence in $\left(L^1(E), d_1\right)$. Then, there is an $f \in L^1(E)$ such that $\{f_k\}_{k=1}^{\infty}$ converges to f in the metric d_1.*

Proof. Let $\{f_k\}_{k=1}^{\infty} \subset L^1(E)$ be a Cauchy sequence. We first show that there is a subsequence $\left\{f_{k_j}\right\}_{j=1}^{\infty} \subset \{f_k\}_{k=1}^{\infty}$ which converges a.e. to an integrable function. Since $\{f_k\}_{k=1}^{\infty}$ is Cauchy, we can pick a subsequence $\left\{f_{k_j}\right\}_{j=1}^{\infty}$ such that

$$d_1\left(f_{k_{j+1}}, f_{k_j}\right) < \frac{1}{2^j}.$$

It follows that

$$\sum_{j=1}^{\infty} d_1\left(f_{k_{j+1}}, f_{k_j}\right) \leq 1.$$

Set $g_j = \sum_{i=1}^{j} \left|f_{k_{i+1}} - f_{k_i}\right|$. Then, $\{g_j\}_{j=1}^{\infty}$ increases to the function g defined by

$$g(x) = \sum_{j=1}^{\infty} \left|f_{k_{j+1}}(x) - f_{k_j}(x)\right|.$$

Since $\int_E g_j \leq 1$, by the Monotone Convergence Theorem, g is Lebesgue integrable and, consequently, finite valued a.e..

Define f by

$$f(x) = \begin{cases} \sum_{j=1}^{\infty} \left\{f_{k_{j+1}}(x) - f_{k_j}(x)\right\} & \text{if the series converges absolutely} \\ 0 & \text{otherwise} \end{cases}.$$

Since the series defining f converges absolutely a.e. in E, the partial sums
$s_j(x) = \sum_{i=1}^{j} \{f_{k_{i+1}}(x) - f_{k_i}(x)\} = f_{k_{j+1}} - f_{k_1}$ converge to f a.e.. Since

$$|s_j(x)| = \left| \sum_{i=1}^{j} \{f_{k_{i+1}}(x) - f_{k_i}(x)\} \right| \le g_j(x) \le g(x)$$

for almost every x, by the Dominated Convergence Theorem, f is absolutely
Lebesgue integrable and

$$d_1(f + f_{k_1}, f_{k_j}) = \int_E |(f + f_{k_1}) - f_{k_{j+1}}|$$

$$= \int_E \left| f - \sum_{i=1}^{j} \{f_{k_{i+1}}(x) - f_{k_i}(x)\} \right| \to 0$$

as $j \to \infty$. Thus, $\{f_{k_j}\}_{j=1}^{\infty}$ converges to $f + f_{k_1}$. Since f and f_{k_1} are
Lebesgue integrable, $f + f_{k_1} \in L^1(E)$. Further, since $f_{k_{j+1}} - f_{k_1} = s_j$
converges to f a.e., it follows that $\{f_{k_j}\}_{j=1}^{\infty}$ converges pointwise to $f + f_{k_1}$
a.e. in E, so that $\{f_k\}_{k=1}^{\infty}$ has a subsequence that converges a.e. to an
integrable function.

It remains to show that $\{f_k\}_{k=1}^{\infty}$ converges to $f + f_{k_1}$. Fix $\epsilon > 0$. Since
$\{f_k\}_{k=1}^{\infty}$ is a Cauchy sequence in $L^1(E)$ and $f_{k_k} \to f + f_{k_1}$, there is a $K > 0$
such that for $k_j, k > K$,

$$\int_E |f_{k_j} - f_k| < \frac{\epsilon}{2} \quad \text{and} \quad \int_E |(f + f_{k_1}) - f_{k_j}| < \frac{\epsilon}{2}.$$

Fix $k_j > K$. If $k > K$, then

$$\int_E |(f + f_{k_1}) - f_k| \le \int_E |(f + f_{k_1}) - f_{k_j}| + \int_E |f_k - f_{k_j}| < \frac{\epsilon}{2} + \frac{\epsilon}{2} = \epsilon,$$

which implies that $\{f_k\}_{k=1}^{\infty}$ converges to $f + f_{k_1}$ in the metric d_1. $\square$

In contrast to the case of the Lebesgue integral, we show that the space
of Riemann integrable functions is not complete under the natural semi-
metric d_1, further justifying that the Lebesgue integral is superior to the
Riemann integral. Let $R([a, b])$ be the space of Riemann integrable functions on $[a, b]$.

Example 3.131   Define $f_k : [0, 1] \to \mathbb{R}$ by setting

$$f_k(x) = \begin{cases} 0 & \text{if } 0 \le x \le \frac{1}{k} \\ x^{-1/2} & \text{if } \frac{1}{k} \le x \le 1 \end{cases}.$$

It is easily checked that $\{f_k\}_{k=1}^\infty$ is a Cauchy sequence in $(R([0,1]), d_1)$. However, $\{f_k\}_{k=1}^\infty$ does not converge to a function in $R([0,1])$. For, suppose that $\{f_k\}_{k=1}^\infty$ converges to f with respect to d_1. It follows from the Monotone Convergence Theorem that $\{f_k\}_{k=1}^\infty$ converges in d_1 to the function $g : [0,1] \to \mathbb{R}$ defined by

$$g(x) = \begin{cases} 0 & \text{if } x = 0 \\ x^{-1/2} & \text{if } 0 < x \le 1 \end{cases}.$$

This implies that $f = g$ a.e. in $[0,1]$ so that the function f does not belong to $R([0,1])$ since f is unbounded.

Note that another counterexample is provided by the functions in Example 3.3.

As an application of the Riesz-Fischer Theorem, we will prove a substitution result for the Lebesgue integral. To do so, we need to extend some earlier results and make an observation about the proof of the Riesz-Fischer Theorem. Recall that a function g is called Lipschitz on an interval $[a, b]$ if there is a constant B such that $|g(x) - g(y)| \le B |x - y|$ for all $x, y \in [a, b]$. Observe first that Lemmas 3.118 and 3.119 are valid with linear replaced by Lipschitz. This follows from the fact that the proof of the first lemma uses only the Lipschitz condition, which is satisfied by all linear functions, and the second lemma follows from the first lemma. The second observation we need is that if $\{f_k\}_{k=1}^\infty$ is Cauchy in L^1 with limit function f, then there is a subsequence $\{f_{k_j}\}_{j=1}^\infty \subset \{f_k\}_{k=1}^\infty$ that converges pointwise to f, which can be found in the proof of the Riesz-Fischer Theorem. We can now prove a change of variables result for the Lebesgue integral.

Theorem 3.132 *Let $\phi : [a, b] \to [\alpha, \beta]$ be continuously differentiable, $\phi' > 0$, $\phi(a) = \alpha$, and $\phi(b) = \beta$. If $f : [\alpha, \beta] \to \mathbb{R}$ is Lebesgue integrable, then $(f \circ \phi) \phi'$ is Lebesgue integrable over $[a, b]$ and*

$$\int_\alpha^\beta f(x)\, dx = \int_a^b f(\phi(t)) \phi'(t)\, dt.$$

Proof. We may assume that $f \ge 0$. Suppose that f is bounded with $|f(t)| \le M$ for all $t \in [\alpha, \beta]$. By Theorem 3.67, there are step functions $\{s_k\}_{k=1}^\infty$ that converge pointwise a.e. to f and such that $|s_k(t)| \le M$ for $t \in [\alpha, \beta]$. By the Dominated Convergence Theorem 3.100, $\int_\alpha^\beta |f - s_k| \to 0$. Since Theorem 2.35 holds for step functions, we have

$$\int_\alpha^\beta s_k(x)\, dx = \int_a^b s_k(\phi(t)) \phi'(t)\, dt$$

for every $k \in \mathbb{N}$. Since $|s_k - s_j|$ is a step function for all $k, j \in \mathbb{N}$, we also have

$$\int_\alpha^\beta |s_k(x) - s_j(x)| \, dx = \int_a^b |s_k(\phi(t)) - s_j(\phi(t))| \, \phi'(t) \, dt$$

which implies that $\{(s_k \circ \phi) \phi'\}_{k=1}^\infty$ is Cauchy in $L^1[a, b]$. Therefore, by the Riesz-Fischer Theorem 3.130, there is $g \in L^1[a, b]$ such that $(s_k \circ \phi) \phi' \to g$ in $L^1[a, b]$. As we observed above, $\{(s_k \circ \phi) \phi'\}_{k=1}^\infty$ has a subsequence that converges to g a.e., so we may assume that the original sequence $\{(s_k \circ \phi) \phi'\}_{k=1}^\infty$ converges to g a.e. Let Z be the null set such that $s_k \to f$ pointwise on $[\alpha, \beta] \setminus Z = Z^c$. Now, $s_k(\phi(t)) \phi'(t) \to f(\phi(t)) \phi'(t)$ for $\phi(t) \in Z^c$, or $(s_k \circ \phi) \phi' \to (f \circ \phi) \phi'$ on $\phi^{-1}(Z^c)$. Since ϕ is continuously differentiable, ϕ satisfies a Lipschitz condition and, therefore, $\phi^{-1}(Z^c)$ is null. Hence, $(s_k \circ \phi) \phi' \to (f \circ \phi) \phi'$ a.e. in $[a, b]$. It now follows that $(f \circ \phi) \phi' = g$ a.e. so that $(f \circ \phi) \phi'$ is Lebesgue integrable and

$$\int_\alpha^\beta s_k(x) \, dx = \int_a^b s_k(\phi(t)) \phi'(t) \, dt \to \int_\alpha^\beta f(x) \, dx = \int_a^b f(\phi(t)) \phi'(t) \, dt$$

so that the result holds for bounded functions.

To complete the proof, assume that f is non-negative and Lebesgue integrable. Set $f_k = f \wedge k$ so that $f_k \uparrow f$ and $(f_k \circ \phi) \phi' \uparrow (f \circ \phi) \phi'$. By the first part of this proof and the Monotone Convergence Theorem 3.77, we get

$$\int_\alpha^\beta f_k(x) \, dx = \int_a^b f_k(\phi(t)) \phi'(t) \, dt \to \int_\alpha^\beta f(x) \, dx = \int_a^b f(\phi(t)) \phi'(t) \, dt$$

which completes the proof. $\qquad\square$

Let (S, d) be a metric space. Let D be a subset of S. As in the case of subsets of $\mathbb{R}$, we define the *closure* of D, denoted $\overline{D}$, to be the set of all $x \in S$ for which there is a sequence $\{s_n\}_{n=1}^\infty \subset D$ that converges to x. A set D in a semi-metric space (S, d) is called *dense* if $\overline{D} = S$. We next give examples of dense subsets of L^1 and then give some applications showing the use of the dense subsets. First we consider the space of step functions.

Theorem 3.133 *Let $E \in \mathcal{M}_n$. The vector space of step functions is dense in $L^1(E)$.*

Proof. Let $f \in L^1(E)$, $f \geq 0$ and $\epsilon > 0$. First assume f is bounded by M and $m_n(E) < \infty$. By Theorem 3.67 there is a sequence of step functions $\{s_k\}_{k=1}^{\infty}$ converging to f a.e. such that $0 \le s_k \le M$. By the Bounded Convergence Theorem (Theorem 3.101), $\int_E |s_k - f| \to 0$.

Suppose that E has infinite measure. Set

$$E_k = E \cap \{t \in E : f(t) \le k\} \cap [-k, k]^n$$

so $E_k \uparrow E$. Since E_k has finite measure, we can use the previous part of the proof to show that for every k there exists a step function s_k such that $\int_{E_k} |s_k - f| < \epsilon/2$ and $s_k = 0$ outside E_k. Then,

$$\int_E |s_k - f| = \int_{E_k} |s_k - f| + \int_{E \setminus E_k} |s_k - f| < \epsilon/2 + \int_{E \setminus E_k} f.$$

Since $\int_{E_k} f \uparrow \int_E f$, the last term on the right hand side above can be made less than $\epsilon/2$ for large enough k. This shows that f can be approximated by step functions on E. Finally, the proof is completed by writing $f = f^+ - f^-$ and applying this result to both f^+ and f^-. □

We next consider spaces of continuous functions which are dense in L^1. Let $C_c(\mathbb{R}^n)$ be the vector space of all real valued continuous functions on $\mathbb{R}^n$ which vanish outside a compact set. We will use the following lemma. In the proof, we define the distance from a point x to a set E by

$$dist\,(x, E) = \inf \{d\,(x, y) : y \in E\}.$$

Note that $dist\,(x, E)$ is a uniformly continuous function of x. See [DS, Proposition 21.13].

Lemma 3.134 *If $K \subset \mathbb{R}^n$ is compact and $V \subset \mathbb{R}^n$ is open with $K \subset V$, then there exists $f \in C_c(\mathbb{R}^n)$ such that $0 \le f \le 1$, $f = 1$ on K and $f = 0$ outside V.*

Proof. Pick an open and bounded set U such that $K \subset U \subset \overline{U} \subset V$. Set $f(t) = dist\,(t, U^c) / \{dist\,(t, K) + dist\,(t, U^c)\}$. It now follows that f is continuous, equal to 1 on K, $0 \le f \le 1$, and since f equals 0 outside of V, $f \in C_c(\mathbb{R}^n)$. □

Using this lemma, we can prove that $C_c(\mathbb{R}^n)$ is dense in $L^1(\mathbb{R}^n)$.

Theorem 3.135 *The space $C_c(\mathbb{R}^n)$ is dense in $L^1(\mathbb{R}^n)$.*

Proof. Since the step functions are dense in $L^1(\mathbb{R}^n)$ by Theorem 3.133, it suffices to show that if B is a brick, then for an $\epsilon > 0$, there exists $f \in C_c(\mathbb{R}^n)$ such that $\int_{\mathbb{R}^n} |\chi_B - f| < \epsilon$. Since the boundary of a brick has measure 0, we may assume B is open. By the n-dimensional analog of Theorem 3.36, there is a closed set $K \subset B$ such that $m_n(B \setminus K) < \epsilon$. Since B is bounded, K is compact. So, by the previous lemma, there exists $f \in C_c(\mathbb{R}^n)$, $0 \leq f \leq 1$, $f = 1$ on K, $f = 0$ on B^c. Then

$$\int_{\mathbb{R}^n} |\chi_B - f| = \int_{B \setminus K} |\chi_B - f| \leq m_n(B \setminus K) < \epsilon.$$

It now follows that the continuous functions with compact support are dense in $L^1(\mathbb{R}^n)$.   □

The two previous theorems are useful for establishing results for elements of L^1. In particular, if one can prove a result about a dense subset of L^1, then one can use the density condition to extend the result to all functions in L^1. We give two examples below. To begin, we establish a version of the Riemann-Lebesgue Lemma.

Lemma 3.136   *(Riemann-Lebesgue) If $f \in L^1[a, b]$, then*

$$\lim_{k \to \infty} \int_a^b f(t) \sin(kt) dt = 0.$$

Proof.   Let $a \leq c \leq d \leq b$ and suppose f has constant value α on $[c, d]$. Then

$$\left| \int_c^d f(t) \sin(kt) dt \right| = \left| \int_c^d \alpha \sin(kt) dt \right|$$

$$= \left| \frac{\alpha}{k} \right| |\cos(kd) - \cos(kc)| \leq 2 \left| \frac{\alpha}{k} \right| \to 0$$

Thus, $\lim_{k \to \infty} \int_c^d f(t) \sin(kt) dt = 0$ and it follows from linearity that $\lim_{k \to \infty} \int_c^d s(t) \sin(kt) dt = 0$ when s is a step function.

Let $f \in L^1[a, b]$ and $\epsilon > 0$. There exists a step function s such that $\int_a^b |f - s| < \epsilon/2$. By the previous paragraph, we can choose $N \in \mathbb{N}$ such that $\left| \int_a^b s(t) \sin(kt) dt \right| < \epsilon/2$ for $k \geq N$. Then, if $k \geq N$,

$$\left| \int_a^b f(t) \sin(kt) \, dt \right| \leq \left| \int_a^b f(t) \sin(kt) \, dt - \int_a^b s(t) \sin(kt) \, dt \right|$$

$$+ \left| \int_a^b s(t) \sin(kt) \, dt \right|$$

$$\leq \int_a^b |f(t) - s(t)| \, dt + \left| \int_a^b s(t) \sin(kt) \, dt \right|$$

$$< \frac{\epsilon}{2} + \frac{\epsilon}{2} = \epsilon.$$

This completes the proof. □

Given a function f, we define the translation of f by $h \in \mathbb{R}^n$ to be the function f_h defined by $f_h(t) = f(t+h)$. Note that f_h is in $L^1(\mathbb{R}^n)$ whenever f is. We give an application of Theorem 3.135 to show that the translations f_h approximate f in $L^1(\mathbb{R}^n)$.

Proposition 3.137 *If $f \in L^1(\mathbb{R}^n)$, then $\lim_{h \to 0} \|f_h - f\|_1 = 0$.*

Proof. Let $\epsilon > 0$. By Theorem 3.135, there exists $g \in C_c(\mathbb{R}^n)$ such that $\|f - g\|_1 < \epsilon/3$. Then, g is uniformly continuous with compact support, so there is an M such that $g(t) = 0$ for $\|t\| > M$. By the uniform continuity, there is a δ, $0 < \delta < 1$, so that

$$|g(t+h) - g(t)| < \frac{\epsilon}{3 \left[2(M+1) \right]^n}$$

for $\|h\| < \delta$. Since the support of g_h is contained in the cube $Q = [-(M+1), M+1]^n$, if $\|h\| < \delta$, then

$$\|g_h - g\|_1 = \int_{\mathbb{R}^n} |g(t+h) - g(t)| \, dt \leq \frac{\epsilon}{3 \left[2(M+1) \right]^n} \int_Q dt = \frac{\epsilon}{3}.$$

Altogether, if $\|h\| < \delta$, then

$$\|f_h - f\|_1 \leq \|f_h - g_h\|_1 + \|g_h - g\|_1 + \|g - f\|_1 < \frac{\epsilon}{3} + \frac{\epsilon}{3} + \frac{\epsilon}{3} = \epsilon,$$

since $\|f_h - g_h\|_1 = \|g - f\|_1$, so the result follows. □

We now use this result to establish some approximation results as was done in Section 3.8. These results use approximate identities which were defined there.

Theorem 3.138   *Let* $\{\varphi_k\}_{k=1}^{\infty}$ *be an approximate identity in* $\mathbb{R}^n$. *If* $f :$
$\mathbb{R}^n \to \mathbb{R}$ *is integrable, then* $\|f * \varphi_k - f\|_1 \to 0$ *as* $k \to \infty$.

Proof.   If $\|f\|_1 = 0$ then $f = 0$ a.e. and there is nothing to prove. Thus,
we may assume that $\|f\|_1 > 0$. Let $\epsilon > 0$. By the previous proposition,
there exists $\delta > 0$ such that $\|h\| < \delta$ implies $\|f_h - f\|_1 < \epsilon/2$. By (3)
of Definition 3.121, there exists N such that $k \geq N$ implies $\int_{\|x\| \geq \delta} \varphi_k <$
$\epsilon/4 \|f\|_1$. Then changing the order of integration and using (2) of the
definition, we have

$$\|f * \varphi_k - f\|_1 = \int_{\mathbb{R}^n} \left| \int_{\mathbb{R}^n} (f(x-y) - f(x)) \varphi_k(y) dy \right| dx$$

$$\leq \int_{\mathbb{R}^n} \varphi_k(y) \int_{\mathbb{R}^n} |f(x-y) - f(x)| dx dy$$

$$= \int_{\mathbb{R}^n} \varphi_k(y) \|f_{-y} - f\|_1 dy$$

$$= \int_{\{y: \|y\| \leq \delta\}} \varphi_k(y) \|f_{-y} - f\|_1 dy$$

$$+ \int_{\{y: \|y\| > \delta\}} \varphi_k(y) \|f_{-y} - f\|_1 dy$$

$$\leq \frac{\epsilon}{2} \int_{\{y: \|y\| \leq \delta\}} \varphi_k(y) dy + 2 \|f\|_1 \int_{\{y: \|y\| > \delta\}} \varphi_k(y) dy$$

$$< \frac{\epsilon}{2} + 2 \|f\|_1 \frac{\epsilon}{4 \|f\|_1} = \epsilon.$$

Thus, $\|f * \varphi_k - f\|_1 \to 0$ as $k \to \infty$ as we wished to show. $\square$

Let $C_c^{\infty}(\mathbb{R}^n)$ be the space of all functions $f : \mathbb{R}^n \to \mathbb{R}$ which are infinitely differentiable and have compact support. We can now show that
$C_c^{\infty}(\mathbb{R}^n)$ is dense in $L^1(\mathbb{R}^n)$.

Theorem 3.139 $C_c^{\infty}(\mathbb{R}^n)$ *is dense in* $L^1(\mathbb{R}^n)$ *with respect to* $\|\cdot\|_1$.

Proof.   By Theorem 3.135, it suffices to show $C_c^{\infty}(\mathbb{R}^n)$ is dense in $C_c(\mathbb{R}^n)$
in the $L^1(\mathbb{R}^n)$ metric. Let $g \in C_c^{\infty}(\mathbb{R}^n)$ and set $K = \text{supp}(g)$. Choose $\{\varphi_k\}$
as in Example 3.125. If $x \in \mathbb{R}^n$ and $dist(x, K) > 1/k$, then $\varphi_k(x-y) = 0$
for $y \in K$. Thus,

$$\varphi_k * g(x) = \int_K \varphi_k(x-y) g(y) dy$$

vanishes for such x; i.e., $\text{supp}(\varphi_k * g) \subset \{x : dist(x, K) \leq 1/k\}$. Hence,
$\varphi_k * g$ has compact support and is infinitely differentiable by Exercise 3.63 so
$\varphi_k * g \in C_c^{\infty}(\mathbb{R}^n)$ and $\varphi_k * g \to g$ in $L^1(\mathbb{R}^n)$ by the previous theorem. $\square$

3.10   Exercises

Measure

Exercise 3.1   Prove that outer measure is translation invariant.

Exercise 3.2   Let $\{I_{i,j}\}_{i,j=1}^{\infty}$ be a doubly indexed collection of intervals. Prove that

$$\sum_{i,j=1}^{\infty} \ell\left(I_{i,j}\right) = \sum_{i=1}^{\infty}\sum_{j=1}^{\infty} \ell\left(I_{i,j}\right).$$

Exercise 3.3   Let $\{a_{jk}\}_{j,k=1}^{\infty}$ be a doubly-indexed sequence of nonnegative terms such that $a_{jk} \leq a_{j(k+1)}$ for all j and k. Prove that

$$\lim_{k\to\infty}\sum_{j} a_{jk} = \sum_{j}\left(\lim_{k\to\infty} a_{jk}\right).$$

Exercise 3.4   Prove that every subset of a null set is a null set and that a countable union of null sets is a null set.

Exercise 3.5   Prove that

$$\mathcal{A} = \{F \subset (0,1) : F \text{ or } (0,1)\setminus F \text{ is a finite or empty set}\}$$

is an algebra.

Exercise 3.6   Let X be a set and $S \subset \mathcal{P}(X)$. Let

$$\mathcal{F} = \{\mathcal{B} : S \subset \mathcal{B} \text{ and } \mathcal{B} \text{ is a } \sigma\text{-algebra}\}.$$

Prove that $C = \cap_{\mathcal{B}\in\mathcal{F}}\mathcal{B}$ is the smallest σ-algebra that contains S.

Exercise 3.7   Suppose that E has finite measure. Show that we can replace "there is a closed set F" in Theorem 3.36 part *(3)* by "there is a compact set F".

Exercise 3.8   A measure μ defined for all elements of $\mathcal{B}(\mathbb{R})$ is called *inner regular* if

$$\mu(E) = \sup\{\mu(K) : K \subset E, K \text{ compact}\}$$

for all $E \in \mathcal{B}(\mathbb{R})$. Prove that Lebesgue measure restricted to the Borel sets is an inner regular measure.

Exercise 3.9   Prove that the complement of the Cantor set is dense in $[0,1]$.

Exercise 3.10   Show that every countable set is a Borel set.

Lebesgue measure in $\mathbb{R}^n$

Exercise 3.11   Prove the Cauchy-Schwarz inequality. That is, if $x, y \in \mathbb{R}^n$, show that $|x \cdot y| = |\sum_{i=1}^{n} x_i y_i| \le \|x\| \, \|y\|$ by expanding the sum

$$\sum_{i=1}^{n} \sum_{j=1}^{n} (x_i y_j - x_j y_i)^2 .$$

Exercise 3.12   Use the Cauchy-Schwarz inequality to prove that $d(x,y) = \|x - y\|$ defines a metric on $\mathbb{R}^n$.

Exercise 3.13   Prove that $(\mathbb{R}^n, d)$ is a complete metric space.

Exercise 3.14   Prove that m_n^* is translation invariant; that is, given $E \subset \mathbb{R}^n$ and $h \in \mathbb{R}^n$, $m_n^*(E + h) = m_n^*(E)$.

Exercise 3.15   Prove that m_n^* is homogeneous of degree n; that is, given $E \subset \mathbb{R}^n$ and $a > 0$, $m_n^*(aE) = a^n m_n^*(E)$, where $aE = \{y \in \mathbb{R}^n : y = ax \text{ for some } x \in E\}$.

Exercise 3.16   Let $E \subset \mathbb{R}^n$.

(1) Prove that E is measurable if, and only if, $E + h$ is measurable for all $h \in \mathbb{R}^n$.
(2) Prove that E is measurable if, and only if, aE is measurable for all $a > 0$.

Exercise 3.17   Suppose that $E \subset \mathbb{R}^j$ is a null set and $F \subset \mathbb{R}^k$. Prove that $E \times F$ is a null set in $\mathbb{R}^{j+k}$.

Exercise 3.18   Either prove or give a counterexample to the following statement: if $E \subset \mathbb{R}$ is measurable and $m(E) > 0$, then E must contain a non-degenerate interval.

Measurable functions

Exercise 3.19   Prove that $E \subset \mathbb{R}$ is a measurable set if, and only if, χ_E is a measurable function.

Exercise 3.20   Prove that the remark following Proposition 3.50 is valid.

Exercise 3.21   Give an example of a nonmeasurable function f on $[0,1]$ such that $|f|$ is measurable.

Exercise 3.22 Let $\{x_i\}_{i=1}^{\infty} \subset \mathbb{R}$. Prove that $\lim_{i \to \infty} x_i$ exists if, and only if, $\limsup x_i = \liminf x_i$.

Exercise 3.23 Suppose that f and g are measurable functions. Prove that $|f|^a$ is measurable for all $a > 0$. Prove that f/g is a measurable function if it is defined a.e..

Exercise 3.24 Suppose that $f : E \to \mathbb{R}^*$ is measurable. Show that there is a sequence of bounded measurable functions $\{f_k\}_{k=1}^{\infty}$ which converges to f pointwise on E.

Exercise 3.25 Show that any derivative is measurable by showing that a derivative is the pointwise limit of a sequence of continuous functions. That is, if $f : [a, b] \to \mathbb{R}$ is differentiable on $[a, b]$, then f' is measurable on $[a, b]$.

Exercise 3.26 Suppose that $f : \mathbb{R}^n \to \mathbb{R}$ and $g : \mathbb{R}^k \to \mathbb{R}$ are measurable. Define $f \otimes g : \mathbb{R}^{n+k} = \mathbb{R}^n \times \mathbb{R}^k \to \mathbb{R}$ by $f \otimes g(x, y) = f(x) g(y)$. Prove that $f \otimes g$ is a measurable function on $\mathbb{R}^{n+k}$.

Exercise 3.27 Let $E \subset \mathbb{R}^n$ be a Lebesgue measurable set. Suppose that $f : E \to \mathbb{R}$ is Lebesgue measurable and $g : \mathbb{R} \to \mathbb{R}$ is continuous. Prove that $g \circ f$ is a Lebesgue measurable function. Note that we cannot conclude that $f \circ g$ is measurable. See [Mu, pages 148-149].

Lebesgue integral

Exercise 3.28 Let $f(0) = 0$ and $f(t) = t^p$ for $t > 0$. For what values of p is f integrable over $[0, 1]$? Over $[1, \infty)$?

Exercise 3.29 Suppose that $f : E \to \mathbb{R}$ is measurable. If E has finite measure and f is bounded, show that f is Lebesgue integrable.

Exercise 3.30 Suppose that g is a bounded, measurable function on E and f is Lebesgue integrable over E. Prove that fg is Lebesgue integrable over E.

Exercise 3.31 Suppose $g : I \to \mathbb{R}$ is measurable and fg is integrable over I for every f integrable on I. Show that g is bounded on the complement of a null set; i.e., g is *essentially bounded*.

Exercise 3.32 Give an example of a pair of integrable functions f and g such that fg is not integrable.

Exercise 3.33 Let $f : E \subset \mathbb{R}^n \to \mathbb{R}^*$ be Lebesgue integrable.

(1) Let $\alpha \in \mathbb{R}^n$ and define $f_\alpha : E - \alpha \to \mathbb{R}^*$ by $f_\alpha(t) = f(t + \alpha)$. Prove that f_α is Lebesgue integrable and satisfies the linear change of variables

$$\int_{E-\alpha} f_\alpha dm_n = \int_E f dm_n.$$

(2) Let $\lambda > 0$. Define $\frac{1}{\lambda} E = \{x \in \mathbb{R}^n : \lambda x \in E\}$ and $g_\lambda : \frac{1}{\lambda} E \to \mathbb{R}^*$ by $g_\lambda(t) = \chi^n g(\lambda t)$. Prove that g_λ is Lebesgue integrable and satisfies the dilation change of variables

$$\int_{\frac{1}{\lambda} E} g_\lambda dm_n = \int_E g dm_n.$$

Exercise 3.34   Let $f : [0, 1] \to \mathbb{R}$ be continuous. Show that the functions $x \to f(x^m)$ are Lebesgue integrable for all m and $\int_0^1 f(x^m) \, dx \to f(0)$.

Exercise 3.35   Evaluate $\lim_n \int_0^n \left(1 + \frac{x}{n}\right)^n e^{-2x} dx$.

Exercise 3.36   Let $f : [0, \infty) \to \mathbb{R}$ be bounded and continuous. Define the *Laplace transform* of f by

$$F(x) = \int_0^\infty e^{-xt} f(t) \, dt.$$

Show that F exists and is continuous and differentiable on (a, ∞) for all $a > 0$.

Exercise 3.37   Let f be Lebesgue integrable on $\mathbb{R}^n$ and define F by $F(E) = \int_E f dm_n$ for all $E \in \mathcal{M}_n$. Show that F is countably additive; that is, $F(\cup_{i=1}^\infty E_i) = \sum_{i=1}^\infty F(E_i)$ for all sequences of pairwise disjoint sets $\{E_i\}_{i=1}^\infty \subset \mathcal{M}_n$.

Exercise 3.38   Suppose that $f : \mathbb{R} \to \mathbb{R}$ is Lebesgue integrable. Show that

$$\lim_{x \to \infty} \int_x^{x+1} f = 0.$$

Exercise 3.39   Suppose that $f \in L^1(\mathbb{R})$. Show that $\lim_{b \to \infty} \int_b^\infty |f| = 0$.

Exercise 3.40   Let $a \in \mathbb{R}$. Prove that $f \in L^1([a, \infty])$ if and only if $f \in L^1([a, b])$ for all $b > a$ and $\left\{ \int_a^b |f(t)| \, dt : b > a \right\}$ is a bounded set.

Exercise 3.41   Suppose that $f_k, h : E \subset \mathbb{R}^n \to \mathbb{R}^*$ are Lebesgue integrable over E and $h \leq f_k$ a.e. for all k. Prove that $\inf_k f_k$ is Lebesgue integrable over E. Can the boundedness condition be deleted?

Exercise 3.42 Prove that Corollary 3.78 implies Theorem 3.77, and hence show that the two are equivalent.

Exercise 3.43 Prove Corollary 3.86.

Exercise 3.44 Prove Corollary 2.23.

Exercise 3.45 Let $I \subset \mathbb{R}$ be an interval. Show that $Z \subset I$ is a null set if and only if there exist an increasing sequence $\{f_k\}_{k=1}^{\infty} \subset L^1(I)$ such that $f_k(t) \to \infty$ for every $t \in Z$ and $\{\int_I f_k\}$ converges in $\mathbb{R}$.

Exercise 3.46 Let $I \subset \mathbb{R}$ be an interval. Show that $Z \subset I$ is a null set if and only if there a sequence $\{f_k\}_{k=1}^{\infty} \subset L^1(I)$ such that $\sum_{k=1}^{\infty} |f_k(t)| = \infty$ for every $t \in Z$ and $\sum_{k=1}^{\infty} \int_I |f_k| < \infty$.

Exercise 3.47 Let $f : [0,1] \to \mathbb{R}$ be Lebesgue integrable. Show that the functions $x \to x^k f(x)$ are Lebesgue integrable for all k and $\int_0^1 x^k f(x)\, dx \to 0$.

Exercise 3.48 If $f : E \subset \mathbb{R}^n \to \mathbb{R}^*$ is Lebesgue integrable and $A_k = \{x : |f(x)| > k\}$, prove that $m_n(A_k) \to 0$ as $k \to \infty$.

Exercise 3.49 Let $A \subset \mathbb{R}^j$ and $B \subset \mathbb{R}^k$ be compact sets. Suppose that $f : A \times B \to \mathbb{R}$ is continuous. Define $F : A \to \mathbb{R}$ by $F(x) = \int_B f(x,y)\, dy$. Show that F is continuous.

Exercise 3.50 If $f : \mathbb{R} \to \mathbb{R}$ is Lebesgue integrable over $\mathbb{R}$ and uniformly continuous on $\mathbb{R}$, show that $\lim_{|x| \to \infty} f(x) = 0$.

Exercise 3.51 Suppose $\{f_k\}_{k=1}^{\infty}$ is a sequence of Lebesgue integrable functions such that $\int_E |f_k| \le M$ for all k. Show that if $\{t_k\}_{k=1}^{\infty}$ satisfies $\sum_{k=1}^{\infty} |t_k| < \infty$, then the series $\sum_{k=1}^{\infty} t_k f_k(x)$ is absolutely convergent for almost all $x \in E$.

Riemann and Lebesgue integrals

Exercise 3.52 Let $\mathcal{R} = \{A \subset [0,1] : \chi_A \text{ is Riemann integrable}\}$. Prove that $\mathcal{R}$ is an algebra which is not a σ-algebra.

Exercise 3.53 Prove that a function which is absolutely Cauchy-Riemann integrable is Lebesgue integrable and the integrals agree.

Fubini's Theorem

Exercise 3.54 Define $f : \mathbb{R}^2 \to \mathbb{R}$ by

$$f(x,y) = \begin{cases} 0 & \text{if } (x,y) = (0,0) \\ xy/(x^2 + y^2) & \text{if } (x,y) \ne (0,0) \end{cases}.$$

Show that

$$\int_{-1}^{1}\int_{-1}^{1} f(x,y)\,dxdy = \int_{-1}^{1}\int_{-1}^{1} f(x,y)\,dydx$$

but f is not integrable over $[-1,1] \times [-1,1]$. [Hint: Consider the integral over the set $[0,1] \times [0,1]$.]

Exercise 3.55   Define $f : \mathbb{R}^2 \to \mathbb{R}$ by

$$f(x,y) = \begin{cases} 0 & \text{if } (x,y) = (0,0) \\ (x^2 - y^2)/(x^2+y^2)^2 & \text{if } (x,y) \neq (0,0) \end{cases}.$$

Compare $\int_{-1}^{1}\int_{-1}^{1} f(x,y)\,dxdy$ and $\int_{-1}^{1}\int_{-1}^{1} f(x,y)\,dydx$.

Exercise 3.56   Find the area inside of a circle of radius r and of an ellipse $\frac{x^2}{a^2} + \frac{y^2}{b^2} = 1$.

Exercise 3.57   Find the volume inside of the ellipsoid $\frac{x^2}{a^2} + \frac{y^2}{b^2} + \frac{z^2}{c^2} = 1$.

Exercise 3.58   Suppose that $f : \mathbb{R}^n \to \mathbb{R}$ and $g : \mathbb{R}^k \to \mathbb{R}$ are Lebesgue integrable. Prove that $f \otimes g$, defined in Exercise 3.26, is a Lebesgue integrable function on $\mathbb{R}^{n+k}$ and

$$\int_{\mathbb{R}^{n+k}} f \otimes g\,dm_{n+k} = \int_{\mathbb{R}^n} f\,dm_n \int_{\mathbb{R}^k} g\,dm_k.$$

Exercise 3.59   Suppose $f, g, h \in L^1(\mathbb{R}^n)$. Show $f * (g * h) = (f * g) * h$ and $f * (g+h) = f * g + f * h$.

Exercise 3.60   Since the convolution product is commutative, Exercise 3.59 implies that $L^1(\mathbb{R}^n)$ is an algebra with convolution as that product. Show $L^1(\mathbb{R}^n)$ has no identity. [Hint: Suppose u is an identity. There exists $a > 0$ such that $\int_{-a}^{a} |u| < 1$. Consider $f = \chi_{[-a,a]}$ and $f * u = f$.]

Exercise 3.61   Suppose $f : \mathbb{R}^n \to \mathbb{R}$ is bounded and uniformly continuous and $g : \mathbb{R}^n \to \mathbb{R}$ is integrable. Show $f * g$ exists everywhere and is bounded and uniformly continuous.

Exercise 3.62   Suppose $f : \mathbb{R}^n \to \mathbb{R}$ is bounded and $g : \mathbb{R}^n \to \mathbb{R}$ is integrable. Show $f * g$ exists everywhere and is bounded and uniformly continuous. [Hint: Use Proposition 3.137 in Section 3.9.]

Exercise 3.63   Suppose $f \in C_c^\infty(\mathbb{R}^n)$ and $g \in L^1(\mathbb{R}^n)$. Show $f * g \in C^\infty(\mathbb{R}^n) \cap L^1(\mathbb{R}^n)$ and give a formula for computing the partial derivatives.

Exercise 3.64 Let $f \in L^1(\mathbb{R}^n)$ and suppose $\int_{\mathbb{R}^n} fg = 0$ for every $g \in C_c^\infty(\mathbb{R}^n)$. Show $f = 0$ a.e..

Exercise 3.65 Let $\{\varphi_k\}$ be an approximate identity. If $g : \mathbb{R}^n \to \mathbb{R}$ is bounded, measurable and continuous at x, show $g * \varphi_k(x) \to g(x)$.

Exercise 3.66 Suppose p is a polynomial in the variables x and y and f is continuous with compact support. Show $\int_{\mathbb{R}} p(x,y) f(y) dy$ defines a polynomial in x.

The Space of Lebesgue integrable functions

Exercise 3.67 If $X \neq \{0\}$ is a vector space, show that the distance-1 metric on X is not induced by a norm.

Exercise 3.68 Let $\| \ \|$ be a (semi-)norm on a vector space X. Prove that $d(x,y) = \|x - y\|$ defines a (semi-)metric.

Exercise 3.69 Show that

$$d_1(x,y) = \sum_{i=1}^n |x_i - y_i|$$

and

$$d_\infty(x,y) = \max\{|x_i - y_i| : 1 \le i \le n\}$$

define metrics on $\mathbb{R}^n$.

Exercise 3.70 Let $(X, \|\cdot\|)$ be a normed linear space and $f : X \to \mathbb{R}$ be a linear functional. Show that the following conditions are equivalent:

(1) f is continuous on X;
(2) f is continuous at 0;
(3) there exists an M such that $|f(x)| \le M \|x\|$ for every $x \in X$.

Exercise 3.71 We say a space X is *separable* if it has a countable, dense subset. Show that $L^1(E)$ is separable. (Use Theorem 3.133 with rational coordinates.)

Exercise 3.72 Suppose $g : I \to \mathbb{R}$ is bounded and measurable. Show $G : L^1(I) \to \mathbb{R}$ defined by $G(f) = \int_I gf$ is linear and continuous.

Chapter 4

Fundamental Theorem of Calculus and the Henstock-Kurzweil integral

In Chapter 2, we gave a brief discussion of the Fundamental Theorem of Calculus for the Riemann integral. In the first part of this chapter, we consider Part I of the Fundamental Theorem of Calculus for the Lebesgue integral and show that the Lebesgue integral suffers from the same defect with respect to Part I of the Fundamental Theorem of Calculus as does the Riemann integral. We then use this result to motivate the discussion of the Henstock-Kurzweil integral for which Part I of the Fundamental Theorem of Calculus holds in full generality.

Recall that Part I of the Fundamental Theorem of Calculus involves the integration of the derivative of a function f and the formula

$$\int_a^b f' = f(b) - f(a). \tag{4.1}$$

In Example 2.31, we gave an example of a derivative which is unbounded and is, therefore, not Riemann integrable, and we showed in Theorem 2.30 that if f' is Riemann integrable, then (4.1) holds. That is, in order for (4.1) to hold, the assumption that the derivative f' is Riemann integrable is required. It would be desirable to have an integration theory for which Part I of the Fundamental Theorem of Calculus holds in full generality. That is, we would like to have an integral which integrates all derivatives and satisfies (4.1). Unfortunately, the example below shows that the general form of Part I of the Fundamental Theorem of Calculus does not hold for the Lebesgue integral.

Example 4.1 In Example 2.31, we considered the function f defined by $f(0) = 0$ and $f(x) = x^2 \cos \frac{\pi}{x^2}$ for $0 < x \leq 1$. The function f is differentiable with derivative f' satisfying $f'(0) = 0$ and $f'(x) = 2x \cos \frac{\pi}{x^2} + \frac{2\pi}{x} \sin \frac{\pi}{x^2}$ for $0 < x \leq 1$. We show that f' is not Lebesgue integrable.

If $0 < a < b < 1$, then f' is continuous on $[a, b]$ and is, therefore, Riemann integrable with

$$\int_a^b f' = b^2 \cos \frac{\pi}{b^2} - a^2 \cos \frac{\pi}{a^2}.$$

Setting $b_k = 1/\sqrt{2k}$ and $a_k = \sqrt{2/(4k+1)}$, we see that $\int_{a_k}^{b_k} f' = 1/2k$. Since the intervals $[a_k, b_k]$ are pairwise disjoint,

$$\int_0^1 |f'| \geq \sum_{k=1}^\infty \int_{a_k}^{b_k} |f'| \geq \sum_{k=1}^\infty \frac{1}{2k} = \infty.$$

Hence, f' is not absolutely integrable on $[0, 1]$ and, therefore, not Lebesgue integrable there.

The most general form of Part I of the Fundamental Theorem of Calculus for the Lebesgue integral is analogous to the result for the Riemann integral; it requires the assumption that the derivative f' be Lebesgue integrable. This result is somewhat difficult to prove, and we do not have the requisite machinery in place at this time to prove it. In order to have a version of the Fundamental Theorem of Calculus for the Lebesgue integral, we prove a special case and later establish the general version for the Lebesgue integral in Theorem 4.93 after we discuss the Henstock-Kurzweil integral and show that it is more general than the Lebesgue integral.

Theorem 4.2 *(Fundamental Theorem of Calculus: Part I) Let f : $[a, b] \to \mathbb{R}$ be differentiable on $[a, b]$ and suppose that f' is bounded. Then, f' is Lebesgue integrable on $[a, b]$ and satisfies (4.1).*

Proof.   Note first that f' is Lebesgue integrable since it is bounded by assumption and measurable by Exercise 3.25. For convenience, extend f to $[a, b+1]$ by setting $f(t) = f(b)$ for $b < t \leq b+1$. Define $f_n : [a, b] \to \mathbb{R}$ by

$$f_n(t) = \frac{f\left(t + \frac{1}{n}\right) - f(t)}{\frac{1}{n}}.$$

By the Mean Value Theorem, for every n, $n > \frac{1}{b-a}$, and $t \in \left[a, b - \frac{1}{n}\right]$, there exists an $s_{n,t} \in [a, b]$ such that $f_n(t) = f'(s_{n,t})$. For $t \in \left[b - \frac{1}{n}, b\right]$, there is an $s_{n,t} \in [a, b]$ such that

$$f_n(t) = \frac{f\left(t + \frac{1}{n}\right) - f(t)}{\frac{1}{n}} = \frac{f(b) - f(t)}{\frac{1}{n}}$$

$$= n(b-t)\frac{f(b) - f(t)}{b-t} = n(b-t)f'(s_{n,t}),$$

where $n(b-t) \leq 1$. Since f' is bounded, it follows that $\{f_n\}_{n=1}^{\infty}$ is uniformly bounded. Since $\{f_n\}_{n=1}^{\infty}$ converges to f' everywhere in $[a, b]$, except possibly b, the Bounded Convergence Theorem shows that

$$\int_a^b f' = \lim_{n \to \infty} \int_a^b f_n = \lim_{n \to \infty} \left\{ \int_a^b \frac{f\left(t + \frac{1}{n}\right)}{\frac{1}{n}} dt - \int_a^b \frac{f(t)}{\frac{1}{n}} dt \right\}.$$

By Exercise 3.33, the linear change of variables $s = t + \frac{1}{n}$ in the next to last integral above shows that

$$\int_a^b f' = \lim_{n \to \infty} \left\{ n \int_{a+\frac{1}{n}}^{b+\frac{1}{n}} f(s)\, ds - n \int_a^b f(t)\, dt \right\} \tag{4.2}$$

$$= \lim_{n \to \infty} \left\{ n \int_b^{b+\frac{1}{n}} f(s)\, ds - n \int_a^{a+\frac{1}{n}} f(t)\, dt \right\}.$$

The function f is continuous and, therefore, Riemann integrable so from the Mean Value Theorem (Exercise 2.18), for every n there are b_n and a_n, $b \leq b_n \leq b + \frac{1}{n}$ and $a \leq a_n \leq a + \frac{1}{n}$, such that $n \int_b^{b+\frac{1}{n}} f = f(b_n)$ and $n \int_a^{a+\frac{1}{n}} f = f(a_n)$. Since $b_n \to b$, $a_n \to a$, and f is continuous on $[a, b+1]$, from (4.2) we obtain

$$\int_a^b f' = \lim_{n \to \infty} \{f(b_n) - f(a_n)\} = f(b) - f(a)$$

as we wished to show. $\qquad\qquad\qquad\qquad\qquad\qquad\qquad\qquad\qquad\square$

4.1 Denjoy and Perron integrals

Upon noting that the general form of Part I of the Fundamental Theorem of Calculus failed to hold for the Lebesgue integral, mathematicians sought a theory of integration for which Part I of the Fundamental Theorem of Calculus holds in full generality, i.e., an integral for which all derivatives are integrable. In 1912, A. Denjoy (1884-1974) introduced such an integration theory. His integral is very technical, and we will make no attempt to define or describe the Denjoy integral. Lusin later gave a more elementary characterization of the Denjoy integral, but this is still quite technical. For a description of the Denjoy integral and references, the reader may consult the text of Gordon [Go].

Later, in 1914, O. Perron (1880-1975) gave another integration theory for which the Part I of the Fundamental Theorem of Calculus holds in full generality. The definition of the Perron integral is quite different from that of the Denjoy integral although later Alexandrov and Looman [Pe, Chap. 9] showed that, in fact, the two integrals are equivalent. We will give a very brief description of the Perron integral since some of the basic ideas will be used later when we show the equivalence of absolute Henstock-Kurzweil integrability and McShane integrability.

Definition 4.3   Let $f : [a, b] \to \mathbb{R}$ and $x \in [a, b]$. The *upper derivative* of f at x is defined to be

$$\overline{D}f(x) = \limsup_{t \to x} \frac{f(t) - f(x)}{t - x}.$$

Similarly, the *lower derivative* is defined to be $\underline{D}f(x) = \liminf_{t \to x} \frac{f(t) - f(x)}{t - x}$.

Thus, f is differentiable at x if, and only if, $\overline{D}f(x) = \underline{D}f(x)$ and both upper and lower derivatives are finite.

Definition 4.4   Let $f : [a, b] \to \mathbb{R}^*$. A function $U : [a, b] \to \mathbb{R}$ is called a *major function* for f if U is continuous on $[a, b]$, $U(a) = 0$, $\underline{D}U(x) > -\infty$ and $\underline{D}U(x) \geq f(x)$ for all $x \in [a, b]$. A function $u : [a, b] \to \mathbb{R}$ is called a *minor function* for f if u is continuous on $[a, b]$, $u(a) = 0$, $\overline{D}u(x) < \infty$ and $\overline{D}u(x) \leq f(x)$ for all $x \in [a, b]$.

It follows that if f is differentiable on $[a, b]$ and has finite-valued derivative, then $f - f(a)$ is both a major and a minor function for f'.

If U is a major function for f and u is a minor function for f, then it can be shown that $U - u$ is increasing. Therefore,

$$-\infty < \sup\{u(b) : u \text{ is a minor function for } f\}$$
$$\leq \inf\{U(b) : U \text{ is a major function for } f\} < \infty.$$

Definition 4.5   A function $f : [a, b] \to \mathbb{R}$ is called *Perron integrable* over $[a, b]$ if, and only if, f has at least one major and one minor function on $[a, b]$ and

$$\sup\{u(b) : u \text{ is a minor function for } f\}$$
$$= \inf\{U(b) : U \text{ is a major function for } f\}. \quad (4.3)$$

The Perron integral of f over $[a, b]$ is defined to be the common value in (4.3).

If a function $f : [a, b] \to \mathbb{R}$ has a finite derivative on $[a, b]$, it then follows from the definition that f' is Perron integrable over $[a, b]$ with Perron integral equal to $f(b) - f(a)$. That is, Part I of the Fundamental Theorem of Calculus holds in full generality for the Perron integral. For a description and development of the Perron integral, see [Go] and [N].

Both the Denjoy and Perron integrals are somewhat technical to define and develop, but in the next section we will use Part I of the Fundamental Theorem of Calculus as motivation to define another integral, called the Henstock-Kurzweil integral, which is just a slight variant of the Riemann integral and for which Part I of the Fundamental Theorem of Calculus holds in full generality. It can be shown that the Henstock-Kurzweil integral is equivalent to the Denjoy and Perron integrals (see [Go]).

4.2 A General Fundamental Theorem of Calculus

Suppose that $f : [a, b] \to \mathbb{R}$ is a differentiable function and we are interested in proving equality (4.1). Let $\mathcal{P} = \{x_0, x_1, \dots, x_n\}$ be a partition of $[a, b]$. By the Mean Value Theorem, there is a $y_i \in (x_{i-1}, x_i)$ such that $f(x_i) - f(x_{i-1}) = f'(y_i)(x_i - x_{i-1})$. Thus, given any partition $\mathcal{P}$, there is a set of sampling points $\{y_1, \dots, y_n\}$ such that

$$
S(f', \mathcal{P}, \{y_i\}_{i=1}^n) = \sum_{i=1}^n f'(y_i)(x_i - x_{i-1})
$$
$$
= \sum_{i=1}^n \{f(x_i) - f(x_{i-1})\} = f(b) - f(a).
$$

The problem is that given a partition $\mathcal{P}$, there may be only one such set of sampling points. However, if we want to show that $\int_a^b f'$ is equal to $f(b) - f(a)$, we do not need the Riemann sums to equal $f(b) - f(a)$, but rather be within some prescribed margin of error. Thus, we are led to consider more closely the relationship between $f'(y_i)(x_{i+1} - x_i)$ and $f(x_{i+1}) - f(x_i)$.

Fix an $\epsilon > 0$ and let $y \in [a, b]$. Since f is differentiable at y, there is a $\delta(y) > 0$ so that if $x \in [a, b]$ and $0 < |x - y| < \delta(y)$ then

$$
\left| \frac{f(x) - f(y)}{x - y} - f'(y) \right| < \epsilon.
$$

Multiplying through by $|x - y|$, we get

$$|f(x) - f(y) - f'(y)(x - y)| \leq \epsilon |x - y|,$$

which is also valid for $x = y$. Now, suppose that $u, v \in [a, b]$ and $y - \delta(y) < u \leq y \leq v < y + \delta(y)$. Then,

$$
\begin{aligned}
&|f(v) - f(u) - f'(y)(v - u)| \\
&= |\{f(v) - f(y) - f'(y)(v - y)\} + \{f(y) - f(u) - f'(y)(y - u)\}| \\
&\leq |f(v) - f(y) - f'(y)(v - y)| + |f(y) - f(u) - f'(y)(y - u)| \\
&\leq \epsilon(v - y) + \epsilon(y - u) = \epsilon(v - u).
\end{aligned}
$$

So, if $y - \delta(y) < u \leq y \leq v < y + \delta(y)$, then $f'(y)(v - u)$ is a good approximation to $f(v) - f(u)$.

This result, known as the *Straddle Lemma*, will be useful to us below.

Lemma 4.6   *(Straddle Lemma) Let $f : [a, b] \to \mathbb{R}$ be differentiable at $y \in [a, b]$. For each $\epsilon > 0$, there is a $\delta > 0$, depending on y, such that*

$$|f(v) - f(u) - f'(y)(v - u)| \leq \epsilon(v - u)$$

whenever $u, v \in [a, b]$ and $y - \delta < u \leq y \leq v < y + \delta$.

The geometric interpretation of the Straddle Lemma is that the slope of the chord between $(u, f(u))$ and $(v, f(v))$ is a good approximation to the slope of the tangent line at $(y, f(y))$. It is important that the values u and v "straddle" y, that is, occur on different sides of y. Consider the function f equal to $x^2 \cos(\pi/x)$ for $x \neq 0$ and $f(0) = 0$. This function has derivative 0 for $x = 0$, but for $u = \frac{1}{2n + \frac{1}{2}}$ and $v = \frac{1}{2n}$, the slope of the chord joining $(u, f(u))$ and $(v, f(v))$ is

$$\frac{\left(\frac{1}{2n}\right)^2 \cos 2n\pi - \left(\left(\frac{1}{2n + \frac{1}{2}}\right)^2 \cos\left(2n\pi + \frac{\pi}{2}\right)\right)}{\frac{1}{2n} - \frac{1}{2n + \frac{1}{2}}} = \frac{\left(\frac{1}{2n}\right)^2}{\frac{1}{2n} - \frac{1}{2n + \frac{1}{2}}} > 2.$$

Thus, if u and v do not straddle 0, then the slope of the chord is not a good approximation to the slope of the tangent line.

This lemma already gives us a hint of how to proceed. When studying Riemann integrals, we chose partitions based on the length of their largest subinterval. This condition does not take into account any of the properties of the function being considered. The Straddle Lemma, on the other hand, assigns a δ to each point where a function is differentiable based on how the function acts near that point. If the function acts smoothly near the point, we would expect the associated δ to be large; if the function oscillates wildly near the point, we would expect δ to be small. This simple change to varying the size of δ from point to point is the key idea behind the Henstock-Kurzweil integral. For the Henstock-Kurzweil integral, we will be interested in partitions $\mathcal{P} = \{x_0, x_1, \ldots, x_n\}$ and sampling points $\{y_i\}_{i=1}^{n}$ such that $[x_{i-1}, x_i] \subset (y_i - \delta(y_i), y_i + \delta(y_i))$, where $\delta : [a, b] \to (0, \infty)$ is a positive function.

There is another point that must be resolved, namely the relationship between the partition and the sampling points. In the Riemann theory, given a partition $\mathcal{P} = \{x_0, x_1, \ldots, x_n\}$, we consider Riemann sums for every set of sampling points $\{y_i\}_{i=1}^{n}$ such that $y_i \in [x_{i-1}, x_i]$. However, if $\mathcal{P}$ is a partition with mesh at most δ, then $[x_{i-1}, x_i] \subset (y_i - \delta, y_i + \delta)$ for every sampling point $y_i \in [x_{i-1}, x_i]$. We use this idea to determine which pairs of partitions and sampling points to consider. In the general case, in which δ is a positive function of y, we will consider only partitions $\mathcal{P}$ and sampling points $\{y_i\}_{i=1}^{n}$ such that $y_i \in [x_{i-1}, x_i] \subset (y_i - \delta(y_i), y_i + \delta(y_i))$.

Fix $[a, b]$. Suppose $\mathcal{P} = \{x_0, x_1, \ldots, x_n\}$ is a partition of $[a, b]$ and $\{t_i\}_{i=1}^{n}$ is a set of sampling points associated to $\mathcal{P}$. Let $I_i = [x_{i-1}, x_i]$, so that $t_i \in I_i$. Thus, we can view a partition together with a set of sampling points as a set of ordered pairs (t, I), where I is a subinterval of $[a, b]$ and t is a point in I.

Definition 4.7 Given an interval $I = [a, b] \subset \mathbb{R}$, a *tagged partition* is a finite set of ordered pairs $\mathcal{D} = \{(t_i, I_i) : i = 1, \ldots, m\}$ such that I_i is a closed subinterval of $[a, b]$, $t_i \in I_i$, $\cup_{i=1}^{m} I_i = [a, b]$ and the intervals have disjoint interiors, $I_i^o \cap I_j^o = \emptyset$ if $i \neq j$. The point t_i is called the *tag* associated to the interval I_i.

In other words, a tagged partition is a partition with a distinguished point (the tag) in each interval.

Given a tagged partition $\mathcal{D}$, a point can be a tag for at most two intervals. This can happen when a tag is an endpoint for two adjacent intervals and is used as the tag for both intervals.

Remark 4.8 *By the preceding argument, a partition with a set of sampling points generates a tagged partition (by setting $I_i = [x_{i-1}, x_i]$). Similarly, a tagged partition generates a partition and a set of associated sampling points. Given a tagged partition $\mathcal{D} = \{(t_i, I_i) : i = 1, \ldots, m\}$, renumber the pairs so the right endpoint of I_{i-1} equals the left endpoint of I_i and set $I_i = [x_{i-1}, x_i]$. Then, $\mathcal{P} = \{x_0, x_1, \ldots, x_m\}$ is a partition of*

$[a, b]$ *and* $t_i \in I_i$. *Note that while a partition is an ordered set of numbers, the intervals in a tagged partition are not ordered (from left to right), so we must first reorder the intervals so that their endpoints create a partition of* $[a, b]$.

Next, we need a way to measure and control the size of a tagged partition. Based on the discussion leading to the Straddle Lemma, we will do this using a positive function, δ, of t.

Definition 4.9 Given an interval $I = [a, b]$, an interval-valued function γ defined on I is called a *gauge* if there is a function $\delta : [a, b] \to (0, \infty)$ such that $\gamma(t) = (t - \delta(t), t + \delta(t))$. If $\mathcal{D} = \{(t_i, I_i) : i = 1, \ldots, m\}$ is a tagged partition of I and γ is a gauge on I, we say that $\mathcal{D}$ is γ-*fine* if $I_i \subset \gamma(t_i)$ for all i. We denote this by writing $\mathcal{D}$ is a γ-*fine tagged partition* of I.

Let $\mathcal{P} = \{x_0, x_1, \ldots, x_n\}$ be a partition of $[a, b]$ with mesh less than δ and let $\{y_i\}_{i=1}^{m}$ be any set of sampling points such that $y_i \in [x_{i-1}, x_i]$. If $\gamma(t) = (t - \delta, t + \delta)$ for all $t \in I$, then $[x_{i-1}, x_i] \subset \gamma(y_i)$ so that $\mathcal{D} = \{(y_i, [x_{i-1}, x_i]) : i = 1, \ldots, m\}$ is a γ-fine tagged partition of $[a, b]$. This is the gauge used for the Riemann integral. Consequently, the constructions used for the Riemann integral are compatible with gauges. The value of changing from a mesh to a gauge is that points where a function behaves nicely can be accentuated by being associated to a large interval, and points where a function acts poorly can be associated to a small interval.

Example 4.10 The Dirichlet function $f : [0, 1] \to \mathbb{R}$,

$$f(x) = \begin{cases} 1 \text{ if } x \in \mathbb{Q} \\ 0 \text{ if } x \notin \mathbb{Q} \end{cases},$$

defined in Example 2.7, is not Riemann integrable. This function is equal to 0 most of the time, so we want a gauge that associates larger intervals to irrational numbers than it does to rational numbers. Let $\{r_i\}_{i=1}^{\infty}$ be an enumeration of the rational numbers in $\mathbb{Q} \cap [0, 1]$. Let $c > 0$ and define $\delta : [0, 1] \to (0, \infty)$ by

$$\delta(x) = \begin{cases} c & \text{if } x \notin \mathbb{Q} \\ 2^{-i}c & \text{if } x = r_i \in \mathbb{Q} \end{cases}.$$

Then,

$$\gamma(x) = \begin{cases} (x - c, x + c) & \text{if } x \notin \mathbb{Q} \\ (x - 2^{-i}c, x + 2^{-i}c) & \text{if } x = r_i \in \mathbb{Q} \end{cases},$$

and every irrational number is associated to an interval of length $2c$ while the rational number r_i is associated to an interval of length $2^{1-i}c$.

After introducing the Henstock-Kurzweil integral, we will use this construction to prove that the Dirichlet function is Henstock-Kurzweil integrable.

If $\mathcal{D} = \{(t_i, I_i) : i = 1, \ldots, m\}$ is a tagged partition of I, we call

$$S(f, \mathcal{D}) = \sum_{i=1}^{m} f(t_i) \ell(I_i)$$

the *Riemann sum* with respect to $\mathcal{D}$.

Let us restate the definition of the Riemann integral in terms of tagged divisions.

Definition 4.11 A function $f : [a, b] \to \mathbb{R}$ is *Riemann integrable* over $[a, b]$ if there is an $A \in \mathbb{R}$ such that for all $\epsilon > 0$ there is a $\delta > 0$ so that if $\mathcal{D} = \{(t_i, [x_{i-1}, x_i]) : 1 \leq i \leq m\}$ is any tagged partition of $[a, b]$ satisfying $[x_{i-1}, x_i] \subset (t_i - \delta, t_i + \delta)$, then $|S(f, \mathcal{D}) - A| < \epsilon$.

Note that the mesh of this partition is at most 2δ.

For the Riemann integral, the partitions are chosen independent of f. Thus, this definition fails to take into account the particular function involved. A major advantage of the Henstock-Kurzweil integral is one only need consider partitions that take the behavior of the function into account.

Definition 4.12 Let $f : [a, b] \to \mathbb{R}$. We call the function f *Henstock-Kurzweil integrable* on $I = [a, b]$ if there is an $A \in \mathbb{R}$ so that for all $\epsilon > 0$ there is a gauge γ on I so that for every γ-fine tagged partition $\mathcal{D}$ of $[a, b]$,

$$|S(f, \mathcal{D}) - A| < \epsilon.$$

The number A is called the *Henstock-Kurzweil integral* of f over $[a, b]$, and we write $A = \int_a^b f = \int_I f$.

The Henstock-Kurzweil integral is also called the *gauge integral* and the *generalized Riemann integral*.

Notation 4.13 *For the remainder of this section, we will use the symbols $\int_a^b f$ and $\int_I f$ to represent the Henstock-Kurzweil integral of f.*

The first question that arises is whether this definition is meaningful. We need to know that, given a gauge γ, there is an associated γ-fine tagged partition, so that we have Riemann sums to define the Henstock-Kurzweil integral, and also that the Henstock-Kurzweil integral is well defined. We will return to both issues at the end of this section.

Observe that every Riemann integrable function is Henstock-Kurzweil integrable. For, suppose that f is Riemann integrable. Let δ correspond to a given ϵ in the definition of the Riemann integral. Set $\gamma(t) = \left(t - \frac{\delta}{2}, t + \frac{\delta}{2}\right)$. Then, any tagged partition that is γ-fine has mesh less than δ. Thus, we have proved

Theorem 4.14 *If* $f : [a, b] \to \mathbb{R}$ *is Riemann integrable then* f *is Henstock-Kurzweil integrable and the two integrals agree.*

However, there are Henstock-Kurzweil integrable functions that are not Riemann integrable. In fact, the Dirichlet function is one such example.

Example 4.15 Let $f : [0, 1] \to \mathbb{R}$ be the Dirichlet function. We will show that $\int_0^1 f = 0$. Let $\epsilon > 0$ and let γ be the gauge defined in Example 4.10 with $c = \frac{\epsilon}{4}$. Let $\mathcal{D} = \{(t_i, I_i) : i = 1, \ldots, m\}$ be a γ-fine tagged partition of $[0, 1]$ and note that

$$|S(f, \mathcal{D}) - 0| = |S(f, \mathcal{D})| = \left| \sum_{i=1}^{m} f(t_i) \ell(I_i) \right|$$

$$= \left| \sum_{\substack{(t_i, I_i) \in \mathcal{D} \\ t_i \notin \mathbb{Q}}} f(t_i) \ell(I_i) + \sum_{\substack{(t_i, I_i) \in \mathcal{D} \\ t_i \in \mathbb{Q}}} f(t_i) \ell(I_i) \right|.$$

The sum for $t_i \notin \mathcal{D} \cap \mathbb{Q}$ equals 0 since $f(t) = 0$ whenever $t \notin \mathbb{Q}$. To estimate the sum for $t_i \in \mathcal{D} \cap \mathbb{Q}$, note that $f(t_i) = 1$ since $t_i \in \mathbb{Q}$ and recall that each tag t_i can be a tag for at most two intervals. Since $t_i \in \mathbb{Q} \cap [0, 1]$, there is an j so that $t_i = r_j$. Thus, if $(t_i, I_i) \in \mathcal{D}$, then $I_i \subset \gamma(t_i)$, so that $\ell(I_i) \leq \ell(\gamma(t_i)) = \ell(\gamma(r_j)) = 2^{1-j}\frac{\epsilon}{4}$. Thus,

$$\left| \sum_{t_i \notin \mathcal{D} \cap \mathbb{Q}} f(t_i) \ell(I_i) + \sum_{t_i \in \mathcal{D} \cap \mathbb{Q}} f(t_i) \ell(I_i) \right| = \left| \sum_{t_i \in \mathcal{D} \cap \mathbb{Q}} f(t_i) \ell(I_i) \right|$$

$$\leq 2 \sum_{j=1}^{\infty} 2^{1-j} \frac{\epsilon}{4} = \epsilon.$$

We have shown that given any $\epsilon > 0$, there is a gauge γ so that for any γ-fine tagged partition, $\mathcal{D}$, $|S(f, \mathcal{D}) - 0| < \epsilon$. In other words, the Dirichlet function is Henstock-Kurzweil integrable over $[0, 1]$ with $\int_0^1 f = 0$.

Notice the use of the variable length intervals in the definition of the gauge. We will give a generalization of this result in Example 4.40; see also Exercise 4.7.

Let us return to the Fundamental Theorem of Calculus. The proof is an easy consequence of the Straddle Lemma.

Theorem 4.16 *(Fundamental Theorem of Calculus: Part I) Suppose that* $f : [a, b] \to \mathbb{R}$ *is differentiable on* $[a, b]$. *Then,* f' *is Henstock-Kurzweil integrable on* $[a, b]$ *and*

$$\int_a^b f' = f(b) - f(a).$$

Proof. Fix an $\epsilon > 0$. For each $t \in [a, b]$, we choose a $\delta(t) > 0$ by the Straddle Lemma (Lemma 4.6) and define a gauge γ on $[a, b]$ by $\gamma(t) = (t - \delta(t), t + \delta(t))$. Suppose that $\mathcal{D} = \{(t_i, I_i) : i = 1, \ldots, m\}$ is a γ-fine tagged partition of $[a, b]$. We reorder the intervals I_i so that the right endpoint of I_{i-1} equals the left endpoint of I_i, and set $I_i = [x_{i-1}, x_i]$ for each i. Then,

$$f(b) - f(a) = \sum_{i=1}^m [f(x_i) - f(x_{i-1})]$$

so, by the Straddle Lemma,

$$|S(f', \mathcal{D}) - (f(b) - f(a))| = \left| \sum_{i=1}^m \{f'(t_i)(x_i - x_{i-1}) - [f(x_i) - f(x_{i-1})]\} \right|$$

$$\leq \sum_{i=1}^m \epsilon(x_i - x_{i-1}) = \epsilon(b - a).$$

Thus, f' is Henstock-Kurzweil integrable and satisfies equation (4.1). $\square$

Thus, every derivative is Henstock-Kurzweil integrable. This is not a surprising coincidence. Kurzweil [K] initiated his study leading to the Henstock-Kurzweil integral in order to study ordinary differential equations. A few years later, working independently, Henstock [He] developed many of the properties of this integral. We will establish a more general version of Theorem 4.16 later in Theorem 4.24.

An immediate consequence of the Fundamental Theorem of Calculus is that the unbounded derivative

$$f'(x) = \begin{cases} 2x \cos \dfrac{\pi}{x^2} + \dfrac{2\pi}{x} \sin \dfrac{\pi}{x^2} & \text{if } 0 < x \le 1 \\ 0 & \text{if } \quad x = 0 \end{cases},$$

defined in Example 2.31, is Henstock-Kurzweil integrable on $[0,1]$ with integral equal to -1. Since f' is unbounded, it is not Riemann integrable and, as we saw in Example 4.1, f' is not Lebesgue integrable.

Before concluding this section, we prove two results which guarantee that the Henstock-Kurzweil integral is well defined. We prove that given a gauge γ, there is a related γ-fine tagged partition, and that the value of the integral is unique.

Theorem 4.17 *Let γ be a gauge on $I = [a,b]$. Then, there is a γ-fine tagged partition of I.*

Proof. Let $E = \{t \in (a,b] : [a,t]$ has a γ-fine tagged partition$\}$. We want to show $b \in E$. First observe that $E \ne \emptyset$ since if $x \in \gamma(a) \cap (a,b)$, then $\{(a,[a,x])\}$ is a γ-fine tagged partition of $[a,x]$. Thus, $x \in E$ and $E \ne \emptyset$.

We next claim that $y = \sup E$ is an element of E. By definition, $y \in [a,b]$, so γ is defined at y. Choose $x \in \gamma(y)$ so that $x < y$ and $x \in E$, and let $\mathcal{D}$ be a γ-fine tagged partition of $[a,x]$. Then, $\mathcal{D}' = \mathcal{D} \cup \{(y,[x,y])\}$ is a γ-fine tagged partition of $[a,y]$. Therefore, $y \in E$.

Finally, we show $y = b$. Suppose $y < b$. Choose $w \in \gamma(y) \cap (y,b)$. Let $\mathcal{D}$ be a γ-fine tagged partition of $[a,y]$. Then, $\mathcal{D}' = \mathcal{D} \cup \{(y,[y,w])\}$ is a γ-fine tagged partition of $[a,w]$. Since $y < w$, this contradicts the definition of y. Thus, $y = b$. $\qquad\square$

Thus, there is a γ-fine tagged partition associated to every gauge γ. In fact, there are many, as we can see by varying the choice of x in the first step of the proof above.

Finally, we prove that the Henstock-Kurzweil integral is unique, justifying our notation in Definition 4.12. The proof employs a very useful technique for working with gauges. Suppose that γ_1 and γ_2 are two gauges defined on an interval $[a,b]$. Then the (interval-valued) function γ defined by $\gamma(t) = \gamma_1(t) \cap \gamma_2(t)$ is also a gauge on $[a,b]$. In fact, if δ_1 and δ_2 are the positive functions used to define γ_1 and γ_2, respectively, and $\delta(t) = \min\{\delta_1(t), \delta_2(t)\}$, then $\gamma(t) = (t - \delta(t), t + \delta(t))$. Further, if $\mathcal{D}$ is a γ-fine tagged partition, then $\mathcal{D}$ is also a γ_1-fine tagged partition and a γ_2-fine tagged partition, since for $(t,I) \in \mathcal{D}$, $I \subset \gamma(t) \subset \gamma_i(t)$, for $i = 1,2$.

Theorem 4.18 *The Henstock-Kurzweil integral of a function is unique.*

Proof. Suppose that f is Henstock-Kurzweil integrable over $[a, b]$ and both A and B satisfy Definition 4.12. Fix $\epsilon > 0$ and choose γ_A and γ_B corresponding to A and B, respectively, in the definition with $\epsilon' = \frac{\epsilon}{2}$. Let $\gamma(t) = \gamma_1(t) \cap \gamma_2(t)$ and suppose that $\mathcal{D}$ be a γ-fine tagged partition, and hence $\mathcal{D}$ is both γ_1-fine and γ_2-fine. Then,

$$|A - B| \leq |A - S(f, \mathcal{D})| + |S(f, \mathcal{D}) - B| < \epsilon' + \epsilon' = \epsilon.$$

Since ϵ was arbitrary, it follows that $A = B$. Thus, the value of the Henstock-Kurzweil integral is unique. $\qquad\square$

Now, review the proof of Proposition 2.3. You will notice that the proof is exactly the same as the one above, replacing positive numbers, δ, with gauges, γ, and partitions and sampling points, $\mathcal{P}$ and $\{t_i\}_{i=1}^n$, with tagged partitions, $\mathcal{D}$. In the following section, in which we establish the basic properties of the Henstock-Kurzweil integral, we will begin with proofs that directly mimic the Riemann proofs. Of course, as we progress with this more advanced theory, we will need to employ more sophisticated proofs.

4.3 Basic properties

We begin with the two most fundamental properties of an integral, linearity and positivity.

Proposition 4.19 *(Linearity) Let $f, g : [a, b] \to \mathbb{R}$ and let $\alpha, \beta \in \mathbb{R}$. If f and g are Henstock-Kurzweil integrable, then $\alpha f + \beta g$ is Henstock-Kurzweil integrable and*

$$\int_a^b (\alpha f + \beta g) = \alpha \int_a^b f + \beta \int_a^b g.$$

Proof. Fix $\epsilon > 0$ and choose $\gamma_f > 0$ so that if $\mathcal{D}$ is a γ_f-fine tagged partition of $[a, b]$, then

$$\left| S(f, \mathcal{D}) - \int_a^b f \right| < \frac{\epsilon}{2(1 + |\alpha|)}.$$

Similarly, choose $\gamma_g > 0$ so that if $\mathcal{D}$ is a γ_g-fine tagged partition of $[a,b]$, then

$$\left| S(g, \mathcal{D}) - \int_a^b g \right| < \frac{\epsilon}{2(1 + |\beta|)}.$$

Now, let $\gamma(t) = \gamma_f(t) \cap \gamma_g(t)$ and suppose that $\mathcal{D}$ is a γ-fine tagged partition of $[a,b]$. Then,

$$\left| S(\alpha f + \beta g, \mathcal{D}) - \left(\alpha \int_a^b f + \beta \int_a^b g \right) \right|$$

$$= \left| (\alpha S(f, \mathcal{D}) + \beta S(g, \mathcal{D})) - \left(\alpha \int_a^b f + \beta \int_a^b g \right) \right|$$

$$= \left| \alpha \left(S(f, \mathcal{D}) - \int_a^b f \right) + \beta \left(S(g, \mathcal{D}) - \int_a^b g \right) \right|$$

$$\leq |\alpha| \left| S(f, \mathcal{D}) - \int_a^b f \right| + |\beta| \left| S(g, \mathcal{D}) - \int_a^b g \right|$$

$$< \frac{\epsilon |\alpha|}{2(1 + |\alpha|)} + \frac{\epsilon |\beta|}{2(1 + |\beta|)} < \epsilon.$$

Since ϵ was arbitrary, it follows that $\alpha f + \beta g$ is Henstock-Kurzweil integrable and

$$\int_a^b (\alpha f + \beta g) = \alpha \int_a^b f + \beta \int_a^b g.$$

$\square$

Proposition 4.20   *(Positivity) Let $f : [a,b] \to \mathbb{R}$. Suppose that f is nonnegative and Henstock-Kurzweil integrable. Then, $\int_a^b f \geq 0$.*

Proof.   Let $\epsilon > 0$ and choose a gauge γ according to Definition 4.12. Then, if $\mathcal{D}$ is a γ-fine tagged partition of $[a,b]$,

$$\left| S(f, \mathcal{D}) - \int_a^b f \right| < \epsilon.$$

Consequently, since $S(f, \mathcal{D}) \geq 0$,

$$\int_a^b f > S(f, \mathcal{D}) - \epsilon > -\epsilon$$

for any positive ϵ. It follows that $\int_a^b f \geq 0$.

$\square$

A comparison of the last two proofs with the corresponding proofs for the Riemann integral immediately shows their similarity.

Remark 4.21   *Suppose that f is a positive function on $[a, b]$. If f is Henstock-Kurzweil integrable, then the best we can conclude is that $\int_a^b f \geq 0$; from our results so far, we cannot conclude that the integral is positive. The Riemann integral has the same defect. However, if f is Lebesgue integrable, then the Lebesgue integral of f is strictly positive. Let $\mathcal{L} \int_a^b f$ be the Lebesgue integral of f. From Tchebyshev's inequality we have*

$$\lambda m\left(\{x \in [a, b] : f(x) > \lambda\}\right) \leq \mathcal{L} \int_a^b f.$$

If f is strictly positive on $[a, b]$, then $[a, b] = \cup_{k=1}^\infty \left\{x \in [a, b] : f(x) > \frac{1}{k}\right\}$, so there must be a k such that $m\left(\left\{x \in [a, b] : f(x) > \frac{1}{k}\right\}\right) > 0$. But, then,

$$\mathcal{L} \int_a^b f \geq \frac{1}{k} m\left(\left\{x \in [a, b] : f(x) > \frac{1}{k}\right\}\right) > 0.$$

Suppose that $f \leq g$. Applying the previous result to $g - f$ yields

Corollary 4.22   *Suppose f and g are Henstock-Kurzweil integrable over $[a, b]$ and $f(x) \leq g(x)$ for all $x \in [a, b]$. Then,*

$$\int_a^b f \leq \int_a^b g.$$

A function f defined on an interval $[a, b]$ is called *absolutely integrable* if both f and $|f|$ are Henstock-Kurzweil integrable over $[a, b]$. A Riemann integrable function is absolutely (Riemann) integrable, and a function is Lebesgue integrable if, and only if, it is absolutely (Lebesgue) integrable. We will see in Section 4.4 that a Henstock-Kurzweil integrable function need not be absolutely integrable. For absolutely integrable functions, we have the following result.

Corollary 4.23   *If f is absolutely integrable over $[a, b]$, then $\left|\int_a^b f\right| \leq \int_a^b |f|$.*

Proof.   Since $-|f| \leq f \leq |f|$, the previous corollary implies that

$$-\int_a^b |f| \leq \int_a^b f \leq \int_a^b |f|.$$

The result follows.   □

While the ability to integrate every derivative is a main feature of the Henstock-Kurzweil integral, the Henstock-Kurzweil integral satisfies an even stronger result. The derivative can fail to exist at a countable number of points and still satisfy equation (4.1).

Theorem 4.24 *(Generalized Fundamental Theorem of Calculus: Part I)*
Let $F, f : [a, b] \to \mathbb{R}$. Suppose that F is continuous and $F' = f$ except for possibly a countable number of points in $[a, b]$. Then, f is Henstock-Kurzweil integrable over $[a, b]$ and

$$\int_a^b f = F(b) - F(a).$$

Proof.   Let $C = \{c_n\}_{n \in \sigma}$ be the points where either F' fails to exist or F' exists but is not equal to f. Let $\epsilon > 0$. If $t \in [a, b] \setminus C$, choose $\delta(t) > 0$ for this ϵ by the Straddle Lemma. If $t \in C$, then $t = c_k$ for some k. Choose $\delta(t) = \delta(c_k) > 0$ so that $|x - c_k| < \delta(c_k)$ implies:

(1) $|F(x) - F(c_k)| < \epsilon 2^{-(k+3)}$;
(2) $|f(c_k)| |x - c_k| < \epsilon 2^{-(k+3)}$.

We can define such a δ since F is continuous on $[a, b]$ and $|x - c_k|$ can be made as small as desired by choosing x sufficiently close to c_k. Define a gauge γ on $[a, b]$ by setting $\gamma(t) = (t - \delta(t), t + \delta(t))$ for all $t \in [a, b]$.

Suppose that $\mathcal{D} = \{(t_i, I_i) : i = 1, \ldots, m\}$ is a γ-fine tagged partition of $[a, b]$, where $I_i = [a_i, b_i]$ for each i. Note that if $a_i \neq a$, then there is a j so that $a_i = b_j$, with a similar statement for each right endpoint $b_i \neq b$. Let $\mathcal{D}_1$ be the set of elements of $\mathcal{D}$ with tags in $[a, b] \setminus C$ and $\mathcal{D}_2$ be the set of elements of $\mathcal{D}$ with tags in C. By the Straddle Lemma,

$$\sum_{(t_i, I_i) \in \mathcal{D}_1} |F(b_i) - F(a_i) - f(t_i)(b_i - a_i)| \leq \sum_{(t_i, I_i) \in \mathcal{D}_1} \epsilon(b_i - a_i) \leq \epsilon(b - a).$$

If $t_i = c_k$ for some k, by (1) and (2)

$$|F(b_i) - F(a_i) - f(t_i)(b_i - a_i)|$$
$$\leq |F(b_i) - F(c_k)| + |F(c_k) - F(a_i)| + |f(c_k)(b_i - a_i)|$$
$$< \frac{\epsilon}{2^{k+3}} + \frac{\epsilon}{2^{k+3}} + \frac{\epsilon}{2^{k+3}} < \frac{\epsilon}{2^{k+1}}.$$

Therefore,

$$\sum_{(t_i, I_i) \in \mathcal{D}_2} |F(b_i) - F(a_i) - f(t_i)(b_i - a_i)| < 2 \sum_{k=1}^{\infty} \frac{\epsilon}{2^{k+1}} = \epsilon$$

since each c_k can be a tag for at most two subintervals of $\mathcal{D}$. Since each endpoint, other than a and b, occurs as both a left and right endpoint,

$$|S(f, \mathcal{D}) - [F(b) - F(a)]| = \left| \sum_{(t_i, I_i) \in \mathcal{D}} \{F(b_i) - F(a_i) - f(t_i)(b_i - a_i)\} \right|$$
$$\leq \epsilon(b - a) + \epsilon = (1 + b - a)\epsilon,$$

and the result is established. $\qquad\square$

The continuity of F in Theorem 4.24 is important; see Exercise 4.17.

Example 4.25 Define F and f on $[0, 1]$ by $F(x) = 2\sqrt{x}$, and $f(0) = 0$ and $f(x) = \frac{1}{\sqrt{x}}$ otherwise. Then, F is continuous on $[0, 1]$ and $F' = f$ except at $x = 0$. Therefore, by Theorem 4.24, f is Henstock-Kurzweil integrable over $[0, 1]$ and $\int_0^1 f = F(1) - F(0) = 2$.

Note that $\int_0^1 f$ is an improper integral in the Riemann sense since f is unbounded, but we were able to show that f is Henstock-Kurzweil integrable directly from Theorem 4.24. We will show in Section 4.5 that there are no improper integrals for the Henstock-Kurzweil integral.

Using Theorem 4.24, we can prove a useful form of the familiar integration by parts formula from calculus. Another more general version will follow.

Theorem 4.26 *(Integration by Parts) Let $F, G, f, g : [a, b] \to \mathbb{R}$. Suppose that F and G are continuous and $F' = f$ and $G' = g$, except for at most a countable number of points. Then, $Fg + fG$ is Henstock-Kurzweil integrable and*

$$\int_a^b (Fg + fG) = F(b)G(b) - F(a)G(a). \tag{4.4}$$

Moreover, Fg is Henstock-Kurzweil integrable if, and only if, fG is Henstock-Kurzweil integrable and, in this case,

$$\int_a^b Fg + \int_a^b fG = F(b)G(b) - F(a)G(a). \tag{4.5}$$

Proof.   Since $(FG)' = Fg + fG$ except possibly at a countable number of points, by Theorem 4.24, $(FG)'$ is Henstock-Kurzweil integrable and (4.4) holds. The last statement follows immediately from (4.4) since, for example, $Fg = (Fg + fG) - fG$.   □

In Example 4.55 below, we give an example in which neither Fg nor fG is Henstock-Kurzweil integrable so that (4.5) makes no sense, even though (4.4) is valid.

We now establish a more general version of the integration by parts formula than Theorem 4.26 and then give an application of this result.

Theorem 4.27   *Let f, g be Henstock-Kurzweil integrable over $[a, b]$ and let $F(t) = \int_a^t f$ and $G(t) = \int_a^t g$ be the indefinite integrals of f and g, respectively. Then $Ff + Gg$ is Henstock-Kurzweil integrable over $[a, b]$ with*

$$\int_a^b (Fg + Gf) = F(b)G(b).$$

Notice that under the hypotheses of Theorem 4.27, F and G are continuous and $F' = f$ (and $G' = g$) almost everywhere, not just off of a countable set.

Proof.   Let $\epsilon > 0$. Since F and G are continuous, there exists an $M \geq b - a > 0$ such that $|F(t)| \leq M$ and $|G(t)| \leq M$ for $a \leq t \leq b$. Further, from the continuity of F and G, there exists a gauge γ on $[a, b]$ such that if $s \in [a, b]$ and $s \in \gamma(t)$, then

$$|f(t)| \, |G(s) - G(t)| < \frac{\epsilon}{4M} \text{ and } |g(t)| \, |F(s) - F(t)| < \frac{\epsilon}{4M}. \qquad (4.6)$$

Since f and g are integrable, we may also assume the gauge γ is such that $\left| S(f, \mathcal{D}) - \int_a^b f \right| < \epsilon/8M$ and $\left| S(g, \mathcal{D}) - \int_a^b g \right| < \epsilon/8M$ for every γ-fine tagged partition $\mathcal{D}$ of $[a, b]$. From Henstock's Lemma if $\mathcal{D} = \{(t_i, [x_{i-1}, x_i]) : i = 1, \ldots n\}$ is a γ-fine tagged partition of I, we have

$$\sum_{i=1}^{n} |f(t_i)(x_i - x_{i-1}) - [F(x_i) - F(x_{i-1})]| \leq \frac{\epsilon}{4M} \qquad (4.7)$$

and

$$\sum_{i=1}^{n} |g(t_i)(x_i - x_{i-1}) - [G(x_i) - G(x_{i-1})]| \leq \frac{\epsilon}{4M}. \qquad (4.8)$$

We can write $F(b)G(b)$ as a telescoping sum

$$F(b)G(b) = F(b)G(b) - F(a)G(a)$$
$$= \sum_{i=1}^{n} [F(x_i)G(x_i) - F(x_{i-1})G(x_{i-1})]$$
$$= \sum_{i=1}^{n} \{F(x_i)[G(x_i) - G(x_{i-1})] + G(x_{i-1})[F(x_i) - F(x_{i-1})]\}$$

and use this expression to obtain

$$|S(Fg + Gf, \mathcal{D}) - F(b)G(b)|$$
$$= \left| \sum_{i=1}^{n} [F(t_i)g(t_i) + f(t_i)G(t_i)](x_i - x_{i-1}) - F(b)G(b) \right|$$
$$\leq \sum_{i=1}^{n} |F(t_i)g(t_i)(x_i - x_{i-1}) - F(x_i)[G(x_i) - G(x_{i-1})]|$$
$$+ \sum_{i=1}^{n} |f(t_i)G(t_i)(x_i - x_{i-1}) - G(x_{i-1})[F(x_i) - F(x_{i-1})]|$$
$$= I + II.$$

Since $F(t_i) = F(x_i) + [F(t_i) - F(x_i)]$, using (4.6) and (4.8), we see that I above is dominated by

$$I \leq \sum_{i=1}^{n} |F(x_i)| \, |g(t_i)(x_i - x_{i-1}) - [G(x_i) - G(x_{i-1})]|$$
$$+ \sum_{i=1}^{n} |g(t_i)| \, |F(t_i) - F(x_i)| (x_i - x_{i-1}) \leq \frac{M\epsilon}{4M} + \frac{(b-a)\epsilon}{4M} < \frac{\epsilon}{2}.$$

Similarly, using (4.7), one can show that $II < \epsilon/2$. Thus, it follows that if $\mathcal{D}$ is a γ-fine partition of $[a, b]$, then

$$|S(Ff + Gg, \mathcal{D}) - F(b)G(b)| < \epsilon.$$

This completes the proof. $\square$

In Section 3.9 we established the Riemann-Lebesgue Lemma 3.136 which states that the Fourier coefficients of a function in $L^1[0, 2\pi]$ converge to 0. However, the Riemann-Lebesgue Lemma does not hold for Henstock-Kurzweil integrable functions. Let $F(t) = x^\alpha \cos(1/x)$ on $[0, 2\pi]$ and $f = F'$. Riemann showed that the Fourier coefficients of the function f do not converge to 0 if $0 < a < 1/2$. (See [Ri1, page 260] or [Z, page 19].) We use Theorem 4.27 to show that the Fourier coefficients of a Henstock-Kurzweil integrable function are $o(1/n)$.

Lemma 4.28   *(Riemann-Lebesgue) Let $f : [0, 2\pi] \to \mathbb{R}$ be Henstock-Kurzweil integrable. Then $\lim_{n \to \infty} \frac{1}{n} \int_0^{2\pi} f(t) \cos(nt) \, dt = 0$.*

Proof.   Set $g(t) = \sin(nt)$ and apply Theorem 4.27 with $F(t) = \int_0^t f(x) \, dx$ and $G(t) = \int_0^t \sin(nx) \, dx$. Since F is continuous, Fg is integrable so fG is integrable from Theorem 4.26 and

$$\int_0^{2\pi} f(t) G(t) \, dt = F(2\pi) G(2\pi) - F(0) G(0) - \int_0^{2\pi} F(t) \sin(nt) \, dt.$$

Thus, $G(t) = (1/n) [1 - \cos(nt)]$ implies

$$\frac{1}{n} \int_0^{2\pi} f(t) [1 - \cos(nt)] \, dt = - \int_0^{2\pi} F(t) \sin(nt) \, dt$$

which yields

$$\frac{1}{n} \int_0^{2\pi} f(t) \cos(nt) \, dt = \frac{1}{n} F(2\pi) + \int_0^{2\pi} F(t) \sin(nt) \, dt.$$

Since F is continuous, the right hand side goes to 0 by the Riemann-Lebesgue Lemma 3.136 and the result follows. $\square$

4.3.1   *Cauchy criterion*

Suppose that $f : [a, b] \to \mathbb{R}$ is Henstock-Kurzweil integrable over $[a, b]$ and $\epsilon > 0$. Then, there is a gauge γ so that if $\mathcal{D}$ is a γ-fine tagged partition of $[a, b]$, then $\left| S(f, \mathcal{D}) - \int_a^b f \right| < \frac{\epsilon}{2}$. Let $\mathcal{D}_1$ and $\mathcal{D}_2$ be two γ-fine tagged partitions of $[a, b]$. Then,

$$|S(f, \mathcal{D}_1) - S(f, \mathcal{D}_2)| \leq \left| S(f, \mathcal{D}_1) - \int_a^b f \right| + \left| \int_a^b f - S(f, \mathcal{D}_2) \right| < \epsilon,$$

which is the Cauchy criterion. As in the case of the Riemann integral, the Henstock-Kurzweil integral is characterized by the Cauchy condition.

Theorem 4.29   *A function $f : [a, b] \to \mathbb{R}$ is Henstock-Kurzweil integrable over $[a, b]$ if, and only if, for every $\epsilon > 0$ there is a gauge γ so that if $\mathcal{D}_1$ and $\mathcal{D}_2$ are two γ-fine tagged partitions of $[a, b]$, then*

$$|S(f, \mathcal{D}_1) - S(f, \mathcal{D}_2)| < \epsilon.$$

Proof. We have already shown that the integrability of f implies the Cauchy criterion. So, assume the Cauchy criterion holds. We will prove that f is Henstock-Kurzweil integrable.

For each $k \in \mathbb{N}$, choose a gauge $\gamma_k > 0$ so that for any two γ_k-fine tagged partitions $\mathcal{D}_1$ and $\mathcal{D}_2$ of $[a, b]$ we have

$$|S(f, \mathcal{D}_1) - S(f, \mathcal{D}_2)| < \frac{1}{k}.$$

Replacing γ_k by $\cap_{j=1}^{k}\gamma_j$, we may assume that $\gamma_{k+1} \subset \gamma_k$. For each k, fix a γ_k-fine tagged partition $\mathcal{D}_k$. Note that for $j > k$, since $\gamma_j \subset \gamma_k$, $\mathcal{D}_j$ is a γ_k-fine tagged partition of $[a, b]$. Thus,

$$|S(f, \mathcal{D}_k) - S(f, \mathcal{D}_j)| < \frac{1}{k},$$

which implies that the sequence $\{S(f, \mathcal{D}_k)\}_{k=1}^{\infty}$ is a Cauchy sequence in $\mathbb{R}$, and hence converges. Let A be the limit of this sequence. It follows from the previous inequality that

$$|S(f, \mathcal{D}_k) - A| \leq \frac{1}{k}.$$

It remains to show that A satisfies Definition 4.12.

Fix $\epsilon > 0$ and choose $K > 2/\epsilon$. Let $\mathcal{D}$ be a γ_K-fine tagged partition of $[a, b]$. Then,

$$|S(f, \mathcal{D}) - A| = |S(f, \mathcal{D}) - S(f, \mathcal{D}_K)| + |S(f, \mathcal{D}_K) - A| < \frac{1}{K} + \frac{1}{K} < \epsilon.$$

It now follows that f is Henstock-Kurzweil integrable on $[a, b]$. $\square$

We will use the Cauchy criterion in the following section.

4.3.2 *The integral as a set function*

Suppose that $f : I = [a, b] \to \mathbb{R}$ is Henstock-Kurzweil integrable over I and J is a subinterval of I. It is reasonable to expect that the Henstock-Kurzweil integral of f over J exists.

Theorem 4.30 *Let $f : [a, b] \to \mathbb{R}$ be Henstock-Kurzweil integrable over $[a, b]$. If $J \subset [a, b]$ is a closed subinterval, then f is Henstock-Kurzweil integrable over J.*

Proof. Let $\epsilon > 0$ and γ be a gauge on $[a, b]$ so that if $\mathcal{D}_1$ and $\mathcal{D}_2$ are two γ-fine tagged partitions of $[a, b]$, then $|S(f, \mathcal{D}_1) - S(f, \mathcal{D}_2)| < \epsilon$. Let

$J = [c, d]$ be a closed subinterval of $[a, b]$. Set $J_1 = [a, c]$ and $J_2 = [d, b]$; if either is degenerate, we need not consider it further. Let $\overline{\gamma} = \gamma|_J$ and $\gamma_i = \gamma|_{J_i}$. Suppose that $\mathcal{D}$ and $\mathcal{E}$ are $\overline{\gamma}$-fine tagged partitions of J, and $\mathcal{D}_i$ is a γ_i-fine tagged partition of J_i, $i = 1, 2$. Then $\mathcal{D}' = \mathcal{D} \cup (\mathcal{D}_1 \cup \mathcal{D}_2)$ and $\mathcal{E}' = \mathcal{E} \cup (\mathcal{D}_1 \cup \mathcal{D}_2)$ are γ-fine tagged partitions of I. Since $\mathcal{D}'$ and $\mathcal{E}'$ contain the same pairs (z_j, I_j) off of J,

$$|S(f, \mathcal{D}) - S(f, \mathcal{E})| = |S(f, \mathcal{D}') - S(f, \mathcal{E}')| < \epsilon.$$

By the Cauchy criterion, f is Henstock-Kurzweil integrable over J. $\square$

Thus, if f is Henstock-Kurzweil integrable over an interval I, then it is Henstock-Kurzweil integrable over every subinterval of I and the set function $F(J) = \int_J f$ is defined for all closed subintervals $J \subset I$. Of course, if f is Henstock-Kurzweil integrable over every closed subinterval $J \subset I$, then f is Henstock-Kurzweil integrable over I, since I is a subinterval of itself. Actually, a much stronger result is true. In order for f to be Henstock-Kurzweil integrable over I, it is enough to know that f is Henstock-Kurzweil integrable over a finite number of closed intervals whose union is I, which is a consequence of the next theorem.

Theorem 4.31 *Let $f : [a, b] \to \mathbb{R}$ and let $\{I_j\}_{j=1}^m$ be a finite set of closed intervals with disjoint interiors such that $[a, b] = \cup_{j=1}^m I_j$. If f is Henstock-Kurzweil integrable over each I_j, then f is Henstock-Kurzweil integrable over $[a, b]$ and*

$$\int_a^b f = \sum_{j=1}^m \int_{I_j} f.$$

Proof. Suppose first that $[a, b]$ is divided into two subintervals, $I_1 = [a, c]$ and $I_2 = [c, b]$, and f is Henstock-Kurzweil integrable over both intervals. Fix $\epsilon > 0$ and, for $i = 1, 2$, choose a gauge γ_i on I_i so that if $\mathcal{D}$ is a γ_i-fine tagged partition of I_i, then $\left| S(f, \mathcal{D}) - \int_{I_i} f \right| < \frac{\epsilon}{2}$. If $x < c$, then the largest interval centered at x that does not contain c is $(x - |x - c|, x + |x - c|) = (x - |x - c|, c)$; similarly, if $x > c$, the largest such interval is $(c, x + |x - c|)$. Define a gauge on all of I as follows:

$$\gamma(x) = \begin{cases} \gamma_1(x) \cap (x - |x - c|, c) & \text{if } x \in [a, c) \\ \gamma_2(x) \cap (c, x + |x - c|) & \text{if } x \in (c, b] \\ \gamma_1(c) \cap \gamma_2(c) & \text{if } \quad x = c \end{cases}.$$

Since $c \in \gamma(x)$ if, and only if, $x = c$, c is a tag for every γ-fine tagged partition. Suppose that $\mathcal{D}$ is a γ-fine tagged partition of $[a, b]$. If $(c, J) \in \mathcal{D}$ and J has a nonempty intersection with both I_1 and I_2, divide J into two intervals $J_i = J \cap I_i$, with $J_i \subset \gamma_i(c)$, $i = 1, 2$. Then, $f(c)\ell(J) = f(c)\ell(J_1) + f(c)\ell(J_2)$. Write $\mathcal{D}$ as $\mathcal{D}_1 \cup \mathcal{D}_2$, where $\mathcal{D}_i = \{(x, J) \in \mathcal{D} : J \subset I_i\}$. By the construction of γ, $\mathcal{D}_i$ is a γ_i-fine tagged partition of I_i. After dividing the interval associated to the tag c, if necessary, we have that $S(f, \mathcal{D}) = S(f, \mathcal{D}_1) + S(f, \mathcal{D}_2)$. Thus,

$$\left| S(f, \mathcal{D}) - \left\{ \int_{I_1} f + \int_{I_2} f \right\} \right| \leq \left| S(f, \mathcal{D}_1) - \int_{I_1} f \right| + \left| S(f, \mathcal{D}_2) - \int_{I_2} f \right|$$
$$< \frac{\epsilon}{2} + \frac{\epsilon}{2} = \epsilon.$$

Thus, f is Henstock-Kurzweil integrable over $[a, b]$ and $\int_a^b f = \int_{I_1} f + \int_{I_2} f$. The proof is now completed by an induction argument. See Exercise 4.18. $\square$

A key point in the previous proof is defining a gauge in which a particular point (c) is always a tag. By iteration, one can design a gauge γ that forces a finite set of points to be tags for every γ-fine tagged partition.

Let φ be a step function defined on $[a, b]$ with canonical form $\sum_{i=1}^{m} a_i \chi_{I_i}$. Since the characteristic function of an interval is Riemann and, hence, Henstock-Kurzweil integrable, by linearity

$$\int_I \varphi = \sum_{i=1}^{m} a_i \int_I \chi_{I_i} = \sum_{i=1}^{m} a_i \ell(I_i).$$

So, every step function defined on an interval is Henstock-Kurzweil integrable there, and the value of the Henstock-Kurzweil integral, $\int_I \varphi$, is the same as the value of the Riemann and Lebesgue integrals of φ.

Lemma 4.32 *Let $f : I = [a, b] \to \mathbb{R}$. Suppose that, for every $\epsilon > 0$, there are Henstock-Kurzweil integrable functions φ_1 and φ_2 such that $\varphi_1 \leq f \leq \varphi_2$ on I and $\int_I \varphi_2 \leq \int_I \varphi_1 + \epsilon$. Then, f is Henstock-Kurzweil integrable on I.*

Proof. Let $\epsilon > 0$ and choose corresponding functions φ_1 and φ_2. There are gauges γ_1 and γ_2 on I so that if $\mathcal{D}$ is a γ_i-fine tagged partition of I, then $\left| S(\varphi_i, \mathcal{D}) - \int_I \varphi_i \right| < \epsilon$ for $i = 1, 2$. Set $\gamma(z) = \gamma_1(z) \cap \gamma_2(z)$. Let $\mathcal{D}$ be a γ-fine tagged partition of I. Then,

$$\int_I \varphi_1 - \epsilon < S(\varphi_1, \mathcal{D}) \leq S(f, \mathcal{D}) \leq S(\varphi_2, \mathcal{D}) < \int_I \varphi_2 + \epsilon < \int_I \varphi_1 + 2\epsilon.$$

Therefore, if $\mathcal{D}_1$ and $\mathcal{D}_2$ are γ-fine tagged partitions of I then

$$S(f, \mathcal{D}_1), S(f, \mathcal{D}_2) \in \left(\int_I \varphi_1 - \epsilon, \int_I \varphi_1 + 2\epsilon \right).$$

This implies that

$$|S(f, \mathcal{D}_1) - S(f, \mathcal{D}_2)| < 3\epsilon.$$

By the Cauchy criterion, f is Henstock-Kurzweil integrable. □

Now, suppose that f is a continuous function on $[a, b]$. Let $\mathcal{P} = \{x_0, x_1, \ldots, x_m\}$ be a partition of $[a, b]$ and recall $m_i = \inf_{x_{i-1} \le t \le x_i} f(t)$ and $M_i = \sup_{x_{i-1} \le t \le x_i} f(t)$. Define step functions φ_1 and φ_2 by

$$\varphi_1(t) = m_1 \chi_{[x_0, x_1]}(t) + \sum_{j=2}^m m_j \chi_{(x_{j-1}, x_j]}(t)$$

and

$$\varphi_2(t) = M_1 \chi_{[x_0, x_1]}(t) + \sum_{j=2}^m M_j \chi_{(x_{j-1}, x_j]}(t).$$

Then, clearly, $\varphi_1 \le f \le \varphi_2$ and φ_1 and φ_2 are Henstock-Kurzweil integrable. Further, since f is uniformly continuous on $[a, b]$, given $\epsilon > 0$, there is a $\delta > 0$ so that $|f(x) - f(y)| < \frac{\epsilon}{b-a}$ for all $x, y \in [a, b]$ such that $|x - y| < \delta$. Suppose we choose a partition $\mathcal{P}$ with mesh less than δ. Then $|M_i - m_i| \le \frac{\epsilon}{b-a}$ for $i = 1, \ldots, m$. It then follows that $\varphi_2(x) \le \varphi_1(x) + \frac{\epsilon}{b-a}$ so that

$$\int_a^b \varphi_2 \le \int_a^b \left(\varphi_1 + \frac{\epsilon}{b-a} \right) = \int_a^b \varphi_1 + \int_a^b \frac{\epsilon}{b-a} = \int_a^b \varphi_1 + \epsilon.$$

By the previous lemma, we have proved that every continuous function defined on a closed interval is Henstock-Kurzweil integrable.

Theorem 4.33 *Let $f : [a, b] \to \mathbb{R}$ be continuous on $[a, b]$. Then, f is Henstock-Kurzweil integrable over $[a, b]$.*

Of course, this result is not surprising. By Theorem 2.27, continuous functions are Riemann integrable and, by Theorem 4.14, Riemann integrable functions are Henstock-Kurzweil integrable.

4.4 Unbounded intervals

We would like to extend the definition of the Henstock-Kurzweil integral to unbounded intervals. Given a function f defined on an interval $I \subset \mathbb{R}$, it is easy to extend f to all of $\mathbb{R}$ by defining f to equal 0 off of I. This 'extension' of f to $\mathbb{R}$ should have the same integral as the original function defined on I. So, we may assume that our function f is defined on $\mathbb{R}$.

To extend the definition of the Henstock-Kurzweil integral to functions on $\mathbb{R}$, we need to define a partition of $\mathbb{R}$. A *partition* of $\mathbb{R}$ is a finite, ordered set of points in $\mathbb{R}^*$, $\mathcal{P} = \{-\infty = x_0, x_1, \ldots, x_n = \infty\}$. Still, if we extend our definition of the Henstock-Kurzweil integral directly to $\mathbb{R}$, we run into problems immediately since any tagged partition of $\mathbb{R}$ will have at least one (and generally two) subintervals of infinite length since, if I is an unbounded interval, we set $\ell(I) = \infty$. Even with the convention that $0 \cdot \infty = 0$, if the value of the function at the tag associated with an interval of infinite length is not 0 then the Riemann sum would not be a finite number. Such a situation arises if we consider a positive function defined on all of $\mathbb{R}$.

Example 4.34 Define $f : \mathbb{R} \to \mathbb{R}$ by $f(x) = \frac{1}{1+x^2}$. Let $P \neq \emptyset$ be a partition of $\mathbb{R}$. Then, P has two unbounded intervals, ones of infinite length, say I_1 and I_n. If $a_1, a_n \in \mathbb{R}$ are the tags, then

$$f(a_1)\ell(I_1) + f(a_n)\ell(I_n) = \infty.$$

If $f(x) = \frac{x}{1+x^4}$ and $a_1 < 0 < a_n$, this expression is not even well defined.

To get around this problem, we consider f to be defined on the extended real line, $\mathbb{R}^* = \mathbb{R} \cup \{-\infty, \infty\}$, and we define $f : \mathbb{R}^* \to \mathbb{R}$ by setting $f(\infty) = f(-\infty) = 0$. We call intervals of the form $[a, \infty]$ and $[-\infty, a]$ *closed intervals* containing ∞ and $-\infty$, and $(a, \infty]$ and $[-\infty, a)$ *open intervals* containing ∞ and $-\infty$. If $a_1 = -\infty$ and $a_n = \infty$, then we avoid the problem above. To handle intervals of infinite length, we will often choose gauges so that the only tag for an interval containing ∞ $(-\infty)$ will be ∞ $(-\infty)$.

Remark 4.35 *Suppose that $f : I \subset \mathbb{R} \to \mathbb{R}$. For the remainder of this chapter, we will always assume that f is extended to $\mathbb{R}^*$ by setting $f(x) = 0$ for $x \notin I$; this, of course, implies that f is equal to 0 at ∞ and $-\infty$.*

Let $I \subset \mathbb{R}^*$ be a closed interval. We define a *partition* of I to be a finite collection of non-overlapping closed intervals $\{I_1, \ldots, I_m\}$ such that

$I = \cup_{i=1}^{m} I_i$. A *tagged partition* of I is a finite set of ordered pairs $\mathcal{D} = \{(t_i, I_i) : i = 1, \ldots, m\}$ such that $\{I_i : i = 1, \ldots, m\}$ is a partition of $[a, b]$ and $t_i \in I_i$, $i = 1, \ldots, m$. The point t_i is called the *tag* associated to the interval I_i.

Let I be a closed subinterval of $\mathbb{R}^*$ and suppose that $f : I \to \mathbb{R}$. Let $\mathcal{D} = \{(t_i, I_i) : i = 1, \ldots, m\}$ be a tagged partition of I. The Riemann sum of f with respect to $\mathcal{D}$ is defined to be

$$S(f, \mathcal{D}) = \sum_{i=1}^{m} f(t_i) \ell(I_i).$$

If ∞ and $-\infty$ are the tags for any intervals of infinite length, then this sum is well defined and finite.

For real numbers t, a gauge at t was defined to be an open interval centered at t, $(t - \delta(t), t + \delta(t))$. This definition does not make sense when $t = \infty$, so we need to revise our definition. It turns out that the important feature of a gauge is that the gauge associates to t an open interval containing t, not that the interval is centered at t. Thus, we can revise the definition of a gauge.

Definition 4.36 Given an interval $I = [a, b]$, an interval-valued function γ defined on I is called a *gauge* if, for all $t \in I$, $\gamma(t)$ is an open interval containing t.

Since $(t - \delta(t), t + \delta(t))$ is an open interval containing t, if a function γ satisfies the Definition 4.9, then it satisfies the Definition 4.36. In fact, the two definitions of a gauge, one defined in terms of a positive function $\delta(t)$ and the other in terms of an open interval containing t, are equivalent. See Exercise 4.19. This new definition extends to elements of $\mathbb{R}^*$ by setting $\gamma(\infty) = (a, \infty]$ and $\gamma(-\infty) = [-\infty, b)$ for some $a, b \in \mathbb{R}$.

Definition 4.37 Given an interval $I \subset \mathbb{R}^*$, an interval-valued function γ defined on I is called a *gauge* if, for all $t \in I$, $\gamma(t)$ is an open interval in $\mathbb{R}^*$ containing t. If $\mathcal{D} = \{(t_i, I_i) : i = 1, \ldots, m\}$ is a tagged partition of I and γ is a gauge on I, we say that $\mathcal{D}$ is *γ-fine* if $I_i \subset \gamma(t_i)$ for all i. We denote this by writing $\mathcal{D}$ is a *γ-fine tagged partition* of I.

We show first that for any gauge γ, there exists a γ-fine tagged partition.

Theorem 4.38 *Let γ be a gauge on a closed interval $I = [a, b] \subset \mathbb{R}^*$. Then, there is a γ-fine tagged partition of I.*

Proof. We will prove the result for $I = [a, \infty]$. The other cases are similar. There is a $b \in \mathbb{R}$ such that $\gamma(\infty) = (b, \infty]$. If $b < a$, then $\mathcal{D} = \{(\infty, I)\}$ is a γ-fine tagged partition of I. If $b \geq a$, let $\mathcal{D}_0$ be a γ-fine tagged partition of $[a, b + 1]$. Then, $\mathcal{D} = \mathcal{D}_0 \cup \{(\infty, [b + 1, \infty])\}$ is a γ-fine tagged partition of I. $\square$

We can now define the Henstock-Kurzweil integral over arbitrary closed subintervals of $\mathbb{R}^*$.

Definition 4.39 Let I be a closed subinterval of $\mathbb{R}^*$ and $f : I \to \mathbb{R}$. We call the function f *Henstock-Kurzweil integrable* over I if there is an $A \in \mathbb{R}$ so that for all $\epsilon > 0$ there is a gauge γ on I so that for every γ-fine tagged partition $\mathcal{D}$ of $[a, b]$,

$$|S(f, \mathcal{D}) - A| < \epsilon.$$

Note that the basic properties of integrals, such as linearity and positivity, are valid for the Henstock-Kurzweil integral. The proof that the value of A is unique is the same as above. Thus, the notation $A = \int_I f$ is well defined. Note that if I is an interval of infinite length, $I = [a, \infty]$, say, we write $\int_I f = \int_a^\infty f$.

Let $I \subset \mathbb{R}$ be an arbitrary interval. Suppose that f and g are Henstock-Kurzweil integrable on I. For all scalars $\alpha, \beta \in \mathbb{R}$,

$$\int_I (\alpha f + \beta g) = \alpha \int_I f + \beta \int_I g;$$

that is the Henstock-Kurzweil integral in linear. It is also positive, so that $f \geq 0$ implies that $\int_I f \geq 0$, and satisfies a Cauchy condition. These results generalize Propositions 4.19 and 4.20 and Theorem 4.29 and follow from the same proofs. Finally, as in Theorem 4.31, the Henstock-Kurzweil integral is additive over disjoint intervals. That is, $f : I \to \mathbb{R}$ is Henstock-Kurzweil integrable over I if, and only if, for every finite set $\{I_j\}_{j=1}^m$ of closed intervals with disjoint interiors such that $I = \cup_{j=1}^m I_j$, f is Henstock-Kurzweil integrable over each I_j. In either case,

$$\int_a^b f = \sum_{j=1}^m \int_{I_j} f.$$

The proof of this result is a little easier than before, since we can use interval gauges. Thus, using the notation of that proof, we can replace the gauge in the proof by

$$\gamma(x) = \begin{cases} \gamma_1(x) \cap (-\infty, c) \text{ if } x \in I_1 \cap (-\infty, c) \\ \gamma_2(x) \cap (c, \infty) \text{ if } x \in I_2 \cap (c, \infty) \\ \gamma_1(c) \cap \gamma_2(c) \text{ if } \qquad x = c \end{cases}.$$

Earlier, we proved that the Dirichlet function is Henstock-Kurzweil integrable over $[0, 1]$ with an integral of 0. It is easy to adapt that proof to show that a function which is 0 except on a countable set has Henstock-Kurzweil integral 0. We now prove a much stronger result, namely that any function which is 0 except on a null set (recall that a set E is null if $m(E) = 0$) is Henstock-Kurzweil integrable with integral 0.

Example 4.40 Let $E \subset \mathbb{R}$ be a null set. Suppose that $f : \mathbb{R} \to \mathbb{R}$ and $f = 0$ except in E; i.e., $f = 0$ a.e. in $\mathbb{R}$. We show that f is Henstock-Kurzweil integrable over $\mathbb{R}$ and $\int_{\mathbb{R}} f = 0$. Fix $\epsilon > 0$. Set $E_m = \{t \in \mathbb{R} : m - 1 < |f(t)| \le m\}$. Note that the sets $\{E_m\}_{m=1}^{\infty}$ are pairwise disjoint and $E_m \subset E$ since f equals 0 off of E, so each E_m is a null set. For all $m \in \mathbb{N}$, there are countably many open intervals $\{I_j^m : j \in \sigma_m\}$ such that $E_m \subset \cup_{j \in \sigma_m} I_j^m$ and $\sum_{j \in \sigma_m} \ell(I_j^m) < \epsilon/2^m m$. If $t \in E_m$, let $m(t)$ be the smallest integer j such that $t \in I_j^m$. Define a gauge γ on $\mathbb{R}$ by setting $\gamma(t) = (t - 1, t + 1)$ for $t \notin E$, $\gamma(t) = I_{m(t)}^m$ for $t \in E_m$, $\gamma(\infty) = (0, \infty]$ and $\gamma(-\infty) = [-\infty, 0)$. (The choice of 0 for an endpoint is arbitrary.)

Suppose that $\mathcal{D} = \{(t_i, J_i) : i = 1, \dots, k\}$ is a γ-fine tagged partition of $\mathbb{R}^*$. Let $\mathcal{D}_0 = \{(t_i, J_i) \in \mathcal{D} : t_i \notin E\}$ and, for $m \in \mathbb{N}$, let $\mathcal{D}_m = \{(t_i, J_i) \in \mathcal{D} : t_i \in E_m\}$. Then, $S(f, \mathcal{D}_0) = 0$ and, since the intervals $\{J_i : (t_i, J_i) \in \mathcal{D}_m\}$ are non-overlapping and $\cup_{(t_i, J_i) \in \mathcal{D}_m} J_i \subset \cup_{j \in \sigma_m} I_j^m$,

$$|S(f, \mathcal{D}_m)| = \left| \sum_{(t_i, J_i) \in \mathcal{D}_m} f(t_i) \ell(J_i) \right| \le m \sum_{(t_i, J_i) \in \mathcal{D}_m} \ell(J_i) < \frac{\epsilon}{2^m}.$$

Thus,

$$|S(f, \mathcal{D})| \le \sum_{m=0}^{\infty} |S(f, \mathcal{D}_m)| < \sum_{m=1}^{\infty} \frac{\epsilon}{2^m} = \epsilon,$$

so f is Henstock-Kurzweil integrable over $\mathbb{R}$ and $\int_{\mathbb{R}} f = 0$. In particular, if E is a null set, then χ_E is Henstock-Kurzweil integrable with $\int_{\mathbb{R}} \chi_E = 0$.

As a consequence of this example, we see that if $f : \mathbb{R} \to \mathbb{R}$ and $f = 0$ except on a null set E, then $\int_I f = 0$ for every interval $I \subset \mathbb{R}$. We will show later, after discussing Part II of the Fundamental Theorem of Calculus, that if $\int_I f = 0$ for every interval $I \subset \mathbb{R}$ then $f = 0$ a.e..

In particular, if $E \subset \mathbb{R}$ is a null set, then χ_E is Henstock-Kurzweil integrable with $\int_I \chi_E = 0$ for any interval $I \subset \mathbb{R}$. We show next that the converse to this statement is true; that is, if $\int_{\mathbb{R}} \chi_E = 0$, then E is measurable with $m(E) = 0$. In order to prove this result, we will use a covering lemma. Suppose we have a set $E \subset \mathbb{R}$ and a collection of sets $\{S\}_{S \in \mathcal{C}}$ such that $E \subset \cup_{S \in \mathcal{C}} S$. This covering lemma will be used to pick a subset of $\mathcal{C}$ so that the union of the members of the subset still cover E and have additional useful properties.

Lemma 4.41 *Let $I \subset \mathbb{R}$ be a closed and bounded interval and $E \subset I$ be nonempty. Let γ be a gauge on I. Then, there is a countable family $\{(t_k, J_k) : k \in \sigma\}$ such that the intervals in $\{J_k : k \in \sigma\}$ are non-overlapping and closed subintervals of I, $t_k \in J_k \cap E$, $J_k \subset \gamma(t_k)$, and $E \subset \cup_{k \in \sigma} J_k \subset I$.*

Proof. Let $\mathcal{D}_k$ be the set of closed subintervals of I obtained by dividing I into 2^k equal subintervals. In other words, $\mathcal{D}_1$ contains the two intervals obtained by bisecting I into two equal parts, $\mathcal{D}_2$ consists of the four intervals obtained by bisecting the two intervals in $\mathcal{D}_1$, and, in general, $\mathcal{D}_k$ is comprised of the 2^k intervals created when the intervals in $\mathcal{D}_{k-1}$ are bisected. Notice that $\cup_{k=1}^{\infty} \mathcal{D}_k$ is a countable set and if $J' \in \mathcal{D}_k$ and $J'' \in \mathcal{D}_l$, then either J' and J'' are non-overlapping or one is contained in the other.

Let $\mathcal{E}_1$ consist of the elements $J \in \mathcal{D}_1$ for which there is a $t \in E \cap J$ with $J \subset \gamma(t)$. Next, let $\mathcal{E}_2$ be the family of intervals $J \in \mathcal{D}_2$ such that there is a $t \in E \cap J$ with $J \subset \gamma(t)$ and J is not contained in any element of $\mathcal{E}_1$, and continue the process. Thus, one gets a sequence of collections of closed subintervals of I, $\{\mathcal{E}_k\}_{k=1}^{\infty}$, some of which may be empty. The set $\mathcal{E} = \cup_{k=1}^{\infty} \mathcal{E}_k$ is a countable collection of non-overlapping, closed intervals in I. By construction, if $J \in \mathcal{E}$, then there is a $t \in E \cap J$ such that $J \subset \gamma(t)$. It remains to show that $E \subset \cup_{J \in \mathcal{E}} J$.

Suppose $t \in E$. Then, there is an integer K so that for $k \geq K$, if $J_{k(t)} \in \mathcal{D}_k$ is the subinterval that contains t, then $J_{k(t)} \subset \gamma(t)$. Either $J_K \in \mathcal{E}_K$ or there is a $J \in \cup_{k=1}^{K-1} \mathcal{E}_k$ such that $J_K \subset J$. Thus, $t \in \cup_{J \in \mathcal{E}} J$, as we wished to prove. $\square$

We are now ready to prove

Theorem 4.42 *Let $E \subset \mathbb{R}$. Then, E is a null set if, and only if, χ_E is Henstock-Kurzweil integrable and $\int_{\mathbb{R}} \chi_E = 0$.*

Proof. The sufficiency is proved in Example 4.40. To prove the necessity, assume that χ_E is Henstock-Kurzweil integrable with integral 0. By Exercise 4.9, it follows that $\chi_{E\cap[-n,n]}$ has integral 0. Thus, we may assume that E is a bounded set, since if we can show that $E \cap [-n,n]$ is a null set for all $n \in \mathbb{N}$, it follows that E is a null set.

Let I be a bounded interval containing E. Fix $\epsilon > 0$ and choose a gauge γ such that $|S(\chi_E, \mathcal{D})| < \frac{\epsilon}{2}$ for every γ-fine tagged partition $\mathcal{D}$ of I. Let $\{(t_k, J_k) : k \in \sigma\}$ be the countable family given by Lemma 4.41. Let $\sigma' \subset \sigma$ be a finite subset. The set $I \setminus \cup_{k\in\sigma'} J_k$ is a union of a finite set of non-overlapping intervals. Let $K_1, \ldots, K_l$ be the closure of these intervals, and let $\mathcal{D}_i$ be a γ-fine tagged partition of K_i, $i = 1, \ldots, l$. Then, $\mathcal{D} = \{(t_k, J_k) : k \in \sigma'\} \cup_{i=1}^l \mathcal{D}_i$ is a γ-fine tagged partition of I. Since $\chi_E \geq 0$,

$$\sum_{k\in\sigma'} \ell(J_k) = \sum_{k\in\sigma'} \chi_E(t_k) \ell(J_k) \leq S(\chi_E, \mathcal{D}) < \frac{\epsilon}{2}.$$

Since this is true for every finite subset of σ, it follows that

$$\sum_{k\in\sigma} \ell(J_k) \leq \frac{\epsilon}{2}.$$

Finally, for each $k \in \sigma$, let I_k be an open interval containing J_k with $\ell(I_k) = \ell(J_k) + \epsilon 2^{-k}$. Since $E \subset \cup_{k\in\sigma} J_k \subset \cup_{k\in\sigma} I_k$, $\{I_k\}_{k\in\sigma}$ is a countable collection of open intervals containing E. Further,

$$\sum_{k\in\sigma} \ell(I_k) = \sum_{k\in\sigma} \left(\ell(J_k) + \frac{\epsilon}{2^k} \right) = \sum_{k\in\sigma} \ell(J_k) + \sum_{k\in\sigma} \frac{\epsilon}{2^k} \leq \frac{\epsilon}{2} + \frac{\epsilon}{2} = \epsilon.$$

Since this holds for all $\epsilon > 0$, we see that E is a null set. $\qquad\square$

In the following example, we relate Henstock-Kurzweil integrals to infinite series.

Example 4.43 Suppose that $\sum_{k=1}^\infty a_k$ is a convergent sequence and set $f(x) = \sum_{k=1}^\infty a_k \chi_{[k,k+1)}(x)$. We claim that f is Henstock-Kurzweil integrable over $[1, \infty)$ and

$$\int_1^\infty f = \sum_{k=1}^\infty a_k.$$

Since the series is convergent, there is a $B > 0$ so that $|a_k| \leq B$ for all $k \in \mathbb{N}$. Let $\epsilon > 0$. Pick a natural number M so that $\left| \sum_{k=j}^\infty a_k \right| < \epsilon$ and $|a_j| < \epsilon$ for $j \geq M$. Define a gauge γ as follows. For $t \in (k, k+1)$, let

$\gamma(t) = (k, k+1)$; for $t = k$, let $\gamma(t) = \left(t - \min\left(\frac{\epsilon}{2^k B}, 1\right), t + \min\left(\frac{\epsilon}{2^k B}, 1\right)\right)$; and, let $\gamma(\infty) = (M, \infty]$. Suppose that $\mathcal{D} = \{(t_i, I_i) : i = 1, \ldots, m\}$ is a γ-fine tagged partition of $[1, \infty]$. Without loss of generality, we may assume that $t_m = \infty$ and $I_m = [b, \infty]$, so that $b > M$ and $f(t_m)\ell(I_m) = 0$. Let K be the largest integer less than or equal to b. Then, $K \geq M$.

Note that for $k \in \mathbb{N}$ and $k \leq b$, k must be a tag. Let $\mathcal{D}_\mathbb{N} = \{(t_i, I_i) \in \mathcal{D} : t_i \in \mathbb{N}\}$. For $k \in \mathbb{N}$, $\cup\{I_i : (t_i, I_i) \in \mathcal{D}_\mathbb{N} \text{ and } t_i = k\} \subset \gamma(k)$. Thus,

$$|S(f, \mathcal{D}_\mathbb{N})| = \left|\sum_{k=1}^{K} a_k \sum_{(t_i, I_i) \in \mathcal{D}_\mathbb{N}; t_i = k} \ell(I_i)\right| \leq \sum_{k=1}^{K} |a_k| \sum_{(t_i, I_i) \in \mathcal{D}_\mathbb{N}; t_i = k} \ell(I_i)$$

$$\leq \sum_{k=1}^{K} |a_k| \ell(\gamma(k)) < \sum_{k=1}^{K} |a_k| \frac{\epsilon}{2^{k-1} B} < \sum_{k=1}^{\infty} \frac{\epsilon}{2^{k-1}} = 2\epsilon.$$

Set $\mathcal{D}_k = \{(t_i, I_i) \in \mathcal{D} : t_i \in (k, k+1)\}$. Note first that

$$|S(f, \mathcal{D}_K) - a_K| = \left|\sum_{(t_i, I_i) \in \mathcal{D}_K} a_K \ell(I_i) - a_K\right|$$

$$= \left|a_K \left(\sum_{(t_i, I_i) \in \mathcal{D}_K} \ell(I_i) - 1\right)\right| \leq |a_K| < \epsilon.$$

For $1 \leq k < K$, by the definition of $\gamma(j)$ for $j \in \mathbb{N}$, $\cup_{(t_i, I_i) \in \mathcal{D}_k} I_i$ is a subinterval of $(k, k+1)$ with length $\ell_k \geq 1 - \frac{\epsilon}{2^k B} - \frac{\epsilon}{2^{k+1} B}$, and

$$S(f, \mathcal{D}_k) = \sum_{(t_i, I_i) \in \mathcal{D}_k} a_k \ell(I_i) = a_k \sum_{(t_i, I_i) \in \mathcal{D}_k} \ell(I_i) = a_k \ell_k.$$

Thus,

$$|S(f, \mathcal{D}_k) - a_k| = |a_k(\ell_k - 1)| \leq B\left(\frac{\epsilon}{2^k B} + \frac{\epsilon}{2^{k+1} B}\right) < \frac{\epsilon}{2^{k-1}}.$$

Therefore,

$$\left| S(f, \mathcal{D}) - \sum_{k=1}^{\infty} a_k \right| = \left| \sum_{k=1}^{\infty} S(f, \mathcal{D}_k) + S(f, \mathcal{D}_\mathbb{N}) - \sum_{k=1}^{\infty} a_k \right|$$

$$\leq \left| \sum_{k=1}^{K-1} \{ S(f, \mathcal{D}_k) - a_k \} \right| + |S(f, \mathcal{D}_K) - a_K|$$

$$+ |S(f, \mathcal{D}_\mathbb{N})| + \left| \sum_{k=K+1}^{\infty} a_k \right|$$

$$< \sum_{k=1}^{\infty} \frac{\epsilon}{2^{k-1}} + \epsilon + 2\epsilon + \epsilon = 6\epsilon.$$

It follows that f is Henstock-Kurzweil integrable over $[1, \infty)$.

Moreover, if the function $f(x) = \sum_{k=1}^{\infty} a_k \chi_{[k,k+1)}(x)$ is Henstock-Kurzweil integrable, then the series $\sum_{k=1}^{\infty} a_k$ converges. See Exercise 4.24.

This example highlights two of the important properties of the Henstock-Kurzweil integral. Note that we have evaluated the integral of a function defined on an interval of infinite length directly from the definition of the Henstock-Kurzweil integral. There is no need to view this as an improper integral. We will discuss this issue in the following section.

We say a function f is *conditionally integrable* if f is Henstock-Kurzweil integrable but $|f|$ is not Henstock-Kurzweil integrable. Using this example, one can now easily construct conditionally integrable functions. If $\sum_{k=1}^{\infty} a_k$ is a conditionally convergent series, then $f(x) = \sum_{k=1}^{\infty} a_k \chi_{(k,k+1)}(x)$ is a Henstock-Kurzweil integrable function by Example 4.43 while $|f(x)| = \sum_{k=1}^{\infty} |a_k| \chi_{(k,k+1)}(x)$ is not Henstock-Kurzweil integrable by Exercise 4.23. Thus, f is a conditionally integrable function. This is in contrast to the Riemann and Lebesgue integrals, for which integrability implies absolute integrability.

Example 4.44 The function $f(x) = \sum_{k=1}^{\infty} \frac{(-1)^k}{k} \chi_{(k,k+1)}(x)$ is a conditionally integrable function on $[1, \infty)$.

4.5 Henstock's Lemma

If f is Henstock-Kurzweil integrable over an interval I, given any $\epsilon > 0$, there is a gauge γ so that if $\mathcal{D} = \{(t_i, I_i) : i = 1, \dots, m\}$ is a γ-fine tagged partition of I, then

$$\left| \sum_{i=1}^{m} f(t_i) \ell(I_i) - \int_I f \right| < \epsilon. \tag{4.9}$$

Since $\int_I f = \sum_{i=1}^{m} \int_{I_i} f$, we can rewrite Equation (4.9) as

$$\left| \sum_{i=1}^{m} \left\{ f(t_i) \ell(I_i) - \int_{I_i} f \right\} \right| = \left| \sum_{i=1}^{m} f(t_i) \ell(I_i) - \sum_{i=1}^{m} \int_{I_i} f \right| < \epsilon.$$

Thus, one is led to consider if, in addition to controlling the difference of sums, one can simultaneously control the estimate for a single interval $\left| f(t_i) \ell(I_i) - \int_{I_i} f \right| < \epsilon$ or, more generally, an estimate of part of the sum; that is, if $\mathcal{D}' \subset \mathcal{D}$, one might expect that

$$\left| \sum_{(t_i, I_i) \in \mathcal{D}'} f(t_i) \ell(I_i) - \int_{\cup_{(t_i, I_i) \in \mathcal{D}'} I_i} f \right|$$

$$= \left| \sum_{(t_i, I_i) \in \mathcal{D}'} f(t_i) \ell(I_i) - \sum_{(t_i, I_i) \in \mathcal{D}'} \int_{I_i} f \right| < \epsilon. \tag{4.10}$$

However, in general, Equation (4.9) holds due to cancellation in the expression on the left hand side. Since the cancellation from one interval may help the estimate for another interval, it is not at all clear that Equation (4.10) will hold, even if $\mathcal{D}'$ contains a single pair (t, I). In this section, we will show that Equation (4.10) (with $<$ replaced by $\leq$) follows from Equation (4.9).

Let $I \subset \mathbb{R}$ be an interval. A *subpartition* of I is a finite set of non-overlapping closed intervals $\{J_i\}_{i=1}^{k}$ such that $J_i \subset I$ for $i = 1, \ldots, k$. A *tagged subpartition* of I is a finite set of ordered pairs $\mathcal{S} = \{(t_i, J_i) : i = 1, \ldots, k\}$ such that $\{J_i\}_{i=1}^{k}$ is a subpartition of I and $t_i \in I_i$. We say that a tagged subpartition is γ-fine if $I_i \subset \gamma(t_i)$ for all i. Note that a γ-fine tagged partition of I is also a γ-fine tagged subpartition of I.

We will now prove *Henstock's Lemma*, which is a valuable tool for deriving results about the Henstock-Kurzweil integral. We will apply Henstock's Lemma to the study of improper integrals and convergence theorems.

Lemma 4.45 *(Henstock's Lemma) Let $f : I \subset \mathbb{R} \to \mathbb{R}$ be Henstock-Kurzweil integrable over I. For $\epsilon > 0$, let γ be a gauge such that if $\mathcal{D}$ is a* *γ-fine tagged partition of I, then*

$$\left| S\left(f, \mathcal{D}\right) - \int_I f \right| < \epsilon.$$

Suppose $\mathcal{D}' = \{(x_1, J_1), \ldots, (x_k, J_k)\}$ is γ-fine tagged subpartition of I. Then

$$\left| \sum_{i=1}^k \left\{ f\left(x_i\right) \ell\left(J_i\right) - \int_{J_i} f \right\} \right| \leq \epsilon \text{ and } \sum_{i=1}^k \left| f\left(x_i\right) \ell\left(J_i\right) - \int_{J_i} f \right| \leq 2\epsilon.$$

Proof. Let $\epsilon > 0$ and γ a gauge satisfying the hypothesis. The set $I \setminus \cup_{i=1}^k J_i$ is a finite union of disjoint intervals. Let $K_1, \ldots, K_m$ be the closure of these intervals. Fix $\eta > 0$. Since f is Henstock-Kurzweil integrable over each K_j, there is a γ-fine tagged partition $\mathcal{D}_j$ of K_j such that

$$\left| S\left(f, \mathcal{D}_j\right) - \int_{K_j} f \right| < \frac{\eta}{m}.$$

One can find such a partition by choosing a gauge γ_j for the interval K_j and the margin of error $\frac{\eta}{m}$, and then choosing a partition which is $\gamma \cap \gamma_j$-fine. Set $\mathcal{D} = \mathcal{D}' \cup \mathcal{D}_1 \cup \cdots \cup \mathcal{D}_m$. Then, $\mathcal{D}$ is a γ-fine tagged partition of I. Since $S\left(f, \mathcal{D}\right) = S\left(f, \mathcal{D}'\right) + \sum_{j=1}^m S\left(f, \mathcal{D}_j\right)$, we have

$$\left| S\left(f, \mathcal{D}'\right) - \sum_{i=1}^k \int_{J_i} f \right| = \left| S\left(f, \mathcal{D}'\right) - \sum_{i=1}^k \int_{J_i} f + \sum_{j=1}^m \left\{ S\left(f, \mathcal{D}_j\right) - \int_{K_j} f \right\} \right.$$

$$\left. - \sum_{j=1}^m \left\{ S\left(f, \mathcal{D}_j\right) - \int_{K_j} f \right\} \right|$$

$$\leq \left| S\left(f, \mathcal{D}\right) - \int_I f \right| + \sum_{j=1}^m \left| S\left(f, \mathcal{D}_j\right) - \int_{K_j} f \right|$$

$$< \epsilon + m\frac{\eta}{m} = \epsilon + \eta.$$

Since $\eta > 0$ was arbitrary, it follows that

$$\left| \sum_{i=1}^k \left\{ f\left(x_i\right) \ell\left(J_i\right) - \int_{J_i} f \right\} \right| = \left| S\left(f, \mathcal{D}'\right) - \sum_{i=1}^k \int_{J_i} f \right| \leq \epsilon.$$

To prove the other estimate, set

$$\mathcal{D}^+ = \left\{ (x_i, J_i) \in \mathcal{D}' : f(x_i)\,\ell(J_i) - \int_{J_i} f \geq 0 \right\}$$

and $\mathcal{D}^- = \mathcal{D}' \backslash \mathcal{D}^+$. Note that both $\mathcal{D}^-$ and $\mathcal{D}^+$ are γ-fine tagged subpartitions of I, so they satisfy the previous estimate. Thus,

$$\sum_{i=1}^{k} \left| f(x_i)\,\ell(J_i) - \int_{J_i} f \right| = \sum_{(x_i, J_i) \in \mathcal{D}^+} \left\{ f(x_i)\,\ell(J_i) - \int_{J_i} f \right\}$$

$$+ \sum_{(x_i, J_i) \in \mathcal{D}^-} \left\{ \int_{J_i} f - f(x_i)\,\ell(J_i) \right\}$$

$$\leq 2\epsilon.$$

This completes the proof of the theorem. □

Suppose that I is a subinterval of $\mathbb{R}$ and $a \in I$. Suppose that $f : I \to \mathbb{R}$ is Henstock-Kurzweil integrable, so that f is integrable over every subinterval of I. Define the *indefinite integral* F of f by $F(x) = \int_a^x f$ for all $x \in I$.

Theorem 4.46 *If $f : I \to \mathbb{R}$ is Henstock-Kurzweil integrable over I, then F is continuous on I.*

Proof. Fix $a \in I$ and $x \in I$. Let $\epsilon > 0$. Choose a gauge γ so that $|S(f, \mathcal{D}) - \int_I f| < \epsilon$ for every γ-fine tagged partition $\mathcal{D}$ of I. If $\gamma(x) = (\alpha, \beta)$, set $\delta = \min\left\{ \beta - x, x - \alpha, \frac{\epsilon}{1+|f(x)|} \right\}$ and suppose that $y \in I$ and $|y - x| < \delta$. Let J be the subinterval of I with endpoints x and y. Applying Henstock's Lemma to the γ-fine tagged subpartition $\{(x, J)\}$ shows that

$$\left| f(x)\,\ell(J) - \int_J f \right| \leq \epsilon.$$

This implies that

$$|F(y) - F(x)| = \left| \int_J f \right| \leq \epsilon + |f(x)|\,\ell(J) < \epsilon + \epsilon = 2\epsilon.$$

Thus, F is continuous at x. Since $x \in I$ was arbitrary, F is continuous on I. □

Thus, Henstock's Lemma implies that the indefinite integral of a Henstock-Kurzweil integrable function is continuous. We apply the second inequality in Henstock's Lemma in the proof of the following corollary.

Corollary 4.47 *Let $f : I = [a, b] \to \mathbb{R}$ be Henstock-Kurzweil integrable over I. If $\int_a^c f = 0$ for every $c \in [a, b]$, then $|f|$ is Henstock-Kurzweil integrable with $\int_I |f| = 0$.*

Proof. By hypothesis, if $a \le c < d \le b$, then $\int_c^d f = \int_a^d f - \int_a^c f = 0$, so that $\int_J f = 0$ for every interval $J \subset I$. Let $\epsilon > 0$ and choose a gauge γ such that

$$\left| S(f, \mathcal{D}) - \int_I f \right| < \epsilon$$

for every γ-fine tagged partition $\mathcal{D}$. Let $\mathcal{D} = \{(t_i, I_i) : i = 1, \ldots, m\}$ be a γ-fine tagged partition. By Henstock's Lemma,

$$\sum_{i=1}^m |f(x_i)| \ell(I_i) = \sum_{i=1}^m \left| f(x_i) \ell(I_i) - \int_{I_i} f \right| \le 2\epsilon,$$

which implies that $|f|$ is Henstock-Kurzweil integrable with $\int_I |f| = 0$. $\square$

We have seen above in Example 4.25 that an unbounded function can be Henstock-Kurzweil integrable, and in Example 4.43 that a function defined on an unbounded interval can be Henstock-Kurzweil integrable. Using Henstock's lemma, we show that there are no improper integrals for the Henstock-Kurzweil integral. We begin by considering a function defined on a bounded interval.

Theorem 4.48 *Let $f : [a, b] \to \mathbb{R}$ be Henstock-Kurzweil integrable over $[c, b]$ for every $a < c < b$. Then, f is Henstock-Kurzweil integrable over $[a, b]$ if, and only if, $\lim_{c \to a^+} \int_c^b f$ exists. In either case,*

$$\int_a^b f = \lim_{c \to a^+} \int_c^b f.$$

Proof. Suppose first that f is Henstock-Kurzweil integrable over $[a, b]$. Let $\epsilon > 0$ and choose a gauge γ so that if $\mathcal{D}$ is a γ-fine tagged partition of $[a, b]$, then

$$\left| S(f, \mathcal{D}) - \int_a^b f \right| < \frac{\epsilon}{3}.$$

For each $c \in (a, b)$, there is a gauge γ_c defined on $[c, b]$ so that if $\mathcal{E}$ is a

γ_c-fine tagged partition of $[c, b]$ then

$$\left| S\left(f, \mathcal{E}\right) - \int_c^b f \right| < \frac{\epsilon}{3}.$$

Without loss of generality, we may assume that $\gamma_c \subset \gamma$, by replacing γ_c by $\gamma_c \cap \gamma$ if necessary. Choose $c \in \gamma(a)$ such that $|f(a)|(c - a) < \epsilon/3$.

Fix $s \in (a, c)$ and let $\mathcal{E}$ be a γ_s-fine tagged partition of $[s, b]$. Set $\mathcal{D} = \{(a, [a, s])\} \cup \mathcal{E}$. Then, $\mathcal{D}$ is a γ-fine tagged partition of $[a, b]$, and

$$\left| \int_a^b f - \int_s^b f \right| \leq \left| \int_a^b f - S\left(f, \mathcal{D}\right) \right| + \left| S\left(f, \mathcal{E}\right) - \int_s^b f \right| + |f(a)|(c - a)$$

$$< \frac{\epsilon}{3} + \frac{\epsilon}{3} + \frac{\epsilon}{3} = \epsilon.$$

Thus, $\lim_{c \to a^+} \int_c^b f = \int_a^b f$.

Next, suppose the limit exists. Choose $\{c_k\}_{k=1}^\infty \subset [a, b]$ so that $c_0 = b$, $c_k > c_{k+1}$ and $c_k \to a$. Define a gauge γ_1 on $[c_1, c_0]$ so that if $\mathcal{D}$ is a γ_1-fine tagged partition of $[c_1, c_0]$, then

$$\left| S\left(f, \mathcal{D}\right) - \int_{c_1}^{c_0} f \right| < \frac{\epsilon}{2}.$$

For $k > 1$, define a gauge γ_k on $[c_k, c_{k-2}]$ so that if $\mathcal{D}$ is a γ_k-fine tagged partition of $[c_k, c_{k-2}]$, then

$$\left| S\left(f, \mathcal{D}\right) - \int_{c_k}^{c_{k-2}} f \right| < \frac{\epsilon}{2^k}.$$

Set $A = \lim_{c \to a^+} \int_c^b f$. Choose K so that $\left| \int_s^{c_0} f - A \right| < \epsilon$ for $a < s \leq c_K$ and $|f(a)|(c_K - a) < \epsilon$. Define a gauge γ on $[a, b]$ by

$$\gamma(t) = \begin{cases} (-\infty, c_K) & \text{if } t = a \\ \gamma_1(t) \cap (c_1, \infty) & \text{if } c_1 < t \leq c_0 \\ \gamma_k(t) \cap (c_k, c_{k-2}) & \text{if } c_k < t \leq c_{k-1} \text{ for } k > 1 \end{cases}.$$

Let $\mathcal{D}$ be a γ-fine tagged partition of $[a, b]$, and $\mathcal{D}_k$ be the subset of $\mathcal{D}$ with tags in $(c_k, c_{k-1}]$. Since $\mathcal{D}$ has a finite number of elements, only finitely many $\mathcal{D}_k \neq \emptyset$ and $\mathcal{D}_i \cap \mathcal{D}_j = \emptyset$ for $i \neq j$. Let J_k be the union of subintervals in $\mathcal{D}_k$. Then, $\mathcal{D}_k$ is γ_k-fine on J_k, and $J_1 \subset (c_1, c_0]$ and $J_k \subset (c_k, c_{k-2})$. By Henstock's Lemma, for $k \geq 1$,

$$\left| \int_{J_k} f - S\left(f, \mathcal{D}_k\right) \right| \leq \frac{\epsilon}{2^k}.$$

Let $(x, [a, d]) \in \mathcal{D}$. By the definition of γ, $a \in \gamma(t)$ if, and only if, $t = a$, so that $x = a$. Since $S(f, \mathcal{D}) = f(a)(d - a) + \sum_{k=1}^{\infty} S(f, \mathcal{D}_k)$ and $\int_d^b f = \sum_{k=1}^{\infty} \int_{J_k} f$, in which both sums have finitely many nonzero terms,

$$|A - S(f, \mathcal{D})| \leq |f(a)|(d - a) + \left| \sum_{k=1}^{\infty} \left\{ \int_{J_k} f - S(f, \mathcal{D}_k) \right\} \right| + \left| A - \int_d^b f \right|$$

$$< \epsilon + \sum_{k=1}^{\infty} \frac{\epsilon}{2^k} + \epsilon = 3\epsilon.$$

Thus, f is Henstock-Kurzweil integrable over $[a, b]$ and $\int_a^b f = A$. $\qquad \square$

This proof can be modified to handle a singularity at b, instead of at a. Further, for a singularity at an interior point $c \in (a, b)$, one may consider the integrals over $[a, c]$ and $[c, b]$ separately.

Suppose that $f : [a, b] \to \mathbb{R}$ is Riemann integrable over $[c, b]$ for all $a < c < b$ and has an improper Riemann integral over $[a, b]$. Then, f is Henstock-Kurzweil integrable over $[c, b]$ and $\lim_{c \to a+} \int_c^b f$ exists. Thus, f is Henstock-Kurzweil integrable over $[a, b]$.

Example 4.49 Let $p \in \mathbb{R}$ and define $f : [0, 1] \to \mathbb{R}$ by $f(t) = t^p$, for $0 < t \leq 1$ and $f(0) = 0$. By Example 2.44, we see that f is Henstock-Kurzweil integrable over $[0, 1]$ with integral $\int_0^1 t^p dt = \dfrac{1}{p + 1}$ if, and only if, $p > -1$.

Suppose, next, that f is defined on an unbounded interval $I = [a, \infty]$. We show that integrals over I exist in the Henstock-Kurzweil sense as proper integrals, demonstrating that there are no Cauchy-Riemann integrals in the Henstock-Kurzweil theory. The proof is similar to the previous one, treating the difficulty at ∞ as the one at a was handled above.

Theorem 4.50 *Let $f : I = [a, \infty] \to \mathbb{R}$ be Henstock-Kurzweil integrable over $[a, b]$ for every $a < b < \infty$. Then, f is Henstock-Kurzweil integrable over $[a, \infty]$ if, and only if, $\lim_{b \to \infty} \int_a^b f$ exists. In either case,*

$$\int_a^{\infty} f = \lim_{b \to \infty} \int_a^b f.$$

Proof. Suppose first that f is Henstock-Kurzweil integrable over I. Let $\epsilon > 0$ and choose a gauge γ so that if $\mathcal{D}$ is a γ-fine tagged partition of I, then

$$\left| S\left(f, \mathcal{D}\right) - \int_I f \right| < \frac{\epsilon}{2}.$$

Suppose $\gamma\left(\infty\right) = \left(T, \infty\right]$. For each $c > \max\left\{T, a\right\}$, there is a gauge γ_c defined on $[a, c]$ so that if $\mathcal{E}$ is a γ_c-fine tagged partition of $[a, c]$ then

$$\left| S\left(f, \mathcal{E}\right) - \int_a^c f \right| < \frac{\epsilon}{2},$$

and such that $\gamma_c\left(z\right) \subset \gamma\left(z\right)$ for all $z \in [a, c]$.

Fix $c > \max\left\{T, a\right\}$ and let $\mathcal{E}$ be a γ_c-fine tagged partition of $[a, c]$. Set $\mathcal{D} = \mathcal{E} \cup \left\{\left(\infty, [c, \infty]\right)\right\}$. Then, $\mathcal{D}$ is a γ-fine tagged partition of I, and

$$\left| \int_I f - \int_a^c f \right| \leq \left| \int_I f - S\left(f, \mathcal{D}\right) \right| + \left| S\left(f, \mathcal{E}\right) - \int_a^c f \right| + \left| f\left(\infty\right) \right| \ell\left([c, \infty]\right)$$

$$< \frac{\epsilon}{2} + \frac{\epsilon}{2} = \epsilon,$$

since $\left| f\left(\infty\right) \right| \ell\left([c, \infty]\right) = 0$ by convention. Thus, $\lim_{b \to \infty} \int_a^b f = \int_I f$.

Next, suppose the limit exists. Choose $\left\{c_k\right\}_{k=1}^\infty \subset [a, \infty)$ so that $c_0 = a$, $c_k < c_{k+1}$ and $c_k \to \infty$. Define a gauge γ_0 on $[c_0, c_1]$ so that

$$\left| S\left(f, \mathcal{D}\right) - \int_{c_0}^{c_1} f \right| < \frac{\epsilon}{2^2}$$

for every γ_0-fine tagged partition $\mathcal{D}$ of $[c_0, c_1]$. For $k \geq 1$, choose a gauge γ_k on $[c_{k-1}, c_{k+1}]$ so that if $\mathcal{D}$ is a γ_k-fine tagged partition of $[c_{k-1}, c_{k+1}]$, then

$$\left| S\left(f, \mathcal{D}\right) - \int_{c_{k-1}}^{c_{k+1}} f \right| < \frac{\epsilon}{2^{k+2}}.$$

Set $A = \lim_{b \to \infty} \int_a^b f$. Choose K so that $\left| \int_a^b f - A \right| < \epsilon/2$ for $b \geq c_K$. Define a gauge γ on I by

$$\gamma\left(t\right) = \begin{cases} \gamma_0\left(t\right) \cap \left(-\infty, c_1\right) & \text{if } c_0 \leq t < c_1 \\ \gamma_k\left(t\right) \cap \left(c_{k-1}, c_{k+1}\right) & \text{if } c_k \leq t < c_{k+1} \text{ for } k \geq 1. \\ \left(c_K, \infty\right] & \text{if } z = \infty \end{cases}$$

Let $\mathcal{D}$ be a γ-fine tagged partition of I. If $I_i = [\alpha, \infty]$ is the unbounded interval of $\mathcal{D}$, then $t_i = \infty$ and $\alpha > c_K$. For $k \geq 0$, let $\mathcal{D}_k$ be the subset of $\mathcal{D}$ with tags in $[c_k, c_{k+1})$. As above, only finitely many $\mathcal{D}_k \neq \emptyset$ and $\mathcal{D}_i \cap \mathcal{D}_j = \emptyset$ for $i \neq j$. Let J_k be the union of subintervals in $\mathcal{D}_k$. Then, $\mathcal{D}_k$ is γ_k-fine on J_k, and $J_0 \subset [c_0, c_1)$ and $J_k \subset (c_{k-1}, c_{k+1})$. By Henstock's Lemma, for $k \geq 0$,

$$\left| \int_{J_k} f - S(f, \mathcal{D}_k) \right| \leq \frac{\epsilon}{2^{k+2}}.$$

Since $\alpha > c_K$, it follows that

$$|A - S(f, \mathcal{D})| \leq \left| A - \int_a^\alpha f \right| + \left| \int_a^\alpha f - S(f, \mathcal{D}) \right|$$

$$< \frac{\epsilon}{2} + \left| \sum_{k=0}^\infty \int_{J_k} f - \sum_{k=0}^\infty S(f, \mathcal{D}_k) + f(\infty) \ell(I_i) \right|$$

$$< \frac{\epsilon}{2} + \sum_{k=0}^\infty \frac{\epsilon}{2^{k+2}} = \epsilon.$$

Thus, f is Henstock-Kurzweil integrable over I and $\int_I f = A$. $\qquad \square$

An analogous result holds for intervals of the form $[-\infty, b]$. A version of this result for $[-\infty, \infty]$ follows by writing $[-\infty, \infty] = [-\infty, a] \cup [a, \infty]$. The value of the integral so obtained does not depend on the choice of a. See Exercise 4.29.

Example 4.51 Let $p \in \mathbb{R}$ and define $f : [1, \infty] \to \mathbb{R}$ by $f(t) = t^{-p}$, for $t \geq 1$. By Example 2.47, we see that f is Henstock-Kurzweil integrable over $[1, \infty]$ with integral $\int_1^\infty t^{-p} dt = \frac{1}{p-1}$ if, and only if, $p > 1$.

Following Example 4.44, we saw that $f(x) = \sum_{k=1}^\infty \frac{(-1)^k}{k} \chi_{(k,k+1)}(x)$ is a conditionally integrable function on $[1, \infty]$. We now give another example of a function that has a conditionally convergent integral.

Example 4.52 It was shown in Example 2.49 that $f(x) = \frac{\sin x}{x}$ has a convergent Cauchy-Riemann integral over $[1, \infty)$, but that $|f|$ is not Cauchy-Riemann integrable there. By Theorem 4.50, f is Henstock-Kurzweil integrable and $|f|$ is not, so f has a conditionally convergent integral.

We now use these theorems to obtain several useful results for guaranteeing absolute integrability. The first result includes a comparison test.

Corollary 4.53 *Let $f : [a, b] \subset \mathbb{R}^* \to \mathbb{R}$. Suppose that f is absolutely integrable over $[a, c]$ for every $a \leq c < b$.*

(1) Suppose f is nonnegative. Then, f is Henstock-Kurzweil integrable over $[a, b]$ if, and only if, $\sup \left\{ \int_a^c f : a \leq c < b \right\} < \infty$.

(2) If there is a Henstock-Kurzweil integrable function $g : [a, b] \to \mathbb{R}$ such that $|f(t)| \leq g(t)$ for all $t \in I$, then f is absolutely integrable over I.

Note that b may be finite or infinite.

Proof. To prove (1), note that the function $F(x) = \int_a^x f$ is increasing on $[a, b]$. Thus, $\sup \left\{ \int_a^c f : a \leq c < b \right\} = \lim_{c \to b} \int_a^c f$, and the result follows from either Theorem 4.48 or 4.50.

For (2), define F as above and set $G(x) = \int_a^x g$. Since g is Henstock-Kurzweil integrable, G satisfies a Cauchy condition near b. We claim that F, too, satisfies a Cauchy condition near b. To see this, note that for $a < x < y < b$,

$$|F(y) - F(x)| = \left| \int_x^y f \right| \leq \int_x^y |f| \leq \int_x^y g = G(y) - G(x).$$

Thus, F is Cauchy near b, and f is Henstock-Kurzweil integrable, by either Theorem 4.48 or 4.50. Applying the same argument to $H(x) = \int_a^x |f|$ shows that $|f|$ is Henstock-Kurzweil integrable, so that f is absolutely integrable. $\qquad \square$

As a consequence of the corollary, we derive the integral test for convergence of series.

Proposition 4.54 *Let $f : [1, \infty] \to \mathbb{R}$ be positive, decreasing and Henstock-Kurzweil integrable over $[1, b]$ for all $1 < b < \infty$. The integral $\int_1^\infty f$ exists if, and only if, the series $\sum_{k=1}^\infty f(k)$ converges. In either case, $\int_1^\infty f \leq \sum_{k=1}^\infty f(k) \leq \int_1^\infty f + f(1)$.*

Proof. Since f is decreasing, $f(i+1) \leq f(x) \leq f(i)$ for $i \leq x \leq i+1$, which implies that $f(i+1) \leq \int_i^{i+1} f \leq f(i)$. Summing in i yields

$$\sum_{i=1}^{n-1} f(i+1) \leq \int_1^n f \leq \sum_{i=1}^{n-1} f(i). \qquad (4.11)$$

By the previous corollary, it now follows that f is Henstock-Kurzweil integrable over $[1, \infty]$ if, and only if, the series converges. Letting $n \to \infty$ in (4.11) shows that $\int_1^\infty f \leq \sum_{k=1}^\infty f(k) \leq \int_1^\infty f + f(1)$. $\qquad \square$

A function φ is called a *multiplier* if the product φf is integrable for every integrable function f. For the Lebesgue integral, every bounded, measurable function is a multiplier. For if φ is measurable and bounded by B, then for any Lebesgue integrable function f, φf is measurable and φf is bounded by the Lebesgue integrable function $B|f|$, so φf is Lebesgue integrable by Proposition 3.94. Surprisingly, for the Henstock-Kurzweil integral, continuous functions need not be multipliers, even on intervals of finite length.

Example 4.55   Define $F, G : [0,1] \to \mathbb{R}$ by $F(0) = G(0) = 0$ and $F(x) = x^2 \sin(x^{-4})$ and $G(x) = x^2 \cos(x^{-4})$ for $0 < x \le 1$ and let $f = F'$ and $g = G'$. Since $(FG)' = Fg + fG$, $Fg + Gf$ is Henstock-Kurzweil integrable by Theorem 4.16. However, $F(x)g(x) - f(x)G(x) = \frac{4}{x}$ for $x \ne 0$, is not Henstock-Kurzweil integrable over $[0,1]$. This implies that neither Fg nor fG is Henstock-Kurzweil integrable over $[0,1]$. Since, for example, F is continuous and g is Henstock-Kurzweil integrable, we see that continuous functions need not be multipliers for the Henstock-Kurzweil integral. See Theorem 4.26.

A function φ is a multiplier for the Henstock-Kurzweil integral if, and only if, it is equal almost everywhere to a function of bounded variation, which we define in the next section. (See [Lee, Theorem 12.9].)

4.6   Absolute integrability

Let f be Henstock-Kurzweil integrable over I. Since f need not be absolutely integrable, we do not know whether or not $|f|$ is Henstock-Kurzweil integrable. We now turn our attention to characterizing when a Henstock-Kurzweil integrable function is absolutely integrable. For this characterization, we will use the concept of *bounded variation*.

4.6.1   *Bounded variation*

The variation of a function is a measure of its oscillation. A function with bounded variation has finite oscillation.

Definition 4.56   Let $\varphi : [a,b] \to \mathbb{R}$. Given a partition $\mathcal{P} = \{x_0, \dots, x_m\}$ of $[a,b]$, define the *variation* of φ with respect to $\mathcal{P}$ by

$$v(\varphi, \mathcal{P}) = \sum_{i=1}^{m} |\varphi(x_i) - \varphi(x_{i-1})|,$$

and the *variation* of φ over $[a, b]$ by

$$Var\left(\varphi, [a, b]\right) = \sup\left\{v\left(\varphi, \mathcal{P}\right) : \mathcal{P} \text{ is a partition of } [a, b]\right\}.$$

We say that φ has *bounded variation* over $[a, b]$ if $Var\left(\varphi, [a, b]\right) < \infty$. In this case, we write $\varphi \in \mathcal{BV}\left([a, b]\right)$.

A constant function has 0 variation, which follows immediately from the definition. A function can have a jump discontinuity and still have bounded variation. For example, the function f defined on $[0, 2]$ by $f(x) = 0$ for $0 \le x < 1$ and $f(x) = 1$ for $1 \le x \le 2$ has a variation of 1, equal to the jump at $x = 1$. Somewhat surprisingly, a continuous function need not have bounded variation.

Example 4.57   The function $\varphi : [0, 1] \to \mathbb{R}$ defined by

$$\varphi(t) = \begin{cases} 0 & \text{if } t = 0 \\ t \sin\left(\frac{1}{t}\right) & \text{if } 0 < t \le 1 \end{cases}$$

is continuous on $[0, 1]$. Set $x_m = \dfrac{1}{\left(m + \frac{1}{2}\right)\pi}$. Then,

$$\varphi(x_m) = \begin{cases} \dfrac{1}{\left(m + \frac{1}{2}\right)\pi} & \text{if } m \text{ is even} \\[4mm] -\dfrac{1}{\left(m + \frac{1}{2}\right)\pi} & \text{if } m \text{ is odd} \end{cases},$$

so that $\left|\varphi(x_m) - \varphi(x_{m-1})\right| = \dfrac{2m}{\left(m + \frac{1}{2}\right)\left(m - \frac{1}{2}\right)\pi} > \dfrac{2}{\pi m}$.   Since $\sum_{m=1}^{\infty} \dfrac{2}{\pi m}$ diverges, it follows that $Var\left(\varphi, [0, 1]\right) = \infty$ and φ does not have bounded variation on $[0, 1]$.

We next develop some of the basic properties of functions with bounded variation. We first show that a function with bounded variation is bounded and that the variation of a function is additive over disjoint intervals.

Proposition 4.58   *If $\varphi \in \mathcal{BV}\left([a, b]\right)$ then φ is bounded on $[a, b]$.*

Proof.   For $x \in [a, b]$, consider the partition $\mathcal{P} = \{a, x, b\}$ of $[a, b]$. Then,

$$\left|\varphi(x) - \varphi(a)\right| + \left|\varphi(b) - \varphi(x)\right| \le Var\left(\varphi, [a, b]\right),$$

which implies

$$|\varphi(x)| \leq \frac{1}{2}\left[|\varphi(a)| + |\varphi(b)| + Var\left(\varphi, [a, b]\right)\right].$$

Thus, φ is bounded on $[a, b]$.   □

If $a < c < b$, by the triangle inequality

$$v\left(\varphi, \{a, b\}\right) = |\varphi(b) - \varphi(a)|$$
$$\leq |\varphi(c) - \varphi(a)| + |\varphi(b) - \varphi(c)| = v\left(\varphi, \{a, c, b\}\right).$$

This inequality is the basic point in the proof that the variation of a function increases as one passes from a partition to one of its refinements. We will use this result to prove that variation is additive.

Proposition 4.59 *Let $\varphi : [a, b] \to \mathbb{R}$. If $\mathcal{P}$ and $\mathcal{P}'$ are partitions of $[a, b]$ and $\mathcal{P}'$ is a refinement of $\mathcal{P}$, then $v\left(\varphi, \mathcal{P}\right) \leq v\left(\varphi, \mathcal{P}'\right)$.*

Proof. Suppose first that $\mathcal{P}'$ has one more element than $\mathcal{P}$; that is, there is an $\hat{x}$ such that $\mathcal{P} = \{x_0, x_1, \ldots, x_n\}$ and $\mathcal{P}' = \{x_0, x_1, \ldots, x_{i-1}, \hat{x}, x_i, \ldots, x_n\}$. Then, all the terms in the sum for $v\left(\varphi, \mathcal{P}\right)$ are the same as those for $v\left(\varphi, \mathcal{P}'\right)$ except for $|\varphi(x_i) - \varphi(x_{i-1})|$ which is bounded by $|\varphi(x_i) - \varphi(\hat{x})| + |\varphi(\hat{x}) - \varphi(x_{i-1})|$. Thus, $v\left(\varphi, \mathcal{P}\right) \leq v\left(\varphi, \mathcal{P}'\right)$. The proof now follows by an induction argument.   □

We can now show that the variation of a function is additive over disjoint intervals.

Proposition 4.60 *Let $\varphi : [a, b] \to \mathbb{R}$ and suppose that $a < c < b$. Then,*

$$Var\left(\varphi, [a, b]\right) = Var\left(\varphi, [a, c]\right) + Var\left(\varphi, [c, b]\right).$$

Proof. Let $\mathcal{P}$ be a partition of $[a, b]$ and set $\mathcal{P}' = \mathcal{P} \cup \{c\}$. Then, $\mathcal{P}_1 = \{x \in \mathcal{P}' : x \leq c\}$ is a partition of $[a, c]$ and $\mathcal{P}_2 = \{x \in \mathcal{P}' : x \geq c\}$ is a partition of $[c, b]$, and $v\left(\varphi, \mathcal{P}'\right) = v\left(\varphi, \mathcal{P}_1\right) + v\left(\varphi, \mathcal{P}_2\right)$. Thus, by the previous proposition,

$$v\left(\varphi, \mathcal{P}\right) \leq v\left(\varphi, \mathcal{P}'\right) = v\left(\varphi, \mathcal{P}_1\right) + v\left(\varphi, \mathcal{P}_2\right) \leq Var\left(\varphi, [a, c]\right) + Var\left(\varphi, [c, b]\right).$$

It follows that $Var\left(\varphi, [a, b]\right) \leq Var\left(\varphi, [a, c]\right) + Var\left(\varphi, [c, b]\right)$.

On the other hand, if $\mathcal{P}_1$ is a partition of $[a, c]$ and $\mathcal{P}_2$ is a partition of $[c, b]$, then $\mathcal{P} = \mathcal{P}_1 \cup \mathcal{P}_2$ is a partition of $[a, b]$. As above,

$$v\left(\varphi, \mathcal{P}_1\right) + v\left(\varphi, \mathcal{P}_2\right) = v\left(\varphi, \mathcal{P}\right) \leq Var\left(\varphi, [a, b]\right).$$

Taking the supremum over all partitions $\mathcal{P}_1$ of $[a, c]$ yields

$$Var\left(\varphi, [a, c]\right) + v\left(\varphi, \mathcal{P}_2\right) \leq Var\left(\varphi, [a, b]\right).$$

Then, taking the supremum over all partitions $\mathcal{P}_2$ of $[c, b]$ shows that

$$Var\left(\varphi, [a, c]\right) + Var\left(\varphi, [c, b]\right) \leq Var\left(\varphi, [a, b]\right),$$

which completes the proof. $\qquad\qquad\square$

Suppose that φ is an increasing function on $[a, b]$. If $a \leq x < z < y \leq b$, then

$$\left|\varphi\left(z\right) - \varphi\left(x\right)\right| + \left|\varphi\left(y\right) - \varphi\left(z\right)\right| = \left|\varphi\left(y\right) - \varphi\left(x\right)\right|.$$

It follows that for any partition $\mathcal{P}$ of $[a, b]$, $v\left(\varphi, \mathcal{P}\right) = \left|\varphi\left(b\right) - \varphi\left(a\right)\right|$ and $\varphi \in \mathcal{BV}\left([a, b]\right)$; moreover, $Var\left(\varphi, [a, b]\right) = \left|\varphi\left(b\right) - \varphi\left(a\right)\right|$. One can argue similarly for a decreasing function, so that every monotone function on a bounded interval has bounded variation there. Another easy consequence of the definition is that

$$v\left(\alpha\varphi + \beta\psi, \mathcal{P}\right) \leq \left|\alpha\right| v\left(\varphi, \mathcal{P}\right) + \left|\beta\right| v\left(\psi, \mathcal{P}\right),$$

which implies that linear combinations of functions of bounded variation have bounded variation. See Exercise 4.39. A surprising fact about functions of bounded variation is that all such functions can be written as the difference of increasing functions.

Theorem 4.61 *A function $\varphi \in \mathcal{BV}\left([a, b]\right)$ if, and only if, there are increasing functions p and q so that $\varphi = p - q$.*

Proof. If p and q are increasing functions on $[a, b]$, by the observations above, $p - q \in \mathcal{BV}\left([a, b]\right)$. So, suppose that $\varphi \in \mathcal{BV}\left[a, b\right]$. Define p by $p\left(x\right) = Var\left(\varphi, [a, x]\right)$, where $Var\left(\varphi, [a, a]\right) = 0$ by definition, and $q = p - \varphi$. From Proposition 4.60, p is increasing. If $a \leq x < y \leq b$, then

$$q\left(y\right) = p\left(y\right) - \varphi\left(y\right) = Var\left(\varphi, [a, y]\right) - \varphi\left(y\right).$$

Thus,

$$\begin{aligned}
q\left(y\right) - q\left(x\right) &= Var\left(\varphi, [a, y]\right) - \varphi\left(y\right) - \left\{Var\left(\varphi, [a, x]\right) - \varphi\left(x\right)\right\} \\
&= \left\{Var\left(\varphi, [a, y]\right) - Var\left(\varphi, [a, x]\right)\right\} - \left\{\varphi\left(y\right) - \varphi\left(x\right)\right\}.
\end{aligned}$$

Since Proposition 4.60 implies that $Var\left(\varphi,[a,y]\right) - Var\left(\varphi,[a,x]\right) =$
$Var\left(\varphi,[x,y]\right),$

$$q\left(y\right) - q\left(x\right) = Var\left(\varphi,[x,y]\right) - \left(\varphi\left(y\right) - \varphi\left(x\right)\right) \geq 0.$$

Therefore, q is increasing and the proof is complete. $\square$

4.6.2   *Absolute integrability and indefinite integrals*

We are now ready to prove that a Henstock-Kurzweil integrable function
is absolutely integrable if, and only if, its indefinite integral has bounded
variation. Recall that we define the indefinite integral of f by $F\left(x\right) = \int_a^x f$.

Theorem 4.62   *Let $f : I = [a,b] \to \mathbb{R}$ be Henstock-Kurzweil integrable
over I. Then, $|f|$ is Henstock-Kurzweil integrable over I if, and only if, the
indefinite integral of f has bounded variation over I. In either case,*

$$Var\left(F,[a,b]\right) = \int_a^b |f|.$$

Proof.   Let $V = Var\left(F,[a,b]\right)$.   Note that for $a \leq x < y \leq b$,
$|F\left(y\right) - F\left(x\right)| = \left|\int_x^y f\right|$. Suppose first that $|f|$ is Henstock-Kurzweil integrable. For any partition $\mathcal{P} = \{x_0, \ldots, x_m\}$ of I,

$$v\left(F,\mathcal{P}\right) = \sum_{i=1}^m |F\left(x_i\right) - F\left(x_{i-1}\right)| = \sum_{i=1}^m \left|\int_{x_{i-1}}^{x_i} f\right| \leq \sum_{i=1}^m \int_{x_{i-1}}^{x_i} |f| = \int_a^b |f|.$$

Thus, $V \leq \int_a^b |f| < \infty$, so $F \in \mathcal{BV}\left([a,b]\right)$.

Next, suppose that $F \in \mathcal{BV}\left([a,b]\right)$ and let $\epsilon > 0$. Choose a partition
$\mathcal{P} = \{x_0, \ldots, x_m\}$ of I such that

$$V - \epsilon < v\left(F,\mathcal{P}\right) \leq V.$$

Since f is Henstock-Kurzweil integrable over I, we can choose a gauge $\bar{\gamma}$
on I so that if $\mathcal{D}$ is a $\bar{\gamma}$-fine tagged partition of I then $\left|S\left(f,\mathcal{P}\right) - \int_I f\right| < \epsilon$.
For convenience, set $x_{-1} = -\infty$ and $x_{m+1} = \infty$. Define a gauge γ on I by:

$$\gamma\left(x\right) = \begin{cases} \bar{\gamma}\left(x\right) \cap \left(x_{i-1}, x_i\right) & \text{if } x \in \left(x_{i-1}, x_i\right) \\ \bar{\gamma}\left(x\right) \cap \left(x_{i-1}, x_{i+1}\right) & \text{if } x = x_i \end{cases}.$$

Note that for $x \notin \mathcal{P}$, $\gamma\left(x\right)$ is an open interval that does not contain any
elements of $\mathcal{P}$. Thus, if $\left(z, J\right) \in \mathcal{D}$ and there is an $x_j \in \mathcal{P}$ such that
$x_j \in J \subset \gamma\left(z\right)$, then $z \in \mathcal{P}$. By the definition of γ for elements of $\mathcal{P}$, it
then follows that $z = x_j$.

Let $\mathcal{D} = \{(z_i, I_i) : i = 1, \ldots, k\}$ be a γ-fine tagged partition of I and, without loss of generality, assume that $\max I_{i-1} = \min I_i$ for $i = 1, \ldots, k$. Let $\mathcal{Q} = \{y_0, \ldots, y_k\}$ be the partition defined by $\mathcal{D}$ so that $I_i = [y_{i-1}, y_i]$. If $x_j \in I_i^o$, the interior of I_i, then x_j is the tag for I_i. We replace I_i by the pair of intervals $I_i^1 = [y_{i-1}, x_j]$ and $I_i^2 = [x_j, y_i]$. Repeating this for all the terms in $\mathcal{P}$ as necessary, one gets a new tagged partition $\mathcal{D}' = \{(z_i', I_i') : i = 1, \ldots, K\}$ in which all such terms $(x_j, I_i) \in \mathcal{D}$ are replaced by the two terms (x_j, I_i^1) and (x_j, I_i^2), and a refinement $\mathcal{P}' = \mathcal{P} \cup \mathcal{Q}$ of $\mathcal{P}$. Note that $\mathcal{D}'$ is γ-fine since $I_i^k \subset I_i \subset \gamma(x_j)$. Finally, since $|f(x_j)| \ell(I_i) = |f(x_j)| \ell(I_i^1) + |f(x_j)| \ell(I_i^2)$ for all $x_j \in \mathcal{P}$, $S(|f|, \mathcal{D}) = S(|f|, \mathcal{D}')$.

Since $\mathcal{P}'$ is a refinement of $\mathcal{P}$, by Proposition 4.59,

$$V - \epsilon < v(F, \mathcal{P}) \leq v(F, \mathcal{P}') = \sum_{i=1}^{K} \left| \int_{I_i'} f \right| \leq V.$$

Since $\mathcal{D}'$ is a γ-fine tagged partition of I, it follows from Henstock's Lemma that

$$\left| \sum_{i=1}^{K} \left\{ |f(z_i')| \ell(I_i') - \left| \int_{I_i'} f \right| \right\} \right| \leq \sum_{i=1}^{K} \left| \left\{ |f(z_i')| \ell(I_i') - \left| \int_{I_i'} f \right| \right\} \right|$$

$$\leq \sum_{i=1}^{K} \left| f(z_i') \ell(I_i') - \int_{I_i'} f \right| \leq 2\epsilon.$$

Thus,

$$|S(|f|, \mathcal{D}) - V| \leq \left| S(|f|, \mathcal{D}') - \sum_{i=1}^{K} \left| \int_{I_i'} f \right| \right| + \left| \sum_{i=1}^{K} \left| \int_{I_i'} f \right| - V \right| < 2\epsilon + \epsilon = 3\epsilon.$$

Therefore, $|f|$ is Henstock-Kurzweil integrable and $\int_I |f| = V$. $\qquad \square$

Since $\mathcal{BV}([a, b])$ is a linear space, the following corollary is immediate.

Corollary 4.63 *If $f, g : I = [a, b] \to \mathbb{R}$ are absolutely integrable over I, then $f + g$ is absolutely integrable over I.*

As a consequence of Theorem 4.62, we obtain the following comparison result for integrals.

Corollary 4.64 *Let $f, g : I = [a, b] \to \mathbb{R}$ be Henstock-Kurzweil integrable over I and suppose that $|f(t)| \leq g(t)$ for all $t \in I$. Then, f is absolutely* *integrable over I and*

$$\int_I |f| \le \int_I g.$$

Proof. Let $\mathcal{P} = \{x_0, \ldots, x_m\}$ be a partition of I. Then

$$\sum_{i=1}^m \left| \int_{x_{i-1}}^{x_i} f \right| \le \sum_{i=1}^m \int_{x_{i-1}}^{x_i} g = \int_I g.$$

Thus, the indefinite integral F of f has bounded variation over $[a, b]$, so by Theorem 4.62 $|f|$ is integrable over I and

$$\int_I |f| = Var\, (F, [a, b]) \le \int_I g.$$

$\square$

Extensions of the three results in this section to functions $f : \mathbb{R} \to \mathbb{R}$ are given in Exercises 4.43 and 4.44.

The examples above of Henstock-Kurzweil integrable functions that are not absolutely integrable involved functions defined on infinite intervals. We conclude this section with an example of such a function on $[0, 1]$.

Example 4.65 In Example 4.1, we exhibited a function f on $[0, 1]$ whose derivative f' is not Lebesgue integrable. The key estimate in that proof is $\int_{a_k}^{b_k} f' = 1/2k$, where $b_k = 1/\sqrt{2k}$ and $a_k = \sqrt{2/(4k+1)}$. By the Fundamental Theorem of Calculus, f' is Henstock-Kurzweil integrable. Since the intervals $[\alpha_k, \beta_k]$ are pairwise disjoint,

$$Var\, (f, [0, 1]) \ge \sum_{k=1}^N \left| \int_{\alpha_k}^{\beta_k} f' \right| = \sum_{k=1}^N \frac{1}{2k}.$$

Thus, $f \notin \mathcal{BV}\,([0, 1])$ so that $|f'|$ is not Henstock-Kurzweil integrable over $[0, 1]$.

4.6.3 *Lattice properties*

We have seen that the sets of Riemann and Lebesgue integrable functions satisfy lattice properties so that, for example, the maximum and minimum of Riemann integrable functions are Riemann integrable. We now study the lattice properties of Henstock-Kurzweil integrable functions.

Proposition 4.66 *Suppose that $f, g : I \to \mathbb{R}$.*

(1) The function f is absolutely integrable over I if, and only if, f^+ and f^- are Henstock-Kurzweil integrable over I.

(2) If f and g are absolutely integrable over I, then $f \vee g$ and $f \wedge g$ are Henstock-Kurzweil integrable over I.

Proof. To prove (1), recall that $f = f^+ - f^-$, $|f| = f^+ + f^-$, $f^+ = \dfrac{|f| + f}{2}$ and $f^- = \dfrac{|f| - f}{2}$. The result now follows from the linearity of the integral. For (2), we observe that $f \vee g = \frac{1}{2}[f + g + |f - g|]$ and $f \wedge g = \frac{1}{2}[f + g - |f - g|]$. By the linearity of the integral and the fact that the sum of absolutely integrable functions is absolutely integrable, the proof is complete. $\qquad\square$

If we only assume that f and g are Henstock-Kurzweil integrable, we need an additional assumption in order to guarantee that the maximum and the minimum of Henstock-Kurzweil integrable functions are Henstock-Kurzweil integrable. For example, if f' is defined as in Example 4.1, then $(f')^+$, the maximum of f' and 0, is not Henstock-Kurzweil integrable while both f' and 0 are Henstock-Kurzweil integrable.

Proposition 4.67 *Suppose that $f, g, h : I \to \mathbb{R}$ are Henstock-Kurzweil integrable over I.*

(1) If $f \leq h$ and $g \leq h$, then $f \vee g$ and $f \wedge g$ are Henstock-Kurzweil integrable over I.

(2) If $h \leq f$ and $h \leq g$, then $f \vee g$ and $f \wedge g$ are Henstock-Kurzweil integrable over I.

Proof. Suppose the conditions of (1) hold. Since $h - f$ and $h - g$ are non-negative and Henstock-Kurzweil integrable, they are absolutely integrable. By the previous proposition, $(h - f) \vee (h - g)$ is Henstock-Kurzweil integrable. Since,

$$
\begin{aligned}
(h - f) \vee (h - g) &= \frac{1}{2}[(h - f) + (h - g) + |(h - f) - (h - g)|] \\
&= \frac{1}{2}[2h - f - g + |-f + g|] \\
&= h - \frac{1}{2}[f + g - |f - g|] \\
&= h - f \wedge g,
\end{aligned}
$$

it follows that $f \wedge g$ is Henstock-Kurzweil integrable. The remaining proofs are similar. $\qquad\square$

We saw in Example 4.55 that the product of a continuous function and a Henstock-Kurzweil integrable function need not be Henstock-Kurzweil integrable, even on a bounded interval, in contrast to the Riemann and Lebesgue integrals. We conclude this section with conditions that guarantee the integrability of the product of two functions.

Proposition 4.68 *(Dedekind's Test) Let $f, g : [a, b] \to \mathbb{R}$ be continuous on $(a, b]$. Suppose that F, defined by $F(x) = \int_x^b f$ for $a < x \le b$, is bounded on $(a, b]$, g' is absolutely integrable over $[a, b]$, and $\lim_{x \to a^+} g(x) = 0$. Then, fg is Henstock-Kurzweil integrable over $[a, b]$.*

Proof. For $a < c \le b$, $(Fg)' = -fg + Fg'$ on the interval $[c, b]$, so that $Fg' = (Fg)' + fg$ and, by the Fundamental Theorem of Calculus and the fact that fg is continuous, Fg' is Henstock-Kurzweil integrable over $[c, b]$. Since F is bounded, there is a $B > 0$ so that $|Fg'| \le B |g'|$. Since g' is absolutely integrable, by Corollary 4.64, Fg' is absolutely integrable over $[c, b]$ for all $a < c < b$. Thus, by Corollary 4.53, Fg' is (absolutely) integrable over $[a, b]$. Since $fg = Fg' - (Fg)'$ from above, by the Fundamental Theorem of Calculus,

$$\int_c^b fg = \int_c^b Fg' + F(c) g(c).$$

Since F is bounded and $\lim_{x \to a^+} g(x) = 0$,

$$\int_a^b fg = \lim_{c \to a^+} \left(\int_c^b Fg' + F(c) g(c) \right) = \lim_{c \to a^+} \int_c^b Fg' = \int_a^b Fg'$$

by Theorem 4.48, so that fg is Henstock-Kurzweil integrable over $[a, b]$. $\square$

See Exercises 4.32, 4.34, and 4.36 for additional examples of integrable products.

4.7 Convergence theorems

The Lebesgue integral is noted for the powerful convergence theorems it satisfies. We now consider their analogs for the Henstock-Kurzweil integral. As we saw in the previous chapter, some restrictions are required for the equation

$$\int_I \lim_{k \to \infty} f_k = \lim_{k \to \infty} \int_I f_k \tag{4.12}$$

to hold.

Example 4.69 Define $f_k : [0, 1] \to \mathbb{R}$ by $f_k(x) = k\chi_{(0,1/k)}(x)$. Then, $\{f_k\}_{k=1}^{\infty}$ converges pointwise to the function f which is identically 0 on $[0, 1]$. Thus, $\int_0^1 f_k = 1$ while $\int_0^1 f = 0$. All the functions are Henstock-Kurzweil integrable, but equation (4.12) does not hold.

A similar problem arises when one considers integrals over unbounded intervals.

Example 4.70 The functions $f_k : [0, \infty) \to \mathbb{R}$ defined by $f_k(x) = \chi_{(k,k+1)}(x)$ converge pointwise to the 0 function, but each f_k has Henstock-Kurzweil integral equal to 1, so equation (4.12) does not hold.

Like the Riemann integral, the simplest condition that allows the interchange of limit and integral on bounded intervals is uniform convergence. See Exercise 4.46. However, such a result does not hold in full generality for the Henstock-Kurzweil integral over unbounded intervals.

Example 4.71 Define $f_k : \mathbb{R} \to \mathbb{R}$ by $f_k(x) = \frac{1}{2k}\chi_{(-k,k)}(x)$. Each f_k is Henstock-Kurzweil integrable and $\int_{\mathbb{R}} f_k = 1$. Further, $\{f_k\}_{k=1}^{\infty}$ converges uniformly to the function f which is identically 0 on $\mathbb{R}$ so that equation (4.12) does not hold.

Equation (4.12) can be viewed as an interchange of limit operations, the operations of integration and limit. In beginning analysis courses it is often shown that a sufficient condition for the interchange of limits in double sequences is uniform convergence in one of the variables and the existence of limits in the other variables. (See [DS, Section 11.5].) For our first result to "pass the limit under the integral sign", we introduce a condition called *uniform integrability*, which is analogous to uniform convergence and is sufficient for interchanging limit and integral for the Henstock-Kurzweil integral. Throughout this section let I be a closed interval (bounded or unbounded) in $\mathbb{R}^*$ and $f_k, f : I \to \mathbb{R}$ for $k \in N$.

Definition 4.72 We say a sequence of functions $\{f_k\}_{k=1}^{\infty}$ is *uniformly integrable* over I if each f_k is integrable over I and for every $\epsilon > 0$ there exists a gauge γ on I such that $\left| S(f_k, \mathcal{D}) - \int_I f_k \right| < \epsilon$ for every $k \in N$ whenever D is a γ-fine tagged partition.

The point of Definition 4.72 is that the same gauge works uniformly for all k. We mention for later use that if one can find a gauge γ that satisfies the condition $\left| S(f_k, \mathcal{D}) - \int_I f_k \right| < \epsilon$ for every $k \geq N$ whenever D is a γ-fine, it will follow that $\{f_k\}_{k=1}^{\infty}$ is uniformly integrable. One need merely choose $\{\gamma_1, \ldots, \gamma_N\}$ so that $\left| S(f_k, \mathcal{D}) - \int_I f_k \right| < \epsilon$ whenever $\mathcal{D}$ is a γ_k-fine tagged partition of I for $k = 1, \ldots, N$ and then set $\gamma' = \gamma \cap (\cap_{k=1}^N \gamma_k)$ to see that $\{f_k\}_{k=1}^\infty$ is uniformly integrable.

For uniformly integrable sequences of integrable functions, we have the following convergence theorem.

Theorem 4.73   *Let $\{f_k\}_{k=1}^\infty$ be a sequence of uniformly integrable functions over I and assume that $f_k \to f$ pointwise. Then f is integrable over I and*

$$\lim_{k \to \infty} \int_I f_k = \int_I \left(\lim_{k \to \infty} f_k \right) = \int_I f.$$

Proof.   Let $\epsilon > 0$. Let γ be a gauge on I such that $\left| S(f_k, \mathcal{D}) - \int_I f_k \right| < \epsilon/3$ for every k whenever $\mathcal{D}$ is γ-fine; we may assume that $\gamma(t)$ is bounded for every $t \in \mathbb{R}$. Fix a γ-fine tagged partition $\mathcal{D}_0$ of I. By the pointwise convergence of $\{f_k\}_{k=1}^\infty$, we can pick an N such that $|S(f_i, \mathcal{D}_0) - S(f_j, \mathcal{D}_0)| < \epsilon/3$ whenever $i, j \geq N$. If $i, j \geq N$, then

$$\left| \int_I f_i - \int_I f_j \right| \leq \left| \int_I f_i - S(f_i, \mathcal{D}_0) \right|$$
$$+ |S(f_i, \mathcal{D}_0) - S(f_j, \mathcal{D}_0)| + \left| S(f_j, \mathcal{D}_0) - \int_I f_j \right| < \epsilon.$$

Hence, $\lim \int_I f_k = L$ exists.

Suppose now that $\mathcal{D}$ is a γ-fine tagged partition of I. As above, pick M such that $|S(f_M, \mathcal{D}) - S(f, \mathcal{D})| < \epsilon/3$ and $\left| L - \int_I f_M \right| < \epsilon/3$. Then,

$$|S(f, \mathcal{D}) - L| \leq |S(f, \mathcal{D}) - S(f_M, \mathcal{D})|$$
$$+ \left| S(f_M, \mathcal{D}) - \int_I f_M \right| + \left| \int_I f_M - L \right| < \epsilon.$$

Hence, f is integrable over I with $\int_I f = L = \lim \int_I f_k$ as desired. $\square$

We give an improvement of the conclusion of Theorem 4.73 in Section 4.11.

We now establish the major convergence theorems for the Henstock-Kurzweil integral. Since we are working over arbitrary subintervals of $\mathbb{R}^*$, we need a preliminary lemma which will enable us to treat the case of unbounded intervals.

Lemma 4.74   *There exist a strictly positive $\varphi : \mathbb{R} \to (0, \infty)$ which is integrable over $\mathbb{R}$ and a gauge $\gamma (= \gamma_\varphi)$ such that $0 \leq S(\varphi, \mathcal{D}) \leq 1$ for every γ-fine partial tagged partition $\mathcal{D}$ of $\mathbb{R}$.*

Proof. Pick any strictly positive function φ on $\mathbb{R}$ with $\int_{\mathbb{R}} \varphi = 1/2$; for example, one such function is $\varphi(t) = 1/(2\pi(1+t))^2$. There is a gauge γ on $\mathbb{R}$ such that $|S(\varphi, \mathcal{D}) - 1/2| < 1/2$ whenever $\mathcal{D}$ is a γ-fine tagged partition of $\mathbb{R}$. Suppose that $\mathcal{D} = \{(t_i, I_i) : 1 \le i \le m\}$ is a γ-fine partial tagged partition of $\mathbb{R}$ and set $I = \bigcup_{i=1}^{m} I_i$. By Henstock's Lemma 4.45, $|S(\varphi, \mathcal{D}) - \int_I \varphi| \le 1/2$ so that

$$0 \le S(\varphi, \mathcal{D}) \le 1/2 + \int_I \varphi \le 1$$

since φ is positive. $\qquad\square$

Using this lemma, we prove a convergence result for series from which we derive the Monotone and Dominated Convergence Theorems.

Theorem 4.75 *Let $f_k : I \to \mathbb{R}$ be nonnegative and integrable over I for each k. Suppose that $f = \sum_{k=1}^{\infty} f_k$ pointwise on I and $\sum_{k=1}^{\infty} \int_I f_k < \infty$. If $s_n = \sum_{k=1}^{n} f_k$, then:*

(1) $\{s_n\}_{k=1}^{\infty}$ is uniformly integrable over I;
(2) f is integrable over I; and
(3) $\sum_{k=1}^{\infty} \int_I f_k = \int_I f = \int_I \sum_{k=1}^{\infty} f_k$.

Proof. In order to avoid treating special cases, we assume in the proof that $I = \mathbb{R}$; for the general case, we can always extend functions from I to $\mathbb{R}$ by setting them equal to 0 on $\mathbb{R} \setminus I$. Let $\epsilon > 0$. For each n, pick a gauge γ_n on $\mathbb{R}^*$ with $\gamma_n(t)$ bounded for each $t \in R$ such that

$$\left| \int_{\mathbb{R}} s_n - S(s_n, \mathcal{D}) \right| < \epsilon/2^n$$

whenever $\mathcal{D}$ is γ_n-fine.

Choose φ and γ_φ satisfying Lemma 4.74. Pick n_0 such that $\sum_{k=n_0}^{\infty} \int_{\mathbb{R}} f_k < \epsilon$. For every $t \in \mathbb{R}$, pick $n(t) \ge n_0$ such that $k \ge j \ge n(t)$ implies

$$\left| \sum_{i=j}^{k} f_i(t) \right| < \epsilon\varphi(t);$$

set $n(\pm\infty) = n_0$. Define a gauge γ on $\mathbb{R}^*$ by

$$\gamma(t) = \left(\bigcap_{j=1}^{n(t)} \gamma_j(t)\right) \cap \gamma_\varphi(t).$$

Suppose $\mathcal{D} = \{(t_i, I_i) : 1 \leq i \leq m\}$ is γ-fine. To establish (1), first note that $\mathcal{D}$ is γ_i-fine for $i = 1, \ldots, n_0$, so that $\left|S(s_i, \mathcal{D}) - \int_{\mathbb{R}} s_i\right| < \epsilon/2^i$. Fix $n > n_0$ and set

$$d_1 = \{i : 1 \leq i \leq m, n(t_i) \geq n\}$$

and

$$d_2 = \{i : 1 \leq i \leq m, n(t_i) < n\}.$$

Set $\mathcal{D}_k = \{(t_i, I_i) : i \in d_k\}$, $k = 1, 2$. Note that $\mathcal{D}_1$ is γ_n-fine by the definition of γ. Set $I = \cup_{i \in d_1} I_i$. Using Henstock's Lemma 4.45,

$$\left|\int_{\mathbb{R}} s_n - S(s_n, \mathcal{D})\right| \leq \left|\int_I s_n - S(s_n, \mathcal{D}_1)\right| + \left|\sum_{i \in d_2} \sum_{j=1}^{n} \left\{\int_{I_i} f_j - f_j(t_i) \ell(I_i)\right\}\right|$$

$$< \frac{\epsilon}{2^n} + \left|\sum_{i \in d_2} \sum_{j=1}^{n(t_i)} \left\{\int_{I_i} f_j - f_j(t_i) \ell(I_i)\right\}\right|$$

$$+ \left|\sum_{i \in d_2} \sum_{j=n(t_i)+1}^{n} \int_{I_i} f_j\right| + \left|\sum_{i \in d_2} \sum_{j=n(t_i)+1}^{n} f_j(t_i) \ell(I_i)\right|$$

$$= \frac{\epsilon}{2^n} + T_1 + T_2 + T_3.$$

We first estimate T_3. From Lemma 4.74,

$$T_3 \leq \sum_{i \in d_2} \left|\sum_{j=n(t_i)+1}^{n} f_j(t_i)\right| \ell(I_i) \leq \sum_{i \in d_2} \epsilon\varphi(t_i) \ell(I_i) = \epsilon S(\varphi, \mathcal{D}_2) \leq \epsilon.$$

Next, from the nonnegativity of the f_j's,

$$T_2 \leq \sum_{i=1}^{m} \sum_{j=n_0}^{n} \int_{I_i} f_j = \sum_{j=n_0}^{n} \int_{\mathbb{R}} f_j \leq \sum_{j=n_0}^{\infty} \int_{\mathbb{R}} f_j < \epsilon.$$

Finally, consider T_1. Set $s = \max\{n(t_i) : i \in d_2\}$ and note that $\{(t_i, I_i) : n(t_i) = k\}$ is γ_k-fine. By Henstock's Lemma 4.45,

$$T_1 = \left| \sum_{i \in d_2} \left\{ \int_{I_i} s_{n(t_i)} - s_{n(t_i)}(t_i)\, \ell(I_i) \right\} \right| \tag{4.13}$$

$$= \left| \sum_{k=1}^{s} \sum_{\{i : n(t_i) = k\}} \left\{ \int_{I_i} s_{n(t_i)} - s_{n(t_i)}(t_i)\, \ell(I_i) \right\} \right| \tag{4.14}$$

$$\leq \sum_{k=1}^{s} \left| \sum_{\{i : n(t_i) = k\}} \left\{ \int_{I_i} s_{n(t_i)} - s_{n(t_i)}(t_i)\, \ell(I_i) \right\} \right| \leq \sum_{k=1}^{s} \epsilon/2^k < \epsilon.$$

Combining these estimates, it follows that $\left| \int_{\mathbb{R}} s_n - S(s_n, \mathcal{D}) \right| < 4\epsilon$ whenever $\mathcal{D}$ is γ-fine and $n > n_0$. This proves (1).

Conditions (2) and (3) follow from Theorem 4.73, which completes the proof. $\qquad\square$

An immediate consequence of this theorem is one of the major convergence theorems for the Henstock-Kurzweil integral, the Monotone Convergence Theorem (sometimes called the Beppo Levi Theorem).

Theorem 4.76 *(Monotone Convergence Theorem) Let $f_k : I \to \mathbb{R}$ be integrable over $\mathbb{R}$ and suppose that $\{f_k(t)\}_{k=1}^{\infty}$ increases to $f(t) \in \mathbb{R}$ for every $t \in \mathbb{R}$. If $\sup_k \int_I f_k < \infty$, then:*

(1) $\{f_k\}_{k=1}^{\infty}$ *is uniformly integrable over I;*
(2) f *is integrable over I; and,*
(3) $\lim_{k \to \infty} \int_I f_k = \int_I (\lim_{k \to \infty} f_k) = \int_I f$.

Proof. Set $f_0 = 0$ and for $k \geq 1$ let $g_k = f_k - f_{k-1}$. Then, $g_k \geq 0$ for all $k \geq 2$ and $s_n = \sum_{k=1}^{n} g_k$ converges pointwise to f. By monotonicity,

$$\sum_{k=1}^{\infty} \int_I g_k = \lim_{n \to \infty} \sum_{k=1}^{n} \int_I (f_k - f_{k-1}) = \lim_{n \to \infty} \int_I f_n = \sup_n \int_I f_n < \infty.$$

Hence, Theorem 4.75 is applicable and completes the proof. $\qquad\square$

Suppose that $\{f_k\}_{k=1}^{\infty}$ is a sequence of Henstock-Kurzweil integrable functions that decreases monotonically for each $x \in I$. Applying the theorem above to $\{-f_k\}_{k=1}^{\infty}$, we get an analogous version of the Monotone Convergence Theorem for a decreasing sequence of functions, under the assumption that $\inf_k \int_I f_k > -\infty$.

The Monotone Convergence Theorem gives a very useful and powerful sufficient condition guaranteeing "passage of the limit under the integral sign", i.e., $\lim \int f_k = \int (\lim f_k)$, but the monotonicity condition is often not satisfied. We next prove another convergence result, the Dominated Convergence Theorem, which relaxes this requirement. For this we require a preliminary result.

Definition 4.77 Suppose that $f_k : I \to \mathbb{R}$. The sequence $\{f_k\}_{k=1}^{\infty}$ is called *uniformly gauge Cauchy* over I if for every $\epsilon > 0$ there is a gauge γ on I and an N such that $i, j \geq N$ implies $|S(f_i, \mathcal{D}) - S(f_j, \mathcal{D})| < \epsilon$ whenever $\mathcal{D}$ is a γ-fine tagged partition of I.

The following proposition shows how the uniform gauge Cauchy condition is related to uniform integrability.

Proposition 4.78 *Let $f_k : I \to \mathbb{R}$ be integrable over I for each k. Then, $\{f_k\}_{k=1}^{\infty}$ is uniformly gauge Cauchy over I if and only if $\{f_k\}_{k=1}^{\infty}$ is uniformly integrable over I and $\lim_{k \to \infty} \int_I f_k$ exists.*

Proof. Let $\epsilon > 0$. Suppose that $\{f_k\}_{k=1}^{\infty}$ is uniformly gauge Cauchy over I. Let γ be a gauge on I and N be such that $|S(f_i, \mathcal{D}) - S(f_j, \mathcal{D})| < \epsilon/3$ whenever $i, j \geq N$ and $\mathcal{D}$ is γ-fine. Fix $i, j \geq N$ and let γ_1 be a gauge on I such that $\left|\int_I f_i - S(f_i, \mathcal{D})\right| < \epsilon/3$ and $\left|\int_I f_j - S(f_j, \mathcal{D})\right| < \epsilon/3$ whenever $\mathcal{D}$ is γ_1-fine. Set $\gamma_2 = \gamma \cap \gamma_1$. If $\mathcal{D}$ is γ_2-fine, then

$$\left|\int_I f_i - \int_I f_j\right| \leq \left|\int_I f_i - S(f_i, \mathcal{D})\right|$$
$$+ |S(f_i, \mathcal{D}) - S(f_j, \mathcal{D})| + \left|S(f_j, \mathcal{D}) - \int_I f_j\right| < \epsilon$$

which implies that $\{\int_I f_k\}_{k=1}^{\infty}$ is Cauchy and therefore converges.

Pick $M \geq N$ such that $\left|\int_I f_i - \int_I f_j\right| < \epsilon/3$ when $i, j \geq M$. Choose a gauge γ_3 on I such that

$$\left|S(f_M, \mathcal{D}) - \int_I f_M\right| < \epsilon/3$$

whenever $\mathcal{D}$ is γ_3-fine. Put $\gamma_4 = \gamma_3 \cap \gamma_2$. If $i \geq M$ and $\mathcal{D}$ is γ_4-fine, then

$$\left|S(f_i, \mathcal{D}) - \int_I f_i\right| \leq |S(f_i, \mathcal{D}) - S(f_M, \mathcal{D})|$$
$$+ \left|S(f_M, \mathcal{D}) - \int_I f_M\right| + \left|\int_I f_M - \int_I f_i\right| < \epsilon.$$

Thus, by the remark above, $\{f_k\}_{k=1}^{\infty}$ is uniformly integrable.

For the converse, let γ' be a gauge on I such that $\left|S\left(f_i, \mathcal{D}\right) - \int_I f_i\right| < \epsilon/3$ whenever $\mathcal{D}$ is γ'-fine and $i \in \mathbb{N}$. Pick N such that $\left|\int_I f_i - \int_I f_j\right| < \epsilon/3$ whenever $i, j \geq N$. If $\mathcal{D}$ is γ'-fine and $i, j \geq N$, then

$$\left|S\left(f_i, \mathcal{D}\right) - S\left(f_j, \mathcal{D}\right)\right| \leq \left|S\left(f_i, \mathcal{D}\right) - \int_I f_i\right|$$
$$+ \left|\int_I f_i - \int_I f_j\right| + \left|\int_I f_j - S\left(f_j, \mathcal{D}\right)\right| < \epsilon$$

so $\{f_k\}_{k=1}^{\infty}$ is uniformly gauge Cauchy. $\qquad\square$

We now have the necessary machinery to prove the Dominated Convergence Theorem.

Theorem 4.79 *(Dominated Convergence Theorem) Let $f_k, f, g : I \to \mathbb{R}$ and assume $\{f_k\}_{k=1}^{\infty}$ and g are integrable over I with $\left|f_i(t) - f_j(t)\right| \leq g(t)$ for all $t \in I$ and $i, j \in \mathbb{N}$. If $\lim_{k \to \infty} f_k = f$ pointwise on I, then*

(1) $\{f_k\}_{k=1}^{\infty}$ is uniformly integrable over I;
(2) f is integrable over I; and
(3) $\lim_{k \to \infty} \int_I f_k = \int_I \left(\lim_{k \to \infty} f_k\right) = \int_I f$.

Proof. Set $t_{j,k} = \vee\{|f_m - f_n| : j \leq m \leq n \leq k\}$. Then each $t_{j,k}$ is integrable by Corollary 4.64 and Proposition 4.67 and for each j, $\{t_{j,k}\}_{k=1}^{\infty}$ is an increasing sequence which converges to the function $t_j = \vee\{|f_m - f_n| : j \leq m \leq n\} \leq g$. The Monotone Convergence Theorem implies that each t_j is integrable and $\int_I t_j \leq \int_I g$. Since $t_{j+1} \leq t_j$ and $t_j \to 0$, the version of the Monotone Convergence Theorem for decreasing functions implies $\int_I t_j \downarrow 0$.

Let $\epsilon > 0$. There exists N such that $\int_I t_N < \epsilon/2$ and there exists a gauge γ on I such that

$$\left|S\left(t_N, \mathcal{D}\right) - \int_I t_N\right| < \epsilon/2$$

whenever $\mathcal{D}$ is γ-fine. If $i, j \geq N$ and $\mathcal{D}$ is γ-fine, then

$$\left|S\left(f_i, \mathcal{D}\right) - S\left(f_j, \mathcal{D}\right)\right| \leq S\left(|f_i - f_j|, \mathcal{D}\right) \leq S\left(t_N, \mathcal{D}\right) \leq \int_I t_N + \epsilon/2 < \epsilon.$$

Hence, $\{f_k\}_{k=1}^{\infty}$ is uniformly gauge Cauchy on I and (1) follows from Proposition 4.78.

As before, (2) and (3) follow from Theorem 4.73 and the proof is complete.   □

Remark 4.80 *The usual "dominating hypothesis" in the Dominated Convergence Theorem for the Lebesgue integral is that there exists an integrable function g such that $|f_j| \leq g$ for all j. Since $|f_i - f_j| \leq |f_i| + |f_j|$, this hypothesis clearly implies the one in Theorem 4.79. On the other hand, the hypothesis $|f_i - f_j| \leq g$ allows for functions f_j that are conditionally integrable. To see this, let $\{f_k\}_{k=1}^{\infty}$ be a sequence of positive functions, let h be any conditionally integrable function h, and consider the sequence $\{f_k + h\}_{k=1}^{\infty}$ of conditionally integrable functions. Thus, such a hypothesis is more appropriate for a conditional integral like the gauge integral. In the more general version of the Dominated Convergence Theorem given later we will use a different, but equivalent, "dominating assumption". Note that if f_k and g are Henstock-Kurzweil integrable and $|f_k| \leq g$, then g is nonnegative and, hence, absolutely integrable and f_k is absolutely integrable by Corollary 4.64. Thus, the dominating condition above is more general than that in Theorem 3.100.*

In the proof of the Monotone Convergence Theorem above, we needed to assume that the limit function was finite on I. In fact, as a consequence of the monotonicity of the sequence of functions and the condition $\sup_k \int_I f_k < \infty$, the limit is finite almost everywhere. We will use this result to derive more general versions of the Monotone and Dominated Convergence Theorems.

Lemma 4.81 *Let $f_k : I \subset \mathbb{R}^* \to \mathbb{R}$ be Henstock-Kurzweil integrable over I and suppose that $\{f_k(x)\}_{k=1}^{\infty}$ increases monotonically for each $x \in I$ and $\sup_k \int_I f_k < \infty$. Then, $\lim_{k \to \infty} f_k(x)$ exists and is finite for almost every $x \in I$.*

Proof. By replacing f_k by $f_k - f_1$, we may assume that each f_k is nonnegative. Then, we may assume that I is a bounded interval, since $I = \cup_{n=1}^{\infty} (I \cap [-n, n])$ and if the conclusion holds on $I \cap [-n, n]$, then it holds almost everywhere in I. Set $M = \sup_k \int_I f_k$ and let $E = \{x \in I : \lim_{k \to \infty} f_k(x) = \infty\}$. Let $f_k^i = 1 \wedge \left(\frac{1}{i}\right) f_k$ and define $h_i : I \to \mathbb{R}$ by

$$h_i(x) = \begin{cases} 1 \wedge \left(\frac{1}{i}\right) \lim_{k \to \infty} f_k(x) & \text{if } x \in I \setminus E \\ 1 & \text{if } x \in E \end{cases}.$$

For each fixed i, $\{f_k^i\}_{k=1}^{\infty}$ increases to h_i pointwise as $k \to \infty$. Since I is a bounded interval, 1 is Henstock-Kurzweil integrable over I. Thus, f_k^i is

Henstock-Kurzweil integrable, since the minimum of absolutely integrable functions is, and

$$\int_I f_k^i \le \frac{1}{i} \int_I f_k \le \frac{M}{i}.$$

By the Monotone Convergence Theorem (for a decreasing sequence of functions), h_i is Henstock-Kurzweil integrable and

$$\int_I \chi_E = \lim_{i \to \infty} \int_I h_i \le \lim_{i \to \infty} \frac{M}{i} = 0.$$

Thus, by Theorem 4.42, E is a null set, so $\lim_{k \to \infty} f_k$ exists and is finite almost everywhere in I. $\qquad\qquad\square$

Using this lemma, we can improve the statement of the Monotone Convergence Theorem by removing the assumption that the pointwise limit is finite everywhere.

Corollary 4.82 *(Monotone Convergence Theorem) Let $f_k : I \subset \mathbb{R}^* \to \mathbb{R}$ and suppose that $\{f_k (x)\}_{k=1}^{\infty}$ increases monotonically for each $x \in I$. Suppose each f_k is Henstock-Kurzweil integrable over I and $\sup_k \int_I f_k < \infty$. Then, $\lim_{k \to \infty} f_k (x)$ is finite for almost every $x \in I$ and the function f, defined by*

$$f (x) = \begin{cases} \lim_{k \to \infty} f_k (x) & \text{if the limit is finite} \\ 0 & \text{otherwise} \end{cases},$$

is Henstock-Kurzweil integrable over I with

$$\int_I f = \lim_{k \to \infty} \int_I f_k.$$

Proof. By the previous lemma, the function f is finite almost everywhere in I. Let $E = \{x \in I : \lim_{k \to \infty} f_k (x) = \infty\}$. Since E is a null set, by Example 4.40

$$\int_E f_k = \int_I \chi_E f_k = 0.$$

Define $\{g_k\}_{k=1}^{\infty}$ by $g_k = \chi_{I \setminus E} f_k$. Since $g_k = f_k - \chi_E f_k$, g_k is Henstock-Kurzweil integrable and

$$\int_I g_k = \int_I f_k - \int_I \chi_E f_k = \int_I f_k.$$

Further, $\{g_k\,(x)\}_{k=1}^{\infty}$ increases pointwise to f on I. By the Monotone Convergence Theorem, f is Henstock-Kurzweil integrable and

$$\int_I f = \lim_{k\to\infty} \int_I g_k = \lim_{k\to\infty} \int_I f_k.$$

$\square$

While this result relaxes the pointwise finiteness condition on the limit function, the sequence of functions need no longer be uniformly integrable.

Example 4.83 Define $f_k : [0,1] \to \mathbb{R}$ by $f_k\,(0) = k$ and $f_k\,(t) = 0$ otherwise. Then the a.e. version of the Monotone Convergence Theorem given in Corollary 4.82 holds, but $\{f_k\}_{k=1}^{\infty}$ is not uniformly integrable over $[0,1]$. See Exercise 4.50.

A similar remark applies to the version of the Dominated Convergence Theorem 4.88 below.

The Monotone Convergence Theorem is equivalent to the following result about infinite series of nonnegative functions.

Theorem 4.84 *Let $f_k : I \subset \mathbb{R}^* \to \mathbb{R}$ be nonnegative and Henstock-Kurzweil integrable over I for each k and define f by*

$$f\,(x) = \begin{cases} \sum_{k=1}^{\infty} f_k\,(x) & \text{if the series converges} \\ 0 & \text{otherwise} \end{cases}.$$

Then, the series converges for almost all $x \in I$ and f is Henstock-Kurzweil integrable over I if, and only if, $\sum_{k=1}^{\infty} \int_I f_k < \infty$. In either case,

$$\int_I f = \sum_{k=1}^{\infty} \int_I f_k.$$

Proof. Suppose first that $\sum_{k=1}^{\infty} \int_I f_k < \infty$. Let $s_m = \sum_{k=1}^{m} f_k$. Since each $f_k \geq 0$, $\{s_m\,(x)\}_{m=1}^{\infty}$ forms an increasing sequence for each $x \in I$ and

$$\sup_m \int_I s_m = \sum_{k=1}^{\infty} \int_I f_k < \infty.$$

By Lemma 4.81, $\sum_{k=1}^{\infty} f_k = \lim_{m\to\infty} s_m$ is finite almost everywhere and, by the Monotone Convergence Theorem (Corollary 4.82), f is Henstock-Kurzweil integrable over I.

On the other hand, suppose that f is Henstock-Kurzweil integrable and $E = \{x \in I : \sum_{k=1}^{\infty} f_k\,(x) = \infty\}$ has measure 0. Then, by the linearity of the integral and the nonnegativity of the functions f_k,

$$\sum_{k=1}^{\infty} \int_I f_k = \sup_m \sum_{k=1}^{m} \int_I f_k = \sup_m \int_I s_m = \sup_m \int_{I \setminus E} s_m \leq \int_I f < \infty.$$

Finally, in either case,

$$\int_I f = \lim_{m \to \infty} \int_I s_m = \lim_{m \to \infty} \sum_{k=1}^{m} \int_I f_k = \sum_{k=1}^{\infty} \int_I f_k. \qquad \square$$

Our next goal is to prove a general version of the Dominated Convergence Theorem for the Henstock-Kurzweil integral. As in the case for the Lebesgue integral, the proof will be based on Fatou's Lemma. We begin with a lemma.

Lemma 4.85 *Let $f_k, \alpha : I \subset \mathbb{R}^* \to \mathbb{R}$ be Henstock-Kurzweil integrable for all k, and suppose that $\alpha \leq f_k$ in I. Then, $\inf_k f_k$ is Henstock-Kurzweil integrable over I.*

Proof. Since $\alpha \leq f_k$, the function $g_k = \inf_{1 \leq j \leq k} f_k$ is Henstock-Kurzweil integrable over I by Proposition 4.67. Since $\alpha \leq g_k$ for all k, $\inf_k \int_I g_k \geq \int_I \alpha > -\infty$. Thus, by the comment in the paragraph following the proof of Theorem 4.76, we can apply the Monotone Convergence Theorem to the decreasing sequence of functions $\{g_k\}_{k=1}^{\infty}$ which converges to $\inf_k f_k$. $\qquad \square$

Note that since $\alpha \leq \inf_k f_k \leq f_1$, $\inf_k f_k$ is finite valued everywhere on I. We can now prove Fatou's Lemma.

Lemma 4.86 *(Fatou's Lemma) Let $f_k, \alpha : I \subset \mathbb{R}^* \to \mathbb{R}$ be Henstock-Kurzweil integrable for all k, and suppose that $\alpha \leq f_k$ in I and $\liminf_{k \to \infty} \int_I f_k < \infty$. Then, $\liminf_{k \to \infty} f_k$ is finite almost everywhere in I and the function f defined by*

$$f(x) = \begin{cases} \liminf_{k \to \infty} f_k(x) & \text{if the limit is finite} \\ 0 & \text{otherwise} \end{cases}$$

is Henstock-Kurzweil integrable over I with

$$\int_I f \leq \liminf_{k \to \infty} \int_I f_k.$$

Proof. By Lemma 4.85, the function Φ_k defined by

$$\Phi_k(x) = \inf \{f_j(x) : j \geq k\}$$

for each $k \in \mathbb{N}$ is Henstock-Kurzweil integrable over I.

Since $\alpha \leq \Phi_k \leq f_k$ on I for all k, it follows that each function Φ_k is finite valued on I and

$$\int_I \alpha \leq \int_I \Phi_k \leq \int_I f_k,$$

which implies

$$\int_I \alpha \leq \liminf_{k \to \infty} \int_I \Phi_k \leq \liminf_{k \to \infty} \int_I f_k. \tag{4.15}$$

Further, by definition, $\{\Phi_k\}_{k=1}^{\infty}$ is an increasing sequence which, by Lemma 4.81, converges pointwise to f almost everywhere in I. Since $\left\{\int_I \Phi_k\right\}_{k=1}^{\infty}$ is monotonic, it then follows from (4.15) that $\left\{\int_I \Phi_k\right\}_{k=1}^{\infty}$ converges and is hence bounded. Thus, by the Monotone Convergence Theorem, f is Henstock-Kurzweil integrable and by (4.15)

$$\int_I f = \lim_{k \to \infty} \int_I \Phi_k = \liminf_{k \to \infty} \int_I \Phi_k \leq \liminf_{k \to \infty} \int_I f_k,$$

which completes the proof. $\square$

As in the case of the Lebesgue integral, the result dual to Fatou's Lemma also holds. For a proof, see Exercise 4.51.

Corollary 4.87 *Let $f_k, \beta : I \subset \mathbb{R}^* \to \mathbb{R}$ be Henstock-Kurzweil integrable for all k, and suppose that $f_k \leq \beta$ in I and $\limsup_{k \to \infty} \int_I f_k > -\infty$. Then, $\limsup_{k \to \infty} f_k$ is finite almost everywhere in I and the function f defined by*

$$f(x) = \begin{cases} \limsup_{k \to \infty} f_k(x) & \text{if the limit is finite} \\ 0 & \text{otherwise} \end{cases}$$

is Henstock-Kurzweil integrable over I with

$$\int_I f \geq \limsup_{k \to \infty} \int_I f_k.$$

We are now prepared to prove the Dominated Convergence Theorem.

Theorem 4.88 *(Dominated Convergence Theorem) Let $f_k : I \subset \mathbb{R}^* \to \mathbb{R}$ be Henstock-Kurzweil integrable over I and suppose that $\{f_k\}_{k=1}^{\infty}$ converges pointwise almost everywhere on I. Define f by*

$$f(x) = \begin{cases} \lim_{k \to \infty} f_k(x) & \text{if the limit is finite} \\ 0 & \text{otherwise} \end{cases}.$$

Suppose that there are Henstock-Kurzweil integrable functions $\alpha, \beta : I \to \mathbb{R}$ such that $\alpha \leq f_k \leq \beta$ almost everywhere in I, for all $k \in \mathbb{N}$. Then, f is Henstock-Kurzweil integrable over I and

$$\int_I f = \int_I \lim_{k \to \infty} f_k = \lim_{k \to \infty} \int_I f_k.$$

Proof. Let $E_k = \{x \in I : f_k(x) < \alpha \text{ or } f_k(x) > \beta(x)\}$. Then, the set

$$E = \left\{ x \in I : \lim_{k \to \infty} f_k(x) \text{ diverges} \right\} \cup \cup_{k \in \mathbb{N}} E_k$$

has measure zero. If $x \notin E$, then $f_k(x) \to f(x)$ and $\alpha(x) \leq f_k(x) \leq \beta(x)$ for all $k \in \mathbb{N}$. Since $\int_I f_k = \int_{I \backslash E} f_k$ and $\int_E f = 0$, we may assume all the hypotheses hold for all $x \in I$.

Since $\alpha \leq f_k$, Fatou's Lemma shows that $\liminf_{k \to \infty} f_k$ is finite almost everywhere in I and

$$\int_I f \leq \liminf_{k \to \infty} \int_I f_k.$$

Similarly, since $f_k \leq \beta$, Corollary 4.87 implies that $\limsup_{k \to \infty} f_k$ is finite almost everywhere in I and

$$\int_I f \geq \limsup_{k \to \infty} \int_I f_k.$$

Combining these results, we see

$$\limsup_{k \to \infty} \int_I f_k \leq \int_I f \leq \liminf_{k \to \infty} \int_I f_k \leq \limsup_{k \to \infty} \int_I f_k$$

so that

$$\int_I f = \lim_{k \to \infty} \int_I f_k.$$

$\square$

It is easy to show the dominating conditions employed in the two versions of the Dominated Convergence Theorem given in Theorems 4.79 and 4.88 are equivalent. See Exercise 4.47.

As an immediate consequence, we get the Bounded Convergence Theorem.

Corollary 4.89 *(Bounded Convergence Theorem) Let $f_k : I \subset \mathbb{R}^* \to \mathbb{R}$ be Henstock-Kurzweil integrable over a bounded interval I and suppose that*

$\{f_k\}_{k=1}^{\infty}$ *converges pointwise almost everywhere on* I. *Define* f *by*

$$f(x) = \begin{cases} \lim_{k \to \infty} f_k(x) & \text{if the limit is finite} \\ 0 & \text{otherwise} \end{cases}.$$

If there is a number M *so that* $|f_k(x)| \le M$ *for all* k *and all* $x \in I$, *then*

$$\int_I f = \lim_{k \to \infty} \int_I f_k.$$

One need only observe that the function $g(x) = M$ for all $x \in I$ is Henstock-Kurzweil integrable over I.

We conclude this section with a mean convergence result for the Henstock-Kurzweil integral.

Corollary 4.90   *(Mean Convergence Theorem) Under the hypotheses of the Dominated Convergence Theorem 4.88,* $f - f_k$ *is absolutely integrable and*

$$\lim_{k \to \infty} \int_I |f - f_k| = 0.$$

Proof.   As in the proof of Theorem 4.88, we may assume the conditions of the hypothesis hold everywhere in I . That theorem shows that f is integrable and, consequently, so is $f - f_k$. Since $\alpha \le f_k \le \beta$, it follows that

$$-(\beta - \alpha) = \alpha - \beta \le f - f_k \le \beta - \alpha$$

so that $|f - f_k| \le \beta - \alpha$. By Corollary 4.64, $f - f_k$ is absolutely integrable and so the Dominated Convergence Theorem implies that $\lim_{k \to \infty} \int_I |f - f_k| = 0$, completing the proof. $\qquad\square$

For the Lebesgue integral, the Dominated Convergence Theorem 3.100 guarantees that both the limit function is absolutely integrable and the mean convergence. For the Henstock-Kurzweil integral, we still have convergence in the mean but we only know that $f - f_k$ is absolutely integrable. The limit function f (and also f_k) may be conditionally integrable.

4.8   Henstock-Kurzweil and Lebesgue integrals

We saw earlier that every Riemann integrable function is Henstock-Kurzweil integrable, by defining the gauge to have constant length. Further, there are Henstock-Kurzweil integrable functions which are not Riemann integrable. The unbounded function $1/\sqrt{x}$ and the Dirichlet function, both defined on $(0, 1]$, provide examples. We now consider the relationship between Lebesgue integrability and Henstock-Kurzweil integrability. Since the Lebesgue integral is an absolute integral (that is, a function is Lebesgue integrable if, and only if, it is absolutely Lebesgue integrable) and the Henstock-Kurzweil integral is a conditional integral, the conditions cannot be equivalent. Further, the function in Example 4.1 is Henstock-Kurzweil integrable and not Lebesgue integrable. We now show that the Henstock-Kurzweil integral is more general than the Lebesgue integral. As above, we will use $\mathcal{L} \int_I f$ to denote the Lebesgue integral of f.

Theorem 4.91 *Suppose that $f : I \to \mathbb{R}$ is nonnegative and measurable. Then, f is Lebesgue integrable if, and only if, f is Henstock-Kurzweil integrable. In either case, $\mathcal{L} \int_I f = \int_I f$.*

Proof. First suppose that f is also bounded, with a bound of M, and $I = [a, b]$ is a bounded interval. Then, by Theorem 3.67, there is a sequence of step functions $\{\varphi_k\}_{k=1}^{\infty}$ such that $\varphi_k \to f$ pointwise a.e. and $|\varphi_k(x)| \le M$ for all $k \in \mathbb{N}$ and $x \in [a, b]$. Since φ_k is a step function, $\mathcal{L} \int_a^b \varphi_k = \int_a^b \varphi_k$, so that by the Bounded Convergence Theorem (which holds for both integrals), $\mathcal{L} \int_a^b f = \int_a^b f$.

Next, suppose that f is an arbitrary nonnegative, measurable, real-valued function defined on an arbitrary interval in $\mathbb{R}$. Define a sequence of functions $\{f_k\}_{k=1}^{\infty}$ by $f_k(x) = \min \{f(x), k\} \chi_{[-k,k]}(x)$. Each f_k is nonnegative, measurable, and bounded so, by the previous case, $\mathcal{L} \int_{-k}^k f_k = \int_{-k}^k f_k$. Since $\{f_k\}_{k=1}^{\infty}$ increases to f pointwise, we can apply the Monotone Convergence Theorem (which, again, holds for both integrals) to conclude that f is Lebesgue integrable if, and only if, f is Henstock-Kurzweil integrable. When either of the integrals is finite, we see that

$$\mathcal{L} \int_I f = \lim_{k \to \infty} \mathcal{L} \int_I f_k = \lim_{k \to \infty} \mathcal{L} \int_{-k}^k f_k = \lim_{k \to \infty} \int_{-k}^k f_k = \lim_{k \to \infty} \int_I f_k = \int_I f.$$

$\square$

If f is Lebesgue integrable, then f^+ and f^- are Lebesgue integrable and, consequently, f^+ and f^- are Henstock-Kurzweil integrable. By linearity, f is absolutely Henstock-Kurzweil integrable. On the other hand, suppose that f is absolutely Henstock-Kurzweil integrable. Then, by linearity, f^+ and f^- are nonnegative and Henstock-Kurzweil integrable. Thus, we have the following corollary.

Corollary 4.92 *Suppose that $f : I \to \mathbb{R}$ is measurable. Then, f is Lebesgue integrable if, and only if, f is absolutely Henstock-Kurzweil integrable. In either case, the integrals agree.*

Thus, Lebesgue integrability implies Henstock-Kurzweil integrability, but the converse is not valid. We will show in Corollary 4.98 that every Henstock-Kurzweil integrable function is measurable, so the measurability condition in Corollary 4.92 can be dropped.

We now have the necessary background to prove a general version of Part I of the Fundamental Theorem of Calculus for the Lebesgue integral.

Theorem 4.93 *(Fundamental Theorem of Calculus: Part I) Suppose that $f : [a, b] \to \mathbb{R}$ is differentiable on $[a, b]$ and f' is Lebesgue integrable on $[a, b]$. Then,*

$$\mathcal{L} \int_a^b f' = f(b) - f(a).$$

Proof. By assumption, f' is Lebesgue integrable, so Corollary 4.92 implies that $\mathcal{L} \int_a^b f' = \int_a^b f'$. By Theorem 4.16, $\int_a^b f' = f(b) - f(a)$, completing the proof. □

For a proof that does not use the Henstock-Kurzweil integral, see [N, Vol. I, IX.17.1] and [Sw1, 4.3.3, page 158].

4.9 Differentiating indefinite integrals

One of the most valuable features of the Henstock-Kurzweil integral is its ability to integrate every derivative. This is the content of Part I of the Fundamental Theorem of Calculus (Theorem 4.16). We now turn our attention to the second part of the Fundamental Theorem of Calculus, that of differentiating integrals. We first observe that if f is continuous at x then its indefinite integral F, $F(x) = \int_a^x f(t) \, dt$, is differentiable at x.

Theorem 4.94 *Let $f : [a, b] \to \mathbb{R}$ be Henstock-Kurzweil integrable on $[a, b]$ and continuous at $x \in [a, b]$. Then, F, the indefinite integral of f, is differentiable at x and $F'(x) = f(x)$.*

Proof. Since f is continuous at x, for $\epsilon > 0$ there is a $\delta > 0$ so that if $t \in [a, b]$ and $|t - x| < \delta$, then

$$-\epsilon < f(t) - f(x) < \epsilon.$$

If $0 < h < \delta$ is such that $x + h \in [a, b]$, then

$$\frac{F(x+h) - F(x)}{h} - f(x) = \frac{1}{h} \int_x^{x+h} f(t)\, dt - f(x)$$

$$= \frac{1}{h} \int_x^{x+h} (f(t) - f(x))\, dt$$

so that

$$-\epsilon \leq \frac{F(x+h) - F(x)}{h} - f(x) \leq \epsilon.$$

Similarly, for $h < 0$,

$$\left| \frac{F(x+h) - F(x)}{h} - f(x) \right| \leq \epsilon.$$

Thus, for $|h| \leq \delta$ and $x + h \in [a, b]$, $\left| \dfrac{F(x+h) - F(x)}{h} - f(x) \right| \leq \epsilon$. Thus, $F'(x) = f(x)$. $\qquad\square$

When f is merely Henstock-Kurzweil integrable, the indefinite integral is still differentiable almost everywhere.

Theorem 4.95 *(Fundamental Theorem of Calculus: Part II) Suppose that $f : [a, b] \to \mathbb{R}$ is Henstock-Kurzweil integrable. Then, F is differentiable at almost all $x \in [a, b]$ and $F'(x) = f(x)$.*

Observe that we cannot do better than a statement which holds almost everywhere. Let $E \subset [0, 1]$ be a null set and consider $f = \chi_E$. Then, f is equal to 0 for almost all $x \in [0, 1]$. It follows that F, and consequently also F', is identically 0 on $[0, 1]$. Thus, $F'(x) \neq f(x)$ if $x \in E$.

In order to prove this theorem, we need another covering lemma. Given an interval I, let $3I$ be the interval concentric with I and having three times the length of I. Recall that if I is an interval in $\mathbb{R}$, then the length of I, $\ell(I)$, is equal to the measure of I, $m(I)$.

Lemma 4.96 *Let $\mathcal{C} = \{I_i : i = 1, \ldots, N\}$ be a finite set of intervals in $\mathbb{R}$. Then, there exists a pairwise disjoint collection $J_1, \ldots, J_k \in \mathcal{C}$ such that*

$$\frac{1}{3} m \left(\cup_{i=1}^N I_i \right) \leq m \left(\cup_{j=1}^k J_j \right).$$

Proof. By reordering the intervals if necessary, we may assume that

$$\ell(I_N) \leq \ell(I_{N-1}) \leq \cdots \leq \ell(I_2) \leq \ell(I_1).$$

Set $J_1 = I_1$. Let $C_1 = \{I \in C : I_1 \cap I = \emptyset\}$ and note that if $I_i \in C$ and $I_i \notin C_1$, then $I_i \subset 3J_1$. Next, we let J_2 be the element of C_1 with the smallest index (and hence the greatest length). Set $C_2 = \{I \in C_1 : I_2 \cap I = \emptyset\}$ and continue as above. Since C is a finite set, the selection of intervals J_j ends after finitely many steps, say k. By construction, the intervals in $\{J_1, \ldots, J_k\}$ are pairwise disjoint, and if $I_i \in C$ is not selected, then there is a j so that $I_i \subset 3J_j$. Thus, $\cup_{i=1}^N I_i = \cup_{j=1}^k \cup_{I_i \cap J_j \neq \emptyset} I_i \subset \cup_{j=1}^k 3J_j$, so that

$$\frac{1}{3}m\left(\cup_{i=1}^N I_i\right) \leq \frac{1}{3}m\left(\cup_{j=1}^k 3J_j\right) \leq \frac{1}{3}\sum_{j=1}^k m\left(3J_j\right) = \sum_{j=1}^k m\left(J_j\right) = m\left(\cup_{j=1}^k J_j\right).$$
$$\square$$

We are now ready to prove Theorem 4.95.

Proof. For a fixed $\mu > 0$, we say that $x \in (a,b)$ satisfies condition $(*_\mu)$ if every neighborhood of x contains an interval $[u,v]$ such that $x \in (u,v)$ and

$$\left|\frac{F(v) - F(u)}{v - u} - f(x)\right| > \mu. \tag{4.16}$$

Let E_μ be the set of all $x \in (a,b)$ that satisfy condition $(*_\mu)$ and set $E = \cup_{n=1}^\infty E_{1/n}$. Suppose that $x \notin E$. Then, for all $n \geq 1$, there is a neighborhood U_n of x such that for any interval $[u,v] \subset U_n$ with $x \in (u,v)$, one has

$$\left|\frac{F(v) - F(u)}{v - u} - f(x)\right| \leq \frac{1}{n}.$$

By the continuity of F (Theorem 4.46), this inequality holds when u is replaced by x. Thus, if $x \notin E$, then F is differentiable at x and $F'(x) = f(x)$.

It suffices to show that E_μ is null for any $\mu > 0$, since then $E = \cup_{n=1}^\infty E_{1/n}$ has measure 0. If $E_\mu = \emptyset$, there is nothing to prove, so assume that $E_\mu \neq \emptyset$. Fix $\epsilon > 0$. Since f is Henstock-Kurzweil integrable, by Henstock's Lemma there is a gauge γ on $[a,b]$ such that

$$\sum_{i=1}^l |F(v_i) - F(u_i) - f(x_i)(v_i - u_i)| < \frac{\epsilon\mu}{6} \tag{4.17}$$

for any γ-fine tagged subpartition $\mathcal{D} = \{(x_i, [u_i, v_i]) : i = 1, \ldots, l\}$ of $[a,b]$. For $x \in E_\mu$, choose an interval $[u_x, v_x]$ such that $x \in [u_x, v_x] \subset \gamma(x)$ and (4.16) holds. Next, choose a gauge γ_1 on E_μ such that $\gamma_1(x) \subset (u_x, v_x)$

for all $x \in E_\mu$. By Lemma 4.41, there exist countably many non-overlapping closed intervals $\{J_k : k \in \sigma\}$ and points $\{x_k : k \in \sigma\}$ such that $x_k \in J_k \cap E_\mu$, $J_k \subset \gamma_1(x_k) \subset (u_{x_k}, v_{x_k})$, and $E_\mu \subset \cup_{k \in \sigma} J_k \subset [a, b]$. Let $\alpha = \sum_{k \in \sigma} \ell(J_k) \leq b - a < \infty$ and pick N such that $\sum_{k=1}^N \ell(J_k) > \dfrac{\alpha}{2}$.

Apply Lemma 4.96 to $\{(u_{x_k}, v_{x_k}) : k = 1, \ldots, N\}$ to get a set of non-overlapping intervals $\{(u_{y_1}, v_{y_1}), \ldots, (u_{y_K}, v_{y_K})\}$ such that

$$\sum_{i=1}^K \ell((u_{y_i}, v_{y_i})) = m\left(\cup_{i=1}^K (u_{y_i}, v_{y_i})\right) \geq \frac{1}{3} m\left(\cup_{k=1}^N (u_{x_k}, v_{x_k})\right) \quad (4.18)$$

$$\geq \frac{1}{3} m\left(\cup_{k=1}^N \ell(J_k)\right) = \frac{1}{3} \sum_{k=1}^N \ell(J_k) > \frac{\alpha}{6}.$$

Since $\{(x_i, [u_{x_i}, v_{x_i}]) : i = 1, \ldots, N\}$ is a γ-fine tagged subpartition of $[a, b]$, by (4.16) and (4.17),

$$\mu \sum_{i=1}^K \ell((u_{y_i}, v_{y_i})) \leq \sum_{i=1}^N |F(v_{x_i}) - F(u_{x_i}) - f(x_i)(v_{x_i} - u_{x_i})| < \frac{\epsilon\mu}{6}.$$

It now follows from (4.18) that $\epsilon > \alpha$. Since $E_\mu \subset \cup_{k \in \sigma} J_k$ and $\sum_{k \in \sigma} \ell(J_k) = \alpha < \epsilon$, it now follows that E_μ is null. $\square$

Since every Lebesgue integrable function is (absolutely) Henstock-Kurzweil integrable, Part II of the Fundamental Theorem of Calculus of Lebesgue integrals follows as an immediate corollary.

Corollary 4.97 *Let $f : [a, b] \to \mathbb{R}$ be Lebesgue integrable. Then, $F' = f$ a.e. in $[a, b]$.*

For a proof that does not use the Henstock-Kurzweil integral, see [N, Vol. I, IX.4.2], [Sw1, 4.1.9, page 150], and [Ro, 5.3.10, page 107].

Suppose f is Henstock-Kurzweil integrable over $[a, b]$. Then, F is continuous on $[a, b]$; extend F to $[a, b+1]$ by setting $F(t) = F(b)$ for $b < t \leq b+1$. Since the extended function is continuous on $[a, b+1]$, it follows that the sequence of functions $\{f_k\}_{k=1}^\infty$ defined by

$$f_k(t) = \frac{F\left(t + \frac{1}{k}\right) - F(t)}{\frac{1}{k}}$$

is measurable on $[a, b]$. By Theorem 4.95, $f(t) = F'(t) = \lim_{k \to \infty} \frac{F\left(t+\frac{1}{k}\right) - F(t)}{\frac{1}{k}}$ for almost all $t \in I$, which implies that f is measurable.

Corollary 4.98   *Let $f : I \subset \mathbb{R} \to \mathbb{R}$ be Henstock-Kurzweil integrable over I. Then, f is (Lebesgue) measurable over I.*

Proof.   The proof when I is a bounded interval is contained in the previous paragraph. If I is unbounded, set $I_n = I \cap [-n, n]$. The function $\chi_{I_n} f$ is measurable since I_n is a bounded interval, and since f is the pointwise limit of $\{\chi_{I_n} f\}_{n=1}^{\infty}$, f is measurable.   □

Other proofs of the measurability of Henstock-Kurzweil integrable functions can be found in [Lee, 5.10] and [Pf, 6.3.3].

Due to this corollary, in Theorem 4.91 and Corollary 4.92, we can drop the assumption that f is measurable, since both Henstock-Kurzweil and Lebesgue integrability imply (Lebesgue) measurability. Thus, we have

Theorem 4.99   *Suppose that $f : I \to \mathbb{R}$. Then, f is Lebesgue integrable if, and only if, f is absolutely Henstock-Kurzweil integrable. In either case, the integrals agree.*

As in the Lebesgue integral case, we define the Henstock-Kurzweil integral of f over a set E in terms of the function $\chi_E f$.

Definition 4.100   Let $f : I \to \mathbb{R}$ and $E \subset I$. We say that f is Henstock-Kurzweil integrable over E if $\chi_E f$ is Henstock-Kurzweil integrable over I and we set

$$\int_E f = \int_I \chi_E f.$$

The next result follows from Theorem 4.99.

Corollary 4.101   *Suppose that $f : I \to \mathbb{R}$ is Henstock-Kurzweil integrable over I. Then, f is Lebesgue integrable over I if, and only if, f is Henstock-Kurzweil integrable over every measurable subset $E \subset I$.*

Proof.   Suppose, first, that f is Lebesgue integrable. Let E be a measurable subset of I. Then, $\chi_E f$ is Lebesgue integrable over I. By Theorem 4.99, $\chi_E f$ is (absolutely) integrable over I, so that f is Henstock-Kurzweil integrable over E.

For the converse, put $E^+ = \{t \in I : f(t) \geq 0\}$ and $E^- = \{t \in I : f(t) < 0\}$. By Corollary 4.98, E^+ and E^- are measurable sets, so that $\int_{E^+} f$ and $\int_{E^-} f$ both exist. Since

$$\mathcal{L}\int_I f^+ = \int_I f^+ = \int_{E^+} f \text{ and } \mathcal{L}\int_I f^- = \int_I f^- = \int_{E^-} (-f),$$

both f^+ and f^- are Lebesgue integrable over I. Thus, $f = f^+ - f^-$ is Lebesgue integrable. $\qquad\square$

4.9.1 Functions with integral 0

We have already seen that sets with measure 0 play an important role in integration. We now investigate some properties of functions with integral 0.

Corollary 4.102 *Let* $f : \mathbb{R} \to \mathbb{R}$ *be Henstock-Kurzweil integrable and suppose that* $\int_I f = 0$ *for every bounded interval* I. *Then,* $f = 0$ *a.e..*

Proof. Let $F(x) = \int_{-n}^{x} f$ for all $x \in [-n, n]$. By definition, $F(x) = 0$ for all $x \in [-n, n]$. By Theorem 4.95, $f(x) = F'(x) = 0$ for almost all $x \in [-n, n]$. It follows that $f = 0$ a.e. in $\mathbb{R}$. $\qquad\square$

Let $E \subset \mathbb{R}$ be a null set. Then, E is measurable and $\mathcal{L} \int \chi_E = 0$. Thus, χ_E is Henstock-Kurzweil integrable and $\int \chi_E = 0$. On the other hand, suppose that χ_E is Henstock-Kurzweil integrable and $\int \chi_E = 0$. Then, χ_E is absolutely Henstock-Kurzweil integrable, so that χ_E is Lebesgue integrable and $m(E) = \mathcal{L} \int \chi_E = 0$. Thus, E is a null set. We have proved

Corollary 4.103 *Let* $E \subset \mathbb{R}$. *Then,* E *is a null set if, and only if,* χ_E *is Henstock-Kurzweil integrable and* $\int \chi_E = 0$.

Of course, this is just Theorem 4.42, proved in Section 4.4. However, the argument above can easily be modified to prove the following result.

Corollary 4.104 *Let* $E \subset \mathbb{R}$. *Then,* E *is a Lebesgue measurable set with finite measure if, and only if,* χ_E *is Henstock-Kurzweil integrable. In either case,* $\int \chi_E = m(E)$.

4.10 Characterizations of indefinite integrals

In this section, we will characterize indefinite integrals for the three integrals considered so far. We begin with the Henstock-Kurzweil integral. Suppose that f is a Henstock-Kurzweil integrable function on an interval $I \subset \mathbb{R}$. Then, the indefinite integral of f, F, is differentiable and $F' = f$ almost everywhere. On the other hand, suppose that a function F is differentiable almost everywhere in an interval I. Does it then follow that the derivative F' is Henstock-Kurzweil integrable? In general the answer is no. As we shall see, in order for F' to be Henstock-Kurzweil integrable, we need to know more about how F acts on the set where it is not differentiable in order to conclude that its derivative is integrable.

Definition 4.105   Let $f : I \to \mathbb{R}$ and $E \subset I$. We say that f has *negligible variation* over E if for every $\epsilon > 0$, there is a gauge γ so that for every γ-fine tagged subpartition of I, $\mathcal{D} = \{(t_i, [x_{i-1}, x_i]) : i = 1, \ldots, m\}$, with $t_i \in E$ for $i = 1, \ldots, m$,

$$\sum_{i=1}^{m} |f(x_i) - f(x_{i-1})| \le \epsilon.$$

Note that the t_i's must be elements of E, in addition to being contained in $[x_{i-1}, x_i]$.

Theorem 4.106   Let $F : I = [a, b] \to \mathbb{R}$. Then, F is the indefinite integral of a Henstock-Kurzweil integrable function $f : I \to \mathbb{R}$ if, and only if, there is a null set $Z \subset I$ such that $F' = f$ on $I \setminus Z$ and F has negligible variation over Z.

Proof.   For the sufficiency, assume F is the indefinite integral of a Henstock-Kurzweil integrable function f. Then, by Theorem 4.95, $F' = f$ almost everywhere on I. Let Z be the set where the equality fails. Define $f_1 : I \to \mathbb{R}$ by $f_1(t) = f(t)$ for $t \in I \setminus Z$ and $f_1(t) = 0$ for $t \in Z$. Then, $F(x) = \int_a^x f_1$ and, consequently, $F(b) = \int_a^b f_1$.

Given $\epsilon > 0$, there is a gauge γ such that $|S(f_1, \mathcal{D}) - F(b)| < \epsilon$ for every γ-fine tagged partition $\mathcal{D}$ of I. Suppose $\mathcal{D} = \{(t_i, [x_{i-1}, x_i]) : i = 1, \ldots, m\}$ is a γ-fine tagged subpartition of I with tags $t_i \in Z$. By Henstock's Lemma,

$$\sum_{i=1}^{m} |f_1(t_i)(x_i - x_{i-1}) - (F(x_i) - F(x_{i-1}))| \le 2\epsilon.$$

But, since $t_i \in Z$, $f(t_i) = 0$ for all $i = 1, \ldots, m$, so that

$$\sum_{i=1}^{m} |F(x_i) - F(x_{i-1})| \le 2\epsilon,$$

as we wished to show.

For the necessity, we assume that $F' = f$ on $I \setminus Z$ and F has negligible variation over Z, for some null set Z. Extend f to all of I by setting $f(t) = 0$ for $t \in Z$. Let $\epsilon > 0$. We claim that the extended function f is Henstock-Kurzweil integrable and $F(x) = \int_a^x f$ for all $x \in I$. Since F has negligible variation over Z, there is a gauge γ_Z satisfying Definition 4.105. Define a gauge γ on I by setting $\gamma(t) = \gamma_Z(t)$ for $t \in Z$ and $\gamma(t) =$

$(t - \delta(t), t + \delta(t))$ for $t \in I \setminus Z$, where $\delta(t)$ is the value corresponding to $\epsilon > 0$ and the function f in the Straddle Lemma (Lemma 4.6).

Suppose that $\mathcal{D} = \{(t_i, [x_{i-1}, x_i]) : i = 1, \ldots, m\}$ is a γ-fine tagged partition of I. Then,

$$\left| F(b) - F(a) - \sum_{i=1}^{m} f(t_i)(x_i - x_{i-1}) \right|$$

$$= \left| \sum_{i=1}^{m} \{F(x_i) - F(x_{i-1}) - f(t_i)(x_i - x_{i-1})\} \right|$$

$$\leq \left| \sum_{\substack{i=1 \\ t_i \in Z}}^{m} \{F(x_i) - F(x_{i-1}) - f(t_i)(x_i - x_{i-1})\} \right|$$

$$+ \left| \sum_{\substack{i=1 \\ t_i \in I \setminus Z}}^{m} \{F(x_i) - F(x_{i-1}) - f(t_i)(x_i - x_{i-1})\} \right|$$

$$= I + II.$$

Since F has negligible variation over Z and $f = 0$ on Z, $I \leq \epsilon$. By the Straddle Lemma,

$$II \leq \sum_{\substack{i=1 \\ t_i \in I \setminus Z}}^{m} \epsilon(x_i - x_{i-1}) \leq \epsilon(b - a).$$

Therefore,

$$\left| F(b) - F(a) - \sum_{i=1}^{m} f(t_i)(x_i - x_{i-1}) \right| \leq \epsilon(1 + b - a),$$

which shows that f is Henstock-Kurzweil integrable over I with

$$F(b) - F(a) = \int_a^b f.$$

Applying the same argument to the interval $[a, x]$ yields $F(x) - F(a) = \int_a^x f$, so that F is an indefinite integral of f. $\qquad \square$

We now turn our attention to characterizing indefinite integrals of Lebesgue integrable functions. To do this, we first study monotone functions and their derivatives.

4.10.1   *Derivatives of monotone functions*

In order to characterize indefinite integrals of Lebesgue integrable functions, we need to know that every increasing function is differentiable almost everywhere. Recall the upper and lower derivatives, $\overline{D}f$ and $\underline{D}f$, discussed in Section 4.1, and that f is differentiable at x if, and only if, $-\infty < \overline{D}f(x) = \underline{D}f(x) < \infty$.

To prove that an increasing function is differentiable almost everywhere, we will use *Vitali covers*. For later use, we will discuss Vitali covers in n-dimensions. Given an interval I in $\mathbb{R}^n$, recall that $v(I)$ represents the volume (and measure) of I.

Definition 4.107   Let $E \subset \mathbb{R}^n$. A family $\mathcal{V}$ of closed, bounded subintervals of $\mathbb{R}^n$ *covers E in the Vitali sense* if for all $x \in E$ and for all $\epsilon > 0$, there is an interval $I \in \mathcal{V}$ so that $x \in I$ and the $v(I) < \epsilon$. If $\mathcal{V}$ covers E in the Vitali sense, we call $\mathcal{V}$ a *Vitali cover* of E.

Given a set $E \subset \mathbb{R}^2$, the set

$$\mathcal{V} = \left\{ \left[x - \frac{1}{n}, x + \frac{1}{n}\right] \times \left[y - \frac{1}{n+1}, y + \frac{1}{2n}\right] : (x,y) \in E \text{ and } n \in \mathbb{N} \right\}$$

is a Vitali cover of E. A typical cover that arises in applications is given in the following example.

Example 4.108   Let $f : [a,b] \to \mathbb{R}$ be differentiable over $[a,b]$. For each $x \in [a,b]$ and $\epsilon > 0$, let $I_{x,\epsilon}$ be a closed interval of length less than ϵ, containing x in its interior, such that

$$\left| \frac{f(y) - f(x)}{y - x} - f'(x) \right| < \epsilon$$

for all $y \in I_{x,\epsilon} \cap [a,b]$. The existence of the intervals $I_{x,\epsilon}$ is guaranteed by the differentiability of f. Since $\ell(I_{x,\epsilon}) < \epsilon$, it follows that $\mathcal{V} = \{I_{x,\epsilon} : x \in [a,b] \text{ and } \epsilon > 0\}$ is a Vitali cover of $[a,b]$.

Suppose a set E of finite outer measure is covered in the Vitali sense by a family of cubes $\mathcal{V}$. The next result, known as the *Vitali Covering Lemma*, shows that a finite set of elements from $\mathcal{V}$ covers all of E except a set of small measure.

Lemma 4.109   *(Vitali Covering Lemma) Let $E \subset \mathbb{R}^n$ have finite outer measure. Suppose that a family of cubes $\mathcal{V}$ is a Vitali cover of E. Given $\epsilon > 0$, there is a finite collection of pairwise disjoint cubes $\{Q_i\}_{i=1}^{N} \subset \mathcal{V}$ such that $m_n^*\left(E \setminus \cup_{i=1}^{N} Q_i\right) < \epsilon$.*

The following proof is due to Banach [Ban].

Proof. Let J be an open set containing E such that $m_n(J) < (1+\epsilon) m_n^*(E)$. We need consider only the $Q \in \mathcal{V}$ such that $Q \subset J$. Given a cube Q, let $e(Q)$ be the length of an edge of Q and note that $v(Q) = m_n(Q) = [e(Q)]^n$.

Define a sequence of cubes by induction. Since

$$\sup \{v(Q) : Q \in \mathcal{V}\} \leq m_n(J) < (1+\epsilon) m_n^*(E) < \infty,$$

we can choose Q_1 so that $e(Q_1) > \frac{1}{2} \sup \{e(Q) : Q \in \mathcal{V}\}$. Assume that $Q_1, \ldots, Q_k$ have been chosen. If $E \subset \cup_{i=1}^k Q_i$, set $N = k$ and $\{Q_i\}_{i=1}^N$ is the desired cover. Otherwise, let

$$S_k = \sup \left\{ e(Q) : Q \in \mathcal{V} \text{ and } Q \cap \left(\cup_{i=1}^k Q_i \right) = \emptyset \right\}.$$

Since $m_n(J) < \infty$, $S_k < \infty$. Since $E \not\subset \cup_{i=1}^k Q_i$, there is an $Q \in \mathcal{V}$ such that $e(Q) > S_k/2$ and $Q \cap \left(\cup_{i=1}^k Q_i \right) = \emptyset$. Set $Q_{k+1} = Q$.

When $E \setminus \cup_{i=1}^k Q_i \neq \emptyset$ for all k, we get a sequence of disjoint cubes such that $\cup_{i=1}^\infty Q_i \subset J$. This implies that

$$\sum_{i=1}^\infty m_n(Q_i) \leq m_n(J) < \infty.$$

Choose an N so that $\sum_{i=N+1}^\infty m_n(Q_i) < \dfrac{\epsilon}{5^n}$.

It remains to show that $m_n^* \left(E \setminus \cup_{i=1}^N Q_i \right) < \epsilon$. Suppose that $x \in E \setminus \cup_{i=1}^N Q_i$. Since $\cup_{i=1}^N Q_i$ is a closed set, there is a $Q \in \mathcal{V}$ such that $x \in Q$ and $Q \cap \left(\cup_{i=1}^N Q_i \right) = \emptyset$. Since $\sum_{i=1}^\infty m_n(Q_i) < \infty$ and $2e(Q_{k+1}) \geq S_k$, it follows that $\lim_{k \to \infty} S_k = 0$. If $Q \cap Q_i = \emptyset$ for all $i \leq k$, then $e(Q) \leq S_k$. Since $e(Q) > 0$, there must be an i such that $Q \cap Q_i \neq \emptyset$. Let j be the smallest such index and let Q_j^* be the cube concentric with Q_j and having edge length 5 times as long. By construction, $j > N$ and $e(Q) \leq S_{j-1} \leq 2e(Q_j)$. Therefore, since $Q \cap Q_j \neq \emptyset$, $Q \subset Q_j^*$. Thus, if $x \in E \setminus \cup_{i=1}^N Q_i$, then $x \in Q_j^*$ for some $j > N$, which implies that $E \setminus \cup_{i=1}^N Q_i \subset \cup_{i=N+1}^\infty Q_i^*$. Since $e(Q_i^*) = 5e(Q_i)$, $m_n(Q_i^*) = 5^n m_n(Q_i)$ and we have

$$m_n^* \left(E \setminus \cup_{i=1}^N Q_i \right) \leq \sum_{i=N+1}^\infty m_n(Q_i^*) = 5^n \sum_{i=N+1}^\infty m_n(Q_i) < \epsilon,$$

as we wished to prove. $\qquad\qquad\qquad\qquad\qquad\qquad\qquad\qquad\qquad\square$

We have actually proved more. Since $\cup_{i=1}^{N} Q_i \subset J$ and the Q_i's are pairwise disjoint, $\sum_{i=1}^{N} m_n(Q_i) < (1+\epsilon) m_n^*(E)$. By iterating the argument, replacing ϵ by $2^{-k}\epsilon$ at the kth iteration, if $\mathcal{V}$ covers E in the Vitali sense, then there is a sequence of pairwise disjoint cubes $\{Q_i\}_{i=1}^{\infty}$ such that $m_n(E \setminus \cup_{i=1}^{\infty} Q_i) = 0$ and $\sum_{i=1}^{\infty} m_n(Q_i) < (1+\epsilon) m_n^*(E)$.

Remark 4.110 *In Theorem 4.128, we will apply the Vitali Covering Lemma to a collection of intervals that are obtained by repeated bisection a fixed interval in $\mathbb{R}^n$. The proof above for cubes applies to this situation since the key geometric estimate, namely $Q \cap Q_j \neq \emptyset$ implies $Q \subset Q_j^*$ (for the smallest j such that $Q \cap Q_i \neq \emptyset$) continues to hold for such a collection of intervals, which has fixed eccentricity.*

We now return to the differentiation of monotone functions.

Theorem 4.111 *Let $f : [a, b] \to \mathbb{R}$ be an increasing function. Then, f' exists almost everywhere in $[a, b]$.*

Proof.   We claim that $E = \{x \in [a, b] : \overline{D}f(x) > \underline{D}f(x)\}$ has measure zero. Since $\overline{D}f(x) \geq \underline{D}f(x)$ for all x, this would imply that $\overline{D}f(x)$ and $\underline{D}f(x)$ are equal almost everywhere.

Set $E_{u,v} = \{x \in [a, b] : \overline{D}f(x) > u > v > \underline{D}f(x)\}$ so that $E = \cup_{u,v \in \mathbb{Q}} E_{u,v}$. It is enough to show that $m^*(E_{u,v}) = 0$ for all $u, v \in \mathbb{Q}$. Let $\tau = m^*(E_{u,v})$, fix $\epsilon > 0$, and choose an open set $I \supset E_{u,v}$ such that $m(I) < \tau + \epsilon$.

Let $x \in E_{u,v}$. Since $\underline{D}f(x) < v$, there are arbitrarily short closed intervals $[\alpha, \beta]$ containing x such that $f(\beta) - f(\alpha) < v(\beta - \alpha)$. Thus, $E_{u,v}$ is covered in the Vitali sense by the collection

$$\mathcal{V} = \{[\alpha, \beta] \subset I : [\alpha, \beta] \cap E_{u,v} \neq \emptyset \text{ and } f(\beta) - f(\alpha) < v(\beta - \alpha)\}.$$

By the Vitali Covering Lemma, there are pairwise disjoint intervals $\{[x_i, y_i]\}_{i=1}^{N} \subset \mathcal{V}$ such that $m^*(E_{u,v} \setminus \cup_{i=1}^{N} [x_i, y_i]) < \epsilon$. This implies that the set $A = E_{u,v} \cap (\cup_{i=1}^{N} [x_i, y_i])$ has outer measure $m^*(A) > \tau - \epsilon$. Further,

$$\sum_{i=1}^{N} [f(y_i) - f(x_i)] < \sum_{i=1}^{N} v(y_i - x_i) < vm(I) < v(\tau + \epsilon).$$

Suppose $s \in A$ is not an endpoint of any $[x_i, y_i]$, $i = 1, \ldots, N$. Since $\overline{D}f(s) > u$, there are arbitrarily short intervals $[\lambda, \mu]$ containing s such that $[\lambda, \mu] \subset [x_i, y_i]$ for some i and $f(\mu) - f(\lambda) > u(\mu - \lambda)$. As above, by the Vitali Covering Lemma, there is a collection of pairwise disjoint intervals $\{[s_j, t_j]\}_{j=1}^{M}$ such that $m^* \left(A \cap \left(\cup_{j=1}^{M} [s_j, t_j] \right) \right) > \tau - 2\epsilon$ and

$$\sum_{j=1}^{M} [f(t_j) - f(s_j)] > \sum_{i=1}^{M} u(t_j - s_j) > u(\tau - 2\epsilon).$$

Since each $[s_j, t_j]$ is contained in $[x_i, y_i]$ for some i, and since f is increasing,

$$\sum_{j=1}^{M} [f(t_j) - f(s_j)] \leq \sum_{i=1}^{N} [f(y_i) - f(x_i)].$$

This implies that $u(\tau - 2\epsilon) < v(\tau + \epsilon)$. Since $\epsilon > 0$ was arbitrary, we have $u\tau \leq v\tau$, and since $u > v$, we see that $\tau = 0$. Thus, $m^*(E_{u,v}) = 0$. Hence, $E_{u,v}$ is measurable with measure 0 and, consequently, E is measurable with measure 0.

It remains to show that $\overline{D}f(x)$ is finite almost everywhere. For if this were the case, then f' exists almost everywhere and the proof is complete.

Fix $k, \epsilon > 0$ and set $E_k = \{x \in [a, b] : \overline{D}f(x) > k\}$. Repeating the argument in the previous paragraph yields

$$f(b) - f(a) \geq \sum_{j=1}^{M} [f(t_j) - f(s_j)] > k(m^*(E_k) - \epsilon).$$

Since $\epsilon > 0$ is arbitrary, $f(b) - f(a) \geq km^*(E_k)$. Finally, since

$$m^* \left(\{x \in [a, b] : \overline{D}f(x) = \infty\} \right) \leq m^* \left(\cap_{k=1}^{\infty} E_k \right) \leq m^*(E_k) \leq \frac{f(b) - f(a)}{k}$$

for all $k > 0$, it follows $m^* \left(\{x \in [a, b] : \overline{D}f(x) = \infty\} \right) = 0$ so that $\overline{D}f(x)$ is finite a.e. in $[a, b]$. Thus, f' exists and is finite almost everywhere in $[a, b]$. $\qquad \square$

We saw in Theorem 4.61 that every function of bounded variation is the difference of two increasing functions. It then follows from Theorem 4.111 that a function of bounded variation is differentiable almost everywhere.

Corollary 4.112 *If $f : [a, b] \to \mathbb{R}$ has bounded variation on $[a, b]$, then f is differentiable a.e. in $[a, b]$.*

In Remark 3.93, we defined a measure Φ to be absolutely continuous with respect to Lebesgue measure if given any $\epsilon > 0$, there is a $\delta > 0$ so that $m(E) < \delta$ implies $\Phi(E) < \epsilon$. Suppose that f is a nonnegative, Lebesgue integrable function on $[a, b]$ and set $F(x) = \mathcal{L} \int_a^x f$. In the same remark, we observed that F is absolutely continuous with respect to Lebesgue measure. Fix $\epsilon > 0$ and choose $\delta > 0$ by absolute continuity. Let $\{[a_i, b_i]\}_{i=1}^{k}$ be a finite set of nonoverlapping intervals in $[a, b]$ and suppose that $\sum_{i=1}^{k} (b_i - a_i) < \delta$. It then follows that

$$\sum_{i=1}^{k} |F(b_i) - F(a_i)| = \sum_{i=1}^{k} \mathcal{L} \int_{a_i}^{b_i} f = \mathcal{L} \int_{\cup_{i=1}^{k} [a_i, b_i]} f < \epsilon.$$

We use this condition to extend the idea of absolute continuity to functions.

Definition 4.113 Let $F : [a, b] \to \mathbb{R}$. We say that F is *absolutely continuous* on $[a, b]$ if for every $\epsilon > 0$, there is a $\delta > 0$ so that $\sum_{i=1}^{k} |F(b_i) - F(a_i)| < \epsilon$ for every finite collection $\{[a_i, b_i]\}_{i=1}^{k}$ of nonoverlapping subintervals of $[a, b]$ such that $\sum_{i=1}^{k} (b_i - a_i) < \delta$.

Clearly, every absolutely continuous function is uniformly continuous, which is seen by considering a single interval $[\alpha, \beta]$ with $\beta - \alpha < \delta$. Further, every such function also has bounded variation.

Proposition 4.114 *Suppose that $F : [a, b] \to \mathbb{R}$ is absolutely continuous on $[a, b]$. Then, F has bounded variation on $[a, b]$.*

Proof. Choose $\delta > 0$ corresponding to $\epsilon = 1$ in the definition of absolute continuity. Thus, if $[c, d] \subset [a, b]$ and $d - c < \delta$, then the variation of F over $[c, d]$ is at most 1. Choose $N \in \mathbb{N}$ such that $N > \dfrac{b - a}{\delta}$. Then, we can divide $[a, b]$ into N nonoverlapping intervals each of length $\dfrac{b - a}{N} < \delta$. It follows that the variation of F over $[a, b]$ is less than or equal to N. $\square$

4.10.2 *Indefinite Lebesgue integrals*

A well-known result for the Lebesgue integral relates absolute continuity and indefinite integration. In the following theorem, we show that these conditions are also equivalent to a condition similar to that of Theorem 4.106.

Theorem 4.115 *Let $F : [a, b] \to \mathbb{R}$. The following statements are equivalent:*

(1) F is the indefinite integral of a Lebesgue integrable function $f : [a, b] \to \mathbb{R}$.

(2) F is absolutely continuous on $[a, b]$.

(3) F has bounded variation on $[a, b]$ and F has negligible variation over Z, where Z is the null set where F' fails to exist.

Condition (3) should be compared with the condition for the Henstock-Kurzweil integral given in Theorem 4.106. In particular, for both integrals, the indefinite integral has negligible variation over the null set where its derivative fails to exist.

Proof. To show that (1) implies (2), note that

$$\sum_{i=1}^{k} |F(b_i) - F(a_i)| = \sum_{i=1}^{k} \left| \mathcal{L} \int_{a_i}^{b_i} f \right| \leq \sum_{i=1}^{k} \mathcal{L} \int_{a_i}^{b_i} |f| = \mathcal{L} \int_{\cup_{i=1}^{k} [a_i, b_i]} |f|,$$

so that the absolute continuity of F follows from the comments above.

Suppose (2) holds. By Proposition 4.114 and Corollary 4.112, F has bounded variation on $[a, b]$ and F' exists almost everywhere. Let Z be the null set where F' fails to exist. We claim that F has negligible variation over Z.

To see this, fix $\epsilon > 0$ and choose $\delta > 0$ by the definition of absolute continuity. Since Z is null, there exists a countable collection of open intervals $\{J_k\}_{k \in \sigma}$ such that $Z \subset \cup_{k \in \sigma} J_k$ and $\sum_{k \in \sigma} \ell(J_k) < \delta$. Define a gauge on $[a, b]$ as follows. If $t \in I \setminus Z$, set $\gamma(t) = \mathbb{R}$; if $t \in Z$, let k_t be the smallest integer such that $t \in J_{k_t}$ and set $\gamma(t) = J_{k_t}$. Suppose that $\mathcal{D} = \{(t_i, [x_{i-1}, x_i]) : i = 1, \ldots, m\}$ is a γ-fine tagged subpartition of $[a, b]$ with tags $t_i \in Z$. Then, $[x_{i-1}, x_i] \subset J_{k_{t_i}}$ so that

$$\sum_{i=1}^{m} (x_i - x_{i-1}) \leq \sum_{k \in \sigma} \ell(J_k) < \delta.$$

By absolute continuity,

$$\sum_{i=1}^{m} |F(x_i) - F(x_{i-1})| < \epsilon$$

so that Z has negligible variation over E. Thus, (2) implies (3).

To show that (3) implies (1), set $f(t) = F'(t)$ for $t \in I \setminus Z$ and $f(t) = 0$ for $t \in Z$. By Theorem 4.106, f is Henstock-Kurzweil integrable and $F(x) - F(a) = \int_a^x f$ for all $x \in [a, b]$. Since F has bounded variation, f is absolutely Henstock-Kurzweil integrable by Theorem 4.62. By Theorem 4.99, f is Lebesgue integrable and F is the indefinite integral of a Lebesgue integrable function. $\qquad \square$

4.10.3   *Indefinite Riemann integrals*

We conclude this section by considering indefinite integrals of Riemann integrable functions. Suppose F is an indefinite integral of a Riemann integrable function $f : [a, b] \to \mathbb{R}$. Since f is bounded, there is an $M > 0$ such that $|f(x)| \leq M$ for all $x \in [a, b]$. We observed above, in Section 2.2.5, that

$$|F(x) - F(y)| \leq M |x - y|,$$

for all $x, y \in [a, b]$. Thus, F satisfies a Lipschitz condition on $[a, b]$. Further, by Corollary 2.42, f is continuous almost everywhere, so that by Part II of the Fundamental Theorem of Calculus for the Riemann integral (Theorem 2.32), $F'(x)$ exists and equals $f(x)$ for almost every $x \in [a, b]$.

Theorem 4.116   *Let $F : [a, b] \to \mathbb{R}$. Then, F is the indefinite integral of a Riemann integrable function $f : [a, b] \to \mathbb{R}$ if, and only if, F satisfies a Lipschitz condition on $[a, b]$, F' exists almost everywhere, and F' is equal almost everywhere to a bounded function f which is continuous a.e..*

Proof.   By the previous remarks, it is enough to prove the necessity of the result. By the Lipschitz condition, F is continuous and F' is bounded whenever it exists. To see this, note that if M is the Lipschitz constant for F, then $F'(x) = \lim_{y \to x} \dfrac{F(y) - F(x)}{y - x}$ and $\left| \dfrac{F(y) - F(x)}{y - x} \right| \leq M$ imply that $|F'(x)| \leq M$ whenever the limit exists. Further, since $F'(x)$ is the almost everywhere limit of the continuous difference quotients $\dfrac{F\left(x + \frac{1}{n}\right) - F(x)}{\frac{1}{n}}$, F' is measurable if we define F' to be 0 where F' fails to exist. Thus, F' is Lebesgue integrable and since $F' = f$ almost everywhere, $\mathcal{L}\int_a^x F' = \mathcal{L}\int_a^x f$ for all $x \in [a, b]$. Since f is continuous a.e., f is Riemann integrable, so that $\mathcal{R}\int_a^x f = \mathcal{L}\int_a^x f = \mathcal{L}\int_a^x F'$.

Since F satisfies a Lipschitz condition, it is absolutely continuous on $[a, b]$; one need only choose $\delta \leq \epsilon/M$, where M is the Lipschitz constant for F. By Theorem 4.115, F is the indefinite integral of a Lebesgue integrable function. As we saw in the proof of that theorem, $F(x) - F(a) = \mathcal{L}\int_a^x f_1$, where $f_1 = F'$ almost everywhere. Thus,

$$F(x) - F(a) = \mathcal{L}\int_a^x f_1 = \mathcal{L}\int_a^x F' = \mathcal{R}\int_a^x f.$$

Thus, F is the indefinite integral of the Riemann integrable function f. $\square$

There is something troubling about this proof. It relies on results for the Lebesgue integral, which in turn are consequences of results for the Henstock-Kurzweil integral, both of which require more sophisticated constructions than the Riemann integral.

4.11 The space of Henstock-Kurzweil integrable functions

In Section 3.9 of Chapter 3, we considered the vector space, $L^1(E)$, of Lebesgue integrable functions on a measurable set E and showed that $L^1(E)$ had a natural norm under which the space was complete. In this section, we consider the space of Henstock-Kurzweil integrable functions. Since the Henstock-Kurzweil integral is a conditional integral, the L^1-norm defined on $L^1(E)$,

$$\|f\|_1 = \int_E |f|,$$

is not meaningful. For example, the function f' defined in Example 2.31 is Henstock-Kurzweil integrable while $|f'|$ is not. However, the space of Henstock-Kurzweil integrable functions does have a natural semi-norm, due to Alexiewicz [A], which we now define.

Definition 4.117 Let $I = [a, b] \subset \mathbb{R}$ and let $\mathcal{HK}(I)$ be the vector space of all Henstock-Kurzweil integrable functions defined on I. If $f \in \mathcal{HK}(I)$, the *Alexiewicz semi-norm* of f is defined to be

$$\|f\| = \sup\left\{ \left| \int_a^x f \right| : a \le x \le b \right\}.$$

(To see that $\|\cdot\|$ defines a semi-norm, see Exercise 4.61.) From Corollary 4.102 (which is also valid for intervals $I \subset \mathbb{R}$) and Exercise 4.65, we have that $\|f\| = 0$ if, and only if, $f = 0$ a.e.. Thus, if functions in $\mathcal{HK}(I)$ which are equal a.e. are identified, then $\|\cdot\|$ is actually a norm on $\mathcal{HK}(I)$.

The Riesz-Fischer Theorem (Theorem 3.130) asserts that the space $L^1(E)$ is complete under the L^1-semi-metric. We show, however, that $\mathcal{HK}(I)$ is not complete under the semi-metric generated by the Alexiewicz semi-norm.

Example 4.118 Let $p : [0, 1] \to \mathbb{R}$ be a continuous and nowhere differentiable function with $p(0) = 0$. (See, for example, [DS, page 137].) By the Weierstrass Approximation Theorem 3.127, there is a sequence of polynomials $\{p_k\}_{k=1}^{\infty}$ that converges uniformly to p. By Part I of the Fundamental

Theorem of Calculus, $p_k(t) = \int_0^t p_k'(x)\, dx - p_k(0)$ for every $t \in [0,1]$. Thus,

$$\|p_k' - p_j'\| = \sup\left\{ \left| \int_0^t (p_k' - p_j') - t(p_k(0) - p_j(0)) \right| : 0 \le t \le 1 \right\}$$
$$= C \sup\{|(p_k - p_j)(t)| : 0 \le t \le 1\},$$

with $C = 2$. Since $\{p_k\}_{k=1}^\infty$ converges uniformly to p, it follows that $\{p_k'\}_{k=1}^\infty$ is a Cauchy sequence in $\mathcal{HK}([0,1])$ with respect to the Alexiewicz semi-norm.

Suppose that there is an $f \in \mathcal{HK}([0,1])$ such that $\|p_k' - f\| \to 0$ as $k \to \infty$. Since $p_k(0) \to p(0) = 0$, $p_k(t) = \int_0^t p_k' - p_k(0) \to \int_0^t f$ uniformly in $t \in [0,1]$ so that $p(t) = \int_0^t f$. By Part II of the Fundamental Theorem of Calculus, p is differentiable a.e. (with derivative f), which is a contradiction to the definition of p. Hence $\mathcal{HK}([0,1])$ is not complete.

Although the space $\mathcal{HK}(I)$ is not complete under the Alexiewicz semi-norm, the space does have other desirable properties. For a discussion of the properties of $\mathcal{HK}([0,1])$, see [Sw2, Chapter 7].

We now show that the conclusion of Theorem 4.73 can be improved to convergence in the Alexiewicz norm. For this we require a uniform version of the Henstock Lemma 4.45. This lemma will also be used to show that the step functions are dense in $\mathcal{HK}(I)$ with respect to the Alexiewicz norm as in Theorem 3.133 for the Lebesgue integral.

Lemma 4.119   *(Uniform Henstock Lemma) Let $f : I \to \mathbb{R}$ be integrable over I and $\epsilon > 0$. Suppose γ is a gauge on I such that $\left| S(f, \mathcal{D}) - \int_I f \right| < \epsilon$ whenever $\mathcal{D}$ is a γ-fine tagged partition of I. If J is any subinterval of I, then*

$$\left| \sum_{i=1}^m \left\{ f(t_i)\, \ell(I_i \cap J) - \int_{I_i \cap J} f \right\} \right| \le 3\epsilon$$

and

$$\sum_{i=1}^m \left| f(t_i)\, \ell(I_i \cap J) - \int_{I_i \cap J} f \right| \le 6\epsilon$$

whenever $\mathcal{D} = \{(t_i, I_i) : 1 \le i \le m\}$ is a γ-fine partial tagged partition of I.

Proof.   Let $\mathcal{D} = \{(t_i, I_i) : 1 \le i \le m\}$ be a γ-fine partial tagged partition of I, set $d_0 = \{i : 1 \le i \le m,\ \ell(I_i \cap J) > 0\}$ and let $\mathcal{D}_0 =$

$\{(t_i, I_i \cap J) : i \in d_0\}$. Then,

$$S = \left| \sum_{i=1}^{m} \left\{ f(t_i)\, \ell\, (I_i \cap J) - \int_{I_i \cap J} f \right\} \right| = \left| \sum_{i \in d_0} \left\{ f(t_i)\, \ell\, (I_i \cap J) - \int_{I_i \cap J} f \right\} \right|,$$

and we wish to show that $S \le 3\epsilon$. Let $d_1 = \{i \in d_0 : t_i \in I_i \cap J\}$ and $\mathcal{D}_1 = \{(t_i, I_i \cap J) : i \in d_1\}$. Then, $\mathcal{D}_1$ is γ-fine so

$$\left| \sum_{i \in d_1} \left\{ f(t_i)\, \ell\, (I_i \cap J) - \int_{I_i \cap J} f \right\} \right| \le \epsilon$$

by Henstock's Lemma 4.45. Setting $d_2 = d_0 \setminus d_1$, to prove the first estimate, it is enough to show that

$$\left| \sum_{i \in d_2} \left\{ f(t_i)\, \ell\, (I_i \cap J) - \int_{I_i \cap J} f \right\} \right| \le 2\epsilon.$$

Note $\{(t_i, I_i \cap J)\}$ is not γ-fine if $i \in d_2$ since $t_i \notin I_i \cap J$. For each $i \in d_2$, we wish to write $I_i \cap J$ as the difference of two intervals whose closures each contain t_i. Suppose, for example, that $J = [c, d]$, $I_i = [a_i, b_i]$, and $a_i \le t_i < c$. Set $B = \min\{b_i, d\}$. Then,

$$I_i \cap J = [c, B] = [a_i, B] \setminus [a_i, c).$$

Setting $A_i = [a_i, B]$ and $B_i = [a_i, c)$ gives the desired decomposition. The other cases are similar. With this decomposition, the sets $\{(t_i, A_i) : i \in d_2\}$ and $\{(t_i, \overline{B_i}) : i \in d_2\}$ are both γ-fine. By Henstock's Lemma, we have

$$\left| \sum_{i \in d_2} \left\{ f(t_i)\, \ell\, (I_i \cap J) - \int_{I_i \cap J} f \right\} \right|$$

$$= \left| \sum_{i \in d_2} \left\{ f(t_i)\, \ell\, (A_i \cap J) - f(t_i)\, \ell\, (B_i \cap J) - \int_{A_i \cap J} f + \int_{B_i \cap J} f \right\} \right|$$

$$\le \left| \sum_{i \in d_2} \left\{ f(t_i)\, \ell\, (A_i \cap J) - \int_{A_i \cap J} f \right\} \right|$$

$$+ \left| \sum_{i \in d_2} \left\{ f(t_i)\, \ell\, (B_i \cap J) - \int_{B_i \cap J} f \right\} \right|$$

$$\le \epsilon + \epsilon = 2\epsilon$$

as desired.

The second inequality is obtained as above by using the second inequality in Henstock's Lemma 4.45. $\qquad\square$

The term "uniform" is used in Lemma 4.119 because the estimates hold uniformly for any subinterval $J \subset I$.

We show that the step functions are dense in $\mathcal{HK}(I)$ with respect to the Alexiewicz norm.

Theorem 4.120 *The step functions are dense in $\mathcal{HK}(I)$ with respect to the Alexiewicz norm.*

Proof. Let $f \in \mathcal{HK}(I)$ and $\epsilon > 0$. Pick a gauge γ on I such that $|S(f, \mathcal{D}) - \int_I f| < \epsilon$ whenever $\mathcal{D}$ is a γ-fine tagged partition of I. Fix a γ-fine tagged partition $\mathcal{D} = \{(t_i, I_i) : 1 \le i \le m\}$ and set $\varphi(t) = \sum_{i=1}^{m} f(t_i)\chi_{I_i}(t)$. By the Uniform Henstock Lemma 4.119 if $J \subset I$ is an interval, then

$$\left| \int_J f - \int_J \varphi \right| = \left| \sum_{i=1}^{m} \left\{ \int_{J \cap I_i} f - \int_{J \cap I_i} \varphi \right\} \right|$$
$$= \left| \sum_{i=1}^{m} \left\{ \int_{J \cap I_i} f - f(t_i)\ell(I_i \cap J) \right\} \right| \le 3\epsilon.$$

Hence, $\|f - \varphi\| = \sup\left\{ \left| \int_a^x (f - \varphi) \right| : a \le x \le b \right\} \le 3\epsilon$. Since $\epsilon > 0$ is arbitrary, we see that the step functions are dense in $\mathcal{HK}(I)$ in the Alexiewicz norm. $\qquad\square$

Finally, we use the Uniform Henstock Lemma to show that the uniform integrability condition in Theorem 4.73 is sufficient for convergence in the Alexiewicz norm.

Theorem 4.121 *If $\{f_k\}_{k=1}^{\infty}$ is uniformly integrable and $\{f_k\}_{k=1}^{\infty}$ converges pointwise to f, then $\{f_k\}_{k=1}^{\infty}$ converges to f in the Alexiewicz norm.*

Proof. By Theorem 4.73, f is integrable so we may assume that $f = 0$. Let $\epsilon > 0$. Let γ be a gauge such that $|S(f_k, \mathcal{D}) - \int_I f_k| < \epsilon$ for every k when $\mathcal{D}$ is a γ-fine tagged partition. Fix a tagged partition $\mathcal{D} = \{(t_i, I_i) : 1 \le i \le m\}$ which is γ-fine. Let $M = \max\{\ell(I_1), \dots, \ell(I_m)\}$. Choose n such that $k \ge n$ implies $|f_k(t_i)| < \epsilon/mM$ for $i = 1, \dots, m$. Suppose J is an arbitrary subinterval of I and $k \ge n$. From the Uniform

Henstock Lemma 4.119, we have

$$\left|\int_J f_k\right| \le \sum_{i=1}^{m}\left|\int_{J\cap I_i} f_k - f_k(t_i)l(I_i\cap J)\right| + \sum_{i=1}^{m}|f_k(t_i)|\,l(I_i\cap J) < 6\epsilon+\epsilon = 7\epsilon.$$

Hence, if $k \ge n$, then $\|f_k\| \le 7\epsilon$ so that $\{f_k\}_{k=1}^{\infty}$ converges to 0 in the Alexiewicz norm. □

Using this theorem, it now follows that the convergence conclusions in the versions of the Monotone Convergence Theorem and Dominated Convergence Theorem given in Theorems 4.76 and 4.79 can be improved to convergence in the Alexiewicz norm.

4.12 Henstock-Kurzweil integrals on $\mathbb{R}^n$

We conclude this chapter by extending the Henstock-Kurzweil integral to functions defined on n-dimensional Euclidean space. Since many of the higher dimensional results follow from proofs analogous to their one-dimensional versions, our presentation will be brief. We begin by laying the groundwork necessary to define the integral.

We define an *interval* I in $(\mathbb{R}^*)^n$ to be a product $I = \prod_{j=1}^{n} I_j$, where each I_j is an interval in $\mathbb{R}^*$. We say that I is open (closed) in $(\mathbb{R}^*)^n$ if, and only if, each I_j is open (closed) in $\mathbb{R}^*$. The *volume* of an interval $I \subset (\mathbb{R}^*)^n$ is defined to be $v(I) = \prod_{j=1}^{n} \ell(I_j)$, with the convention $0 \cdot \infty = 0$.

Definition 4.122 A *partition* of a closed interval $I \subset (\mathbb{R}^*)^n$ is a finite collection of closed, nonoverlapping subintervals $\{J_j : j = 1, \ldots, k\}$ of I with $I = \cup_{j=1}^{k} J_j$. A *tagged partition* of I is a finite set of ordered pairs $\mathcal{D} = \{(x_j, J_j) : j = 1, \ldots, k\}$ such that $\{J_j : j = 1, \ldots, k\}$ is a partition of I and $x_j \in J_j$ for all j. The point x_j is called the *tag* associated to the interval J_j.

As in the one-dimensional case, a gauge on $I \subset (\mathbb{R}^*)^n$ associates open intervals to points in I.

Definition 4.123 A *gauge* γ on an interval $I \subset (\mathbb{R}^*)^n$ is a mapping defined on I that associates to each $x \in I$ an open interval J_x containing x. A tagged partition $\mathcal{D} = \{(x_j, J_j) : j = 1, \ldots, k\}$ is called γ-*fine* if $x_j \in J_j \subset \gamma(J_j)$ for $j = 1, \ldots, k$.

If $f : I \subset (\mathbb{R}^*)^n \to \mathbb{R}$ and $\mathcal{D} = \{(x_j, J_j) : j = 1, \ldots, k\}$ is a tagged partition of the interval I, the Riemann sum of f with respect to $\mathcal{D}$ is defined to be

$$S(f, \mathcal{D}) = \sum_{j=1}^{k} f(x_j) v(J_j).$$

We assume, as before, that the function f has value 0 at all infinite points (that is, any point with at least one coordinate equal to ∞) and $0 \cdot \infty = 0$. In order to use these sums to define a multi-dimensional integral, we need to know that every gauge γ has at least one γ-fine tagged partition associated to it.

Theorem 4.124 *Let I be a closed interval in $(\mathbb{R}^*)^n$ and γ be a gauge on I. Then, there is a γ-fine tagged partition of I.*

Proof.   First, suppose that $I = I_1 \times \cdots \times I_n$ is closed and bounded. Assume that there is no γ-fine tagged partition of I. Bisect each I_j and consider all the products of the n bisected intervals. This partitions I into 2^n nonoverlapping closed subintervals. At least one of these subintervals must not have any γ-fine tagged partitions, for if each of the 2^n subintervals had a γ-fine tagged partition, then the union of these partitions would be a γ-fine tagged partition of I. Let J_1 be one of the subintervals without a γ-fine tagged partition. Continuing this bisection procedure produces a decreasing sequence of subintervals $\{J_i\}_{i=1}^{\infty}$ of I such that the diameters of the J_i's approach 0 and no J_i has a γ-fine tagged partition. Let $\{x\} = \cap_{i=1}^{\infty} J_i$. Since the diameters decrease to 0, there is a i_0 such that $J_{i_0} \subset \gamma(x)$, which implies that $\mathcal{D} = \{(x, J_{i_0})\}$ is a γ-fine tagged partition of J_{i_0}. This contradiction shows that I has a γ-fine tagged partition.

Next, suppose that $I \subset (\mathbb{R}^*)^n$ is a closed, unbounded interval. Define $h : \mathbb{R}^* \to \left[-\frac{\pi}{2}, \frac{\pi}{2}\right]$ by

$$h(x) = \begin{cases} -\dfrac{\pi}{2} & \text{if} \quad x = -\infty \\ \arctan x & \text{if} \; -\infty < x < \infty \\ \dfrac{\pi}{2} & \text{if} \quad x = \infty \end{cases}$$

and $\vec{h} : (\mathbb{R}^*)^n \to \left[-\frac{\pi}{2}, \frac{\pi}{2}\right]^n$ by $\vec{h}(x) = \vec{h}(x_1, \ldots, x_n) = (h(x_1), \ldots, h(x_n))$. Note that $\vec{h}$ is one-to-one and let $\vec{g}$ be the inverse function of $\vec{h}$. Then, $\vec{h}$ and $\vec{g}$ map closed intervals onto closed intervals and open intervals onto open intervals. Consequently, $\vec{h}(I)$ is a closed and bounded interval and

$\vec{h} \circ \gamma \circ g$ is a gauge on $\vec{h}(I)$. By the previous case, $\vec{h}(I)$ has an $\vec{h} \circ \gamma \circ g$-fine tagged partition $\mathcal{D} = \{(x_j, J_j) : j = 1, \ldots, k\}$. It then follows that $\vec{g}(\mathcal{D}) = \{(\vec{g}(x_j), \vec{g}(J_j)) : j = 1, \ldots, k\}$ is a γ-fine tagged partition of I. $\square$

We can now define the Henstock-Kurzweil integral for functions defined on intervals in $(\mathbb{R}^*)^n$.

Definition 4.125 Let $f : I \subset (\mathbb{R}^*)^n \to \mathbb{R}$. We call the function f *Henstock-Kurzweil integrable* on I if there is an $A \in \mathbb{R}$ so that for all $\epsilon > 0$ there is a gauge γ on I so that for every γ-fine tagged partition $\mathcal{D}$ of I,

$$|S(f, \mathcal{D}) - A| < \epsilon.$$

The number A is called the *Henstock-Kurzweil integral* of f over I, and we write $A = \int_I f$.

The basic properties of the integral, such as linearity, positivity and additivity, and the Cauchy condition, carry over to subintervals of $(\mathbb{R}^*)^n$ as before; we do not repeat the statements or proofs. In $\mathbb{R}^*$, a tag can be associated to one or two intervals in a tagged partition; each tag in a tagged partition in $(\mathbb{R}^*)^n$ can appear as the tag for up to 2^n different subintervals in the partition.

We first show that the Henstock-Kurzweil integral of the characteristic function of a bounded interval equals its volume.

Example 4.126 Let $I \subset \mathbb{R}^n$ be a bounded interval. Then, $\int_{\mathbb{R}^n} \chi_I = v(I)$. Without loss of generality, assume $n = 2$. Let $I \subset \mathbb{R}^2$ and $\bar{I} = [a, b] \times [c, d]$. Fix $\epsilon > 0$ and choose $\delta > 0$ so that the sum of the areas of the four strips surrounding the boundary of I, $(a - \delta, a + \delta) \times (c - \delta, d + \delta)$, $(b - \delta, b + \delta) \times (c - \delta, d + \delta)$, $(a - \delta, b + \delta) \times (c - \delta, c + \delta)$, and $(a - \delta, b + \delta) \times (d - \delta, d + \delta)$, is less than ϵ. Let S be the union of these four intervals. Define a gauge γ on $\mathbb{R}^2$ so that $\gamma(x) = I^\circ$ for $x \in I^\circ$, $\gamma(x) \subseteq S$ for $x \in \partial I$, the boundary of I, and $\gamma(x) \cap \bar{I} = \emptyset$ for $x \notin \bar{I}$. If $\mathcal{D}$ is a γ-fine tagged partition of $\mathbb{R}^2$, then

$$|S(\chi_I, \mathcal{D}) - v(I)| = \left| \sum_{(x,J) \in \mathcal{D}, x \in I^\circ} v(J) + \sum_{(x,J) \in \mathcal{D}, x \in \partial I} \chi_I(x) v(J) - v(I) \right|$$

$$\leq \left| \sum_{(x,J) \in \mathcal{D}, x \in \partial I} v(J) \right|$$

$$\leq v(S) < \epsilon,$$

since $I \setminus \cup_{(x,J) \in \mathcal{D}, x \in I^\circ} J \subset \cup_{(x,J) \in \mathcal{D}, J \subset S} J$.

Since the Henstock-Kurzweil integral is linear, it follows from Example 4.126 that step functions are Henstock-Kurzweil integrable. Further, if $\varphi(x) = \sum_{i=1}^{k} a_i \chi_{I_i}(x)$ is a step function, then

$$\int_{\mathbb{R}^n} \varphi = \int_{\mathbb{R}^n} \sum_{i=1}^{k} a_i \chi_{I_i} = \sum_{i=1}^{k} a_i \int_{\mathbb{R}^n} \chi_{I_i} = \sum_{i=1}^{k} a_i v(I_i).$$

The following example generalizes Example 4.43 to higher dimensions.

Example 4.127 Suppose that $\sum_{k=1}^{\infty} a_k$ is a convergent sequence. Set $J_k = (k, k+1) \times (k, k+1)$ and, for $x \in \mathbb{R}^2$, set $f(x) = \sum_{k=1}^{\infty} a_k \chi_{J_k}(x)$. We claim that f is Henstock-Kurzweil integrable over $\mathbb{R}^2$ and

$$\int_{\mathbb{R}^2} f = \sum_{k=1}^{\infty} a_k.$$

Since the series is convergent, there is a $B > 0$ so that $|a_k| \leq B$ for all $k \in \mathbb{N}$. Let $\epsilon > 0$. Pick a natural number M so that $\left| \sum_{k=j}^{\infty} a_k \right| < \epsilon$ and $|a_j| < \epsilon$ for $j \geq M$. For each $k \in \mathbb{N}$, let O_k be an open interval containing $\overline{J_k}$ such that $v(O_k \setminus J_k) < \min \left\{ \frac{\epsilon}{2^k B}, 1 \right\}$. If $x \notin \cup_{k=1}^{\infty} \overline{J_k} \cup \{(\infty, \infty)\}$, let I_x be an open interval disjoint from $\cup_{k=1}^{\infty} \overline{J_k} \cup \{(\infty, \infty)\}$. Define a gauge γ as follows:

$$\gamma(x) = \begin{cases} O_k & \text{if} & x \in \overline{J_k} \\ I_x & \text{if } x \notin \cup_{k=1}^{\infty} \overline{J_k} \cup \{(\infty, \infty)\} \\ (M, \infty] \times (M, \infty] & \text{if} & x = (\infty, \infty) \end{cases}.$$

Suppose that $\mathcal{D} = \{(t_i, I_i) : i = 1, \ldots, m\}$ is a γ-fine tagged partition of $\mathbb{R}^2$. Without loss of generality, we may assume that $t_m = (\infty, \infty)$ and $I_m = [a, \infty] \times [b, \infty]$, so that $a, b > M$ and $f(t_m) v(I_m) = 0$. Let K be the largest integer less than or equal to $\max \{a, b\}$. Then, $K \geq M$.

Set $\mathcal{D}_j = \{(t_i, I_i) \in \mathcal{D} : t_i \in J_j\}$. Note that $v(J_j) = 1$ and, by the definition of γ, $\cup_{(t_i, I_i) \in \mathcal{D}_j} I_i \subset O_j$. For $j = K$, we have

$$|S(f, \mathcal{D}_K) - a_K| = \left| \sum_{(t_i, I_i) \in \mathcal{D}_K} a_K v(I_i) - a_K \right|$$

$$= \left| a_K \left(\sum_{(t_i, I_i) \in \mathcal{D}_K} v(I_i) - v(J_K) \right) \right|$$

$$\leq |a_K| \, v(O_k \setminus J_k) < \epsilon,$$

while for $1 \leq j < K$,

$$|S(f, \mathcal{D}_j) - a_j| = \left| \sum_{(t_i, I_i) \in \mathcal{D}_j} a_j v(I_i) - a_j \right| = \left| a_j \left(\sum_{(t_i, I_i) \in \mathcal{D}_j} v(I_i) - v(J_j) \right) \right|$$

$$\leq |a_j| \, v(O_j \setminus J_j) < |a_j| \frac{\epsilon}{2^j B} \leq \frac{\epsilon}{2^j}.$$

Therefore,

$$\left| S(f, \mathcal{D}) - \sum_{k=1}^{\infty} a_k \right| = \left| \sum_{k=1}^{\infty} S(f, \mathcal{D}_k) - \sum_{k=1}^{\infty} a_k \right|$$

$$\leq \left| \sum_{k=1}^{K-1} \{ S(f, \mathcal{D}_k) - a_k \} \right| + |S(f, \mathcal{D}_K) - a_K|$$

$$+ \left| \sum_{k=K+1}^{\infty} a_k \right|$$

$$< \sum_{k=1}^{\infty} \frac{\epsilon}{2^k} + \epsilon + \epsilon = 3\epsilon.$$

It follows that f is Henstock-Kurzweil integrable over $\mathbb{R}^2$ with integral equal to $\sum_{k=1}^{\infty} a_k$.

Henstock's Lemma holds for functions defined on intervals in $(\mathbb{R}^*)^n$ and, hence, the Monotone Convergence Theorem, Fatou's Lemma and the Dominated Convergence Theorem are also valid for the Henstock-Kurzweil integral in $(\mathbb{R}^*)^n$. Given the validity of the Dominated Convergence Theorem, Corollary 4.92 extends to $\mathbb{R}^n$ and we have that absolute Henstock-Kurzweil integrability in $\mathbb{R}^n$ is equivalent to Lebesgue integrability, once we know that every Henstock-Kurzweil integrable function on $\mathbb{R}^n$ is measurable. We now prove this latter result, which implies a generalization of Corollary 4.99.

Theorem 4.128 *Suppose that $f : I \subset \mathbb{R}^n \to \mathbb{R}$ is Henstock-Kurzweil integrable. Then, f is measurable.*

Proof. Without loss of generality, we may assume $n = 2$. Consider first the case in which $I = [a_1, b_1] \times [a_2, b_2]$ is a bounded interval. Set

$$I_{j,k}^{(l)} = \left[a_1 + (j-1)(b_1 - a_1) 2^{-l}, a_1 + j(b_1 - a_1) 2^{-l} \right]$$
$$\times \left[a_2 + (k-1)(b_2 - a_2) 2^{-l}, a_2 + k(b_2 - a_2) 2^{-l} \right]$$

and let $E_l = \left\{ I_{j,k}^{(l)} : j, k = 1, \ldots, 2^l \right\}$. Then,

(1) $I = \bigcup_{I_{j,k}^{(l)} \in E_l} I_{j,k}^{(l)}$;

(2) $\left(I_{j,k}^{(l)} \right)^o \cap \left(I_{j',k'}^{(l)} \right)^o = \emptyset$ unless $(j, k) = (j', k')$;

(3) $m_2 \left(I_{j,k}^{(l)} \right) = 2^{-2l} m_2(I)$.

Define $f_l : I \to \mathbb{R}$ by

$$f_l(x) = \sum_{I_{j,k}^{(l)} \in E_l} \left(\frac{1}{m_2 \left(I_{j,k}^{(l)} \right)} \int_{I_{j,k}^{(l)}} f \right) \chi_{\left(I_{j,k}^{(l)} \right)^o}(x).$$

Let $\Delta I = I^o \setminus \left(\bigcup_{l=1}^{\infty} \bigcup_{I_{j,k}^{(l)} \in E_l} \partial I_{j,k}^{(l)} \right)$, where ∂J represents the boundary of the interval J. Since $m_2 \left(\partial I_{j,k}^{(l)} \right) = 0$ (see the comment on page 83 preceding Definition 3.43), it follows that $m_2(\Delta I) = m_2(I^o) = m_2(I)$. Thus, if $\{ f_l(x) \}_{l=1}^{\infty}$ converges to $f(x)$ a.e. in ΔI, then $\{ f_l(x) \}_{l=1}^{\infty}$ converges to $f(x)$ a.e. in I.

We next show that $\{ f_l(x) \}_{l=1}^{\infty}$ converges to $f(x)$ a.e. in ΔI. Let $X = \{ x \in \Delta I : \{ f_l(x) \}_{l=1}^{\infty}$ does not converge to $f(x) \}$. If $x \in X$, then there is a $M \in \mathbb{N}$ and a sequence $\{ l(x) \} \subset \mathbb{N}$ such that $\left| f_{l(x)}(x) - f(x) \right| > \frac{1}{M}$, for all $l(x)$. Let $J_{l(x)}$ be the interval $I_{j,k}^{(l(x))} \in E_{l(x)}$ that contains x in its interior. Then, by the definition of f_m,

$$\left| \int_{J_{l(x)}} f - f(x) m_2 \left(J_{l(x)} \right) \right| > \frac{1}{M} m_2 \left(J_{l(x)} \right). \tag{4.19}$$

Let $X_M = \left\{ x \in \Delta I : \left| f_{l(x)}(x) - f(x) \right| > \frac{1}{M}, \text{ for all } l(x) \right\}$, so that $X = \cup_{M=1}^{\infty} X_M$. It is enough to show that $m_2(X_M) = 0$ for all $M \in \mathbb{N}$ to prove the claim.

Fix $\epsilon > 0$. Since f is Henstock-Kurzweil integrable over I, there is a gauge γ on I such that $\left| \int_I f - S(f, \mathcal{D}) \right| < \epsilon$ for every γ-fine tagged partition $\mathcal{D}$ of I.

Let $\mathcal{V}_M = \left\{ J_{l(x)} : x \in X_M \text{ and } J_{l(x)} \subset \gamma(x) \right\}$ and note that $\mathcal{V}_M$ is a Vitali cover of X_M. By the Vitali Covering Lemma, we can choose a finite set of pairwise disjoint intervals $J_{l(x_1)}, J_{l(x_2)}, \ldots, J_{l(x_K)}$ such that $m_2^* \left(X_M \setminus \cup_{i=1}^K J_{l(x_i)} \right) < \epsilon$. Further, $\mathcal{D}' = \left\{ \left(x_1, J_{l(x_1)} \right), \ldots, \left(x_K, J_{l(x_K)} \right) \right\}$ is a γ-fine partial tagged subpartition of I. Thus, by (4.19) and Henstock's Lemma,

$$m_2^* (X_M) < \sum_{i=1}^K m_2 \left(J_{l(x_i)} \right) + \epsilon$$

$$< M \sum_{i=1}^K \left| \int_{J_{l(x_i)}} f - f(x_i) \, m_2 \left(J_{l(x_i)} \right) \right| + \epsilon$$

$$\leq 2M\epsilon + \epsilon.$$

Since $\epsilon > 0$ is arbitrary, it follows that X_M is measurable with measure 0. Consequently, $m_2(X) = 0$ and $\{ f_l(x) \}_{l=1}^{\infty}$ converges to $f(x)$ a.e. in I. Since every step function is measurable and the pointwise (a.e.) limit of measurable functions is measurable, it follows that f is measurable.

Suppose I is an unbounded interval, and set $I_k = I \cap ([-k, k] \times [-k, k])$. By the n-dimensional analog of Theorem 4.30, f is integrable over I_k and hence measurable on I_k by the first part of the proof. Let $f_k = f \chi_{I_k}$. Since $I \setminus I_k$ is a measurable set, it follows that f_k is measurable on I for all k. Thus, since $\{ f_k \}_{k=1}^{\infty}$ converges pointwise to f on I, f is measurable on I. $\square$

Since a function is absolutely Henstock-Kurzweil integrable if, and only if, it is Lebesgue integrable, this implies that the Fubini and Tonelli Theorems (Theorems 3.111 and 3.112) hold for absolutely Henstock-Kurzweil integrable functions in $\mathbb{R}^n$.

Theorem 4.129 *(Fubini's Theorem) Let* $f : \mathbb{R} \times \mathbb{R} \to \mathbb{R}$ *be absolutely Henstock-Kurzweil integrable. Then:*

(1) f_x *is absolutely Henstock-Kurzweil integrable in* $\mathbb{R}$ *for almost every* $x \in \mathbb{R}$;

(2) the function $x \longmapsto \int_{\mathbb{R}} f_x = \int_{\mathbb{R}} f(x,y) \, dy$ is absolutely Henstock-Kurzweil integrable over $\mathbb{R}$;

(3) the following equality holds:

$$\int_{\mathbb{R} \times \mathbb{R}} f = \int_{\mathbb{R}} \left(\int_{\mathbb{R}} f_x \right) dx = \int_{\mathbb{R}} \int_{\mathbb{R}} f(x,y) \, dy dx.$$

Theorem 4.130   *(Tonelli's Theorem) Let $f : \mathbb{R} \times \mathbb{R} \to \mathbb{R}$ be nonnegative and measurable. Then:*

(1) f_x is measurable on $\mathbb{R}$ for almost every $x \in \mathbb{R}$;

(2) the function $x \longmapsto \int_{\mathbb{R}} f_x = \int_{\mathbb{R}} f(x,y) \, dy$ is measurable on $\mathbb{R}$;

(3) the following equality holds:

$$\int_{\mathbb{R} \times \mathbb{R}} f = \int_{\mathbb{R}} \left(\int_{\mathbb{R}} f_x \right) dx = \int_{\mathbb{R}} \int_{\mathbb{R}} f(x,y) \, dy dx.$$

It should be pointed out that there are versions of the Fubini Theorem for (Henstock-Kurzweil) conditionally integrable functions in $\mathbb{R}^n$, but as the proofs are somewhat long and technical, we do not give them. We refer the reader to [Ma], [McL, 6.1] and [Sw2, 8.13].

4.13   Exercises

Denjoy and Perron integrals

Exercise 4.1   Let $f(x) = |x|$. Find $\overline{D}f(0)$ and $\underline{D}f(0)$.

Exercise 4.2   Suppose that $f : [a,b] \to \mathbb{R}$ is increasing. Prove that $\overline{D}f(x) \geq \underline{D}f(x) \geq 0$ for all $x \in [a,b]$.

A General Fundamental Theorem of Calculus

Exercise 4.3   Define positive functions $\delta_1, \delta_2 : [0,1] \to (0,\infty)$ by $\delta_1(t) = \frac{1}{8}$ for all $t \in [0,1]$ and $\delta_2(0) = \delta_2(1) = \frac{1}{4}$ and $\delta_2(t) = t$ for $0 < t < 1$. Let γ_i be the gauge on $[0,1]$ defined by δ_i, for $i = 1,2$; that is, $\gamma_i(t) = (t - \delta_i(t), t + \delta_i(t))$. Give examples of γ_i-fine tagged partitions of $[0,1]$.

Exercise 4.4   Suppose that γ_1 and γ_2 are gauges on an interval I such that $\gamma_1(t) \subset \gamma_2(t)$ for all $t \in I$. Show that any γ_1-fine tagged partition of I is also γ_2-fine.

Exercise 4.5 Suppose that γ_1 and γ_2 are gauges on an interval I and set $\gamma(t) = \gamma_1(t) \cap \gamma_2(t)$. Show that γ is a gauge on I such that any γ-fine tagged partition of I is also γ_i-fine, for $i = 1, 2$.

Exercise 4.6 Let $I = [a, b]$ and let γ be a gauge on I. Fix $c \in (a, b)$ and set $I_1 = [a, c]$ and $I_2 = [c, b]$. Suppose that $\mathcal{D}_i$ is a γ-fine tagged partition of I_i, for $i = 1, 2$. Show that $\mathcal{D} = \mathcal{D}_1 \cup \mathcal{D}_2$ is a γ-fine tagged partition of I.

Exercise 4.7 Let $f : [a, b] \to \mathbb{R}$ and let $C = \{c_i\}_{i \in \sigma} \subset [a, b]$ be a countable set. Suppose that $f(x) = 0$ except for $x \in C$. Using only the definition, prove that f is Henstock-Kurzweil integrable and $\int_{[a,b]} f = 0$. Note that f may take on a different value at each $c_i \in C$.

Exercise 4.8 Use the following outline to give an alternate proof of Theorem 4.17:

Assume that the theorem is false. Use bisection and Exercise 4.6 to construct intervals $I_0 = I \supset I_1 \supset I_2 \supset \cdots$ such that $\ell(I_k) \leq \ell(I_{k-1})/2$ and no γ-fine tagged partition of I_k exists. Use the fact that $\cap_{k=1}^{\infty} I_k = \{x\}$ to obtain a contradiction.

Exercise 4.9 Suppose that $f, g : I = [a, b] \subset \mathbb{R} \to \mathbb{R}$, g is nonnegative and Henstock-Kurzweil integrable, and $|f(t)| \leq g(t)$ for all $t \in I$. If $\int_I g = 0$, show that f is Henstock-Kurzweil integrable over I and $\int_I f = 0$.

Exercise 4.10 Let $f : I \to \mathbb{R}$. If $|f|$ is Henstock-Kurzweil integrable over I and $\int_I |f| = 0$, show that f is Henstock-Kurzweil integrable over I and $\int_I f = 0$.

Exercise 4.11 Let $a \leq x_0 \leq b$. Show that there is a gauge γ on $[a, b]$ such that if $\mathcal{D}$ is a γ-fine tagged partition of $[a, b]$ and $(t, J) \in \mathcal{D}$ with $x_0 \in J$, then $t = x_0$; that is, x_0 must be the tag for J. Generalize this result to a finite number of points $\{x_1, \ldots, x_n\}$.

Basic properties

Exercise 4.12 Let $f : I \to \mathbb{R}$. Suppose there is a real number A such that for every $\epsilon > 0$, there are Henstock-Kurzweil integrable functions g and h satisfying $g \leq f \leq h$ and $A - \epsilon < \int_I g \leq \int_I h < A + \epsilon$. Prove that f is Henstock-Kurzweil integrable with $\int_I f = A$.

Exercise 4.13 Let $f, g : I \to \mathbb{R}$. Suppose that f is Henstock-Kurzweil integrable over I and g is equal to f except at countably many points in I. Show that g is Henstock-Kurzweil integrable with $\int_I g = \int_I f$.

Exercise 4.14 Suppose that $\int_I |f - g| = 0$. Prove that f is Henstock-Kurzweil integrable over I if, and only if, g is Henstock-Kurzweil integrable over I and $\int_I f = \int_I g$.

Exercise 4.15 Give an example of a pair of Henstock-Kurzweil integrable functions f and g whose product fg is not Henstock-Kurzweil integrable.

Exercise 4.16 This example studies the relationships between the Henstock-Kurzweil integral and translations or dilations. Assume that $f : [a, b] \to \mathbb{R}$ is Henstock-Kurzweil integrable over $[a, b]$.

(1) (Translation) Let $h \in \mathbb{R}$. Define $f_h : [a + h, b + h] \to \mathbb{R}$ by $f_h(t) = f(t - h)$. Show that f_h is Henstock-Kurzweil integrable over $[a + h, b + h]$ with $\int_{a+h}^{b+h} f_h = \int_a^b f$.
(2) (Dilation) Let $c > 0$ and define $g : [a/c, b/c] \to \mathbb{R}$ by $g(t) = f(ct)$. Show that g is Henstock-Kurzweil integrable over $[a/c, b/c]$ with $c \int_{a/c}^{b/c} g(t) \, dt = \int_a^b f(t) \, dt$.

Exercise 4.17 Give an example which shows the importance of the continuity assumption in the Generalized Fundamental Theorem of Calculus, Theorem 4.24.

Exercise 4.18 Complete the induction proof of Theorem 4.31.

Unbounded intervals

Exercise 4.19 Prove that Definitions 4.9 and 4.36 of a gauge are equivalent. That is, given a gauge γ satisfying the Definition 4.36, prove that there is a gauge γ' satisfying Definition 4.9 so that $\gamma'(t) \subset \gamma(t)$. This implies that every γ'-fine tagged partition is also a γ-fine tagged partition.

Exercise 4.20 Suppose $f, g : \mathbb{R} \to \mathbb{R}$. If f is Henstock-Kurzweil integrable over $\mathbb{R}$ and $g = f$ a.e., show that g is Henstock-Kurzweil integrable over $\mathbb{R}$ with $\int_{\mathbb{R}} f = \int_{\mathbb{R}} g$.

Exercise 4.21 Let $f : \mathbb{R} \to \mathbb{R}$. Show that f is Henstock-Kurzweil integrable over $\mathbb{R}$ if, and only if, there is an $A \in \mathbb{R}$ such that for every $\epsilon > 0$ there exist $a, b \in \mathbb{R}$, $a < b$, and a gauge γ on $[a, b]$ such that $|S(f, \mathcal{D}) - A| < \epsilon$ for every γ-fine tagged partition $\mathcal{D}$ of $[a, b]$.

Exercise 4.22 Let $f : \mathbb{R} \to \mathbb{R}$. Show that f is Henstock-Kurzweil integrable over $\mathbb{R}$ if, and only if, there is an $A \in \mathbb{R}$ such that for every $\epsilon > 0$ there exist $r > 0$ and a gauge γ on $\mathbb{R}$ such that if $a \leq -r$ and $b \geq r$, then $|S(f, \mathcal{D}) - A| < \epsilon$ for every γ-fine tagged partition $\mathcal{D}$ of $[a, b]$.

Exercise 4.23 Suppose that $a_k \geq 0$ for all k and $\sum_{k=1}^{\infty} a_k = \infty$. Prove that the function f defined by $f(x) = \sum_{k=1}^{\infty} a_k \chi_{[k,k+1)}(x)$ is not Henstock-Kurzweil integrable over $[1, \infty)$.

Exercise 4.24 Suppose $\{a_k\}_{k=1}^{\infty} \subset \mathbb{R}$ and set $f(x) = \sum_{k=1}^{\infty} a_k \chi_{[k,k+1)}(x)$. Show that if f is Henstock-Kurzweil integrable over $[1, \infty)$, then the series $\sum_{k=1}^{\infty} a_k$ converges. For the converse, see Example 4.43.

Henstock's Lemma

Exercise 4.25 Using the notation of Henstock's Lemma (Lemma 4.45), show that

$$\left| \sum_{i=1}^{k} \left\{ |f(x_i)| \, \ell(J_i) - \left| \int_{J_i} f \right| \right\} \right| \leq 2\epsilon.$$

Exercise 4.26 Suppose that $f : [a, b] \to \mathbb{R}$ is bounded on $[a, b]$ and Henstock-Kurzweil integrable over $[c, b]$ for every $a < c \leq b$. Show that f is Henstock-Kurzweil integrable over $[a, b]$.

Exercise 4.27 Use Example 4.49 to show that the product of Henstock-Kurzweil integrable functions need not be Henstock-Kurzweil integrable.

Exercise 4.28 Recall that a function f has a Cauchy principal value integral over $[a, b]$ if, for some $a < c < b$, f is Henstock-Kurzweil integrable over $[a, c - \epsilon]$ and $[c + \epsilon, b]$ for every (sufficiently small) $\epsilon > 0$, and the limit

$$\lim_{\epsilon \to 0^+} \left(\int_a^{c-\epsilon} f + \int_{c+\epsilon}^b f \right)$$

exists and is finite. Give an example of a function f whose principal value integral over $[a, b]$ exists but such that f is not Henstock-Kurzweil integrable over $[a, b]$.

Exercise 4.29 Suppose that $f : [-\infty, \infty] \to \mathbb{R}$ is Henstock-Kurzweil integrable over $[-\infty, \infty]$. Prove $\int_{-\infty}^{\infty} f = \int_{-\infty}^a f + \int_a^{\infty} f$ for every choice of $a \in \mathbb{R}$.

Exercise 4.30 Let $f : [a, \infty) \to \mathbb{R}$ be differentiable. Give necessary and sufficient conditions for f' to be Henstock-Kurzweil integrable over $[a, \infty)$.

Exercise 4.31 Show that the *Fresnel integral*, $\int_0^{\infty} \sin(x^2) \, dx$, exists in the Henstock-Kurzweil sense. Is the integral absolutely convergent? [Hint: try the substitution $t = x^2$.]

Exercise 4.32   Let $f, g : I \to \mathbb{R}$. Suppose that fg and f are Henstock-Kurzweil integrable over $[a, c]$ for every $a \leq c < b$, g is differentiable and g' is absolutely integrable over $[a, b]$. Set $F(t) = \int_a^t f$ for $a \leq t < b$ and assume that $\lim_{t \to b^-} F(t)$ exists. Prove that fg is Henstock-Kurzweil integrable over $[a, b]$. [Hint: integrate by parts.]

Exercise 4.33   Prove the following limit form of the Comparison Test:
Suppose that $f, g : [a, b] \to \mathbb{R}$ are Henstock-Kurzweil integrable over $[a, c]$ for all $a \leq c < b$, and $f(t) \geq 0$ and $g(t) > 0$ for all $t \in [a, b]$. Assume $\lim_{t \to b^-} \dfrac{f(t)}{g(t)} = L \in \mathbb{R}^*$.

(1) If $L = 0$ and g is Henstock-Kurzweil integrable over $[a, b]$, then f is Henstock-Kurzweil integrable over $[a, b]$.
(2) If $0 < L < \infty$, then g is Henstock-Kurzweil integrable over $[a, b]$ if, and only if, f is Henstock-Kurzweil integrable over $[a, b]$.
(3) If $L = \infty$ and f is Henstock-Kurzweil integrable over $[a, b]$, then g is Henstock-Kurzweil integrable over $[a, b]$.

Exercise 4.34   (Abel's Test) Prove the following result:
Let $f, g : [a, \infty) \to \mathbb{R}$. Suppose that f is continuous on $[a, \infty)$. Assume that $F(t) = \int_a^t f$ is bounded and assume that g is nonnegative, differentiable and decreasing. If either (a) $\lim_{t \to \infty} g(t) = 0$ or (b) $\int_a^\infty f$ exists, then $\int_a^\infty fg$ exists. [Hint: integrate by parts.]

Exercise 4.35   Use Abel's Test in Exercise 4.34 to show that $\int_1^\infty \dfrac{\sin t}{t^p} dt$ exists for $p > 0$. Show that the integral is conditionally convergent for $0 < p \leq 1$. It may help to review Example 4.52.

Exercise 4.36   Suppose $f : [a, \infty) \to \mathbb{R}$ is continuous and $F(t) = \int_a^t f$ is bounded on $[a, \infty)$. Assume $g : [a, \infty) \to \mathbb{R}$ with $\lim_{t \to \infty} g(t) = 0$ and that g' is nonpositive and continuous on $[a, \infty)$. Prove that $\int_a^\infty fg$ exists.

Exercise 4.37   Use Exercise 4.36 to show that $\int_3^\infty \dfrac{\sin t}{\log t} dt$ exists.

Exercise 4.38   Suppose that $f, g : [a, b] \to \mathbb{R}$ are continuous on $(a, b]$ and g' is absolutely integrable over $[a, b]$. Assume $F(t) = \int_t^b f$ is bounded. Show that fg is Henstock-Kurzweil integrable over $[a, b]$ if, and only if, $\lim_{c \to a^+} F(c) g(c)$ exists.

Absolute integrability

Exercise 4.39 Suppose that $\varphi, \psi \in \mathcal{BV}([a,b])$ and $\alpha, \beta \in \mathbb{R}$. Prove that $\alpha\varphi + \beta\psi \in \mathcal{BV}([a,b])$ and

$$Var\,(\alpha\varphi + \beta\psi, [a,b]) \le |\alpha|\,Var\,(\varphi, [a,b]) + |\beta|\,Var\,(\psi, [a,b]).$$

Exercise 4.40 Suppose that $\varphi \in \mathcal{BV}([a,b])$. Prove that $|\varphi| \in \mathcal{BV}([a,b])$. Is the converse true? Either prove or give a counterexample.

Exercise 4.41 Prove that $Var\,(\varphi, [a,b]) = 0$ if, and only if, φ is constant on $[a,b]$.

Exercise 4.42 We say a function $\varphi \in \mathcal{BV}(\mathbb{R})$ if $\varphi \in \mathcal{BV}([-a,a])$ for all $a > 0$ and $Var\,(\varphi, \mathbb{R}) \equiv \lim_{a \to \infty} Var\,(\varphi, [-a,a])$ exists and is finite.

(1) Prove that $\varphi \in \mathcal{BV}(\mathbb{R})$ implies $\varphi \in \mathcal{BV}([a,b])$ for all $[a,b] \subset \mathbb{R}$.
(2) Give an example of a function $\varphi \in \mathcal{BV}([a,b])$ for all $[a,b] \subset \mathbb{R}$ such that $\varphi \notin \mathcal{BV}(\mathbb{R})$.
(3) Prove that if $\varphi, \psi \in \mathcal{BV}(\mathbb{R})$ and $\alpha, \beta \in \mathbb{R}$, then $\alpha\varphi + \beta\psi \in \mathcal{BV}(\mathbb{R})$.

Exercise 4.43 Prove Theorem 4.62 for $I = \mathbb{R}$.

Exercise 4.44 Extend Corollaries 4.63 and 4.64 to $I = \mathbb{R}$.

Exercise 4.45 Suppose that $f : I \to \mathbb{R}$ is absolutely integrable over I and let $c > 0$. Define f_c by

$$f_c(t) = \begin{cases} f(t) & \text{if } |f(t)| \le c \\ 0 & \text{if } |f(t)| > c \end{cases}.$$

Show that f_c is absolutely integrable over I.

Convergence theorems

Exercise 4.46 State and prove a uniform convergence theorem for the Henstock-Kurzweil integral.

Exercise 4.47 Let $f_k : I \to \mathbb{R}$ be Henstock-Kurzweil integrable over I. Show that there is a Henstock-Kurzweil integrable function $g : I \to \mathbb{R}$ such that $|f_k - f_j| \le g$ for all $k, j \in \mathbb{N}$ if, and only if, there are Henstock-Kurzweil integrable functions α and β satisfying $\alpha \le f_k \le \beta$ for all $k \in \mathbb{N}$.

Exercise 4.48 Suppose that $f, g, h : I \to \mathbb{R}$ are Henstock-Kurzweil integrable. If $|f - h| \le g$ and h is conditionally integrable, prove that f is conditionally integrable.

Exercise 4.49   Suppose that $f_k, \beta : I \to \mathbb{R}$ are Henstock-Kurzweil integrable over I and $f_k \leq \beta$ for all k. Prove that $\sup_k f_k$ is Henstock-Kurzweil integrable over I.

Exercise 4.50   Recall the functions $f_k : [0, 1] \to \mathbb{R}$ defined in Example 4.83 by $f_k(0) = k$ and $f_k(t) = 0$ otherwise. Show that $\{f_k\}_{k=1}^{\infty}$ is not uniformly integrable over $[0, 1]$.

Exercise 4.51   Prove Corollary 4.87, the dual to Fatou's Lemma.

Exercise 4.52   Suppose that $f_k : I \to \mathbb{R}$ is Henstock-Kurzweil integrable over I and $\{f_k\}_{k=1}^{\infty}$ converges to f pointwise. Suppose there exists a Henstock-Kurzweil integrable function $g : I \to \mathbb{R}$ such that $|f_k| \leq g$ for all $k \in \mathbb{N}$. Show that the conclusion of the Dominated Convergence Theorem can be improved to include $\int_I |f_k - f| \to 0$.

Exercise 4.53   Let $f : \mathbb{R} \to \mathbb{R}$ and suppose $A \subset B$ and both sets are measurable. Show that if f is absolutely integrable over B then f is absolutely integrable over A. Show that the result fails if we replace "absolutely integrable" with "Henstock-Kurzweil integrable".

Exercise 4.54   Suppose $f, g : I \to [0, \infty)$, f is Henstock-Kurzweil integrable over I and g is Henstock-Kurzweil integrable over every bounded subinterval of I. Show that $f \wedge g$ is Henstock-Kurzweil integrable over I. In particular, for every $k \in \mathbb{N}$, $f \wedge k$ is Henstock-Kurzweil integrable over I. [Hint: Use the Monotone Convergence Theorem.]

Exercise 4.55   Let $f : I \to \mathbb{R}$ be absolutely integrable over I. For $k \in \mathbb{N}$, define f^k, the truncation of f at height k, by

$$f^k(t) = \begin{cases} -k & \text{if } f(t) < -k \\ f(t) & \text{if } |f(t)| \leq k \\ k & \text{if } f(t) > k \end{cases}.$$

Show that f^k is absolutely integrable over I. [Hint: consider $g = f \wedge k$ and $h = (-k) \vee g$.]

Henstock-Kurzweil and Lebesgue integrals

Exercise 4.56   Let $f : I \to \mathbb{R}$. Prove that f is absolutely integrable over I if, and only if, $\chi_E f$ is Henstock-Kurzweil integrable over I for all measurable $E \subset I$.

Exercise 4.57   Suppose $f : I \to \mathbb{R}$ is Henstock-Kurzweil integrable over I. Show f is Lebesgue integrable over I if and only if fg is Henstock-Kurzweil integrable for every bounded, measurable function g.

Characterizations of indefinite integrals

Exercise 4.58   Show that $V = \left\{ \left[x - \frac{1}{n}, x + \frac{1}{n} \right] : x \in [0,1] \cap \mathbb{Q} \text{ and } n \in \mathbb{N} \right\}$ is a Vitali cover of $[0,1]$.

Exercise 4.59   Show that the set of intervals with rational endpoints is a Vitali cover of $\mathbb{R}$.

Exercise 4.60   Let $E = [0,1] \times [0,1]$ and set $\|x\|_\infty = \max \{|x_1|, |x_2|\}$.

(1) If $E_r = [-r,r] \times \left[-r^2, r^2 \right]$, show that

$$V = \{x + E_r : \|x\|_\infty \le 1 \text{ and } 0 < r \le 1\}$$

is a Vitali cover of E.

(2) Fix $\alpha > 0$. If $F_r = [-r,r] \times [-\alpha r, \alpha r]$, show that

$$V = \{x + F_r : \|x\|_\infty \le 1 \text{ and } 0 < r \le 1\}$$

is a Vitali cover of E.

The space of Henstock-Kurzweil integrable functions

Exercise 4.61   Show that the function $\|\cdot\|$ defined in Definition 4.117 is a semi-norm. That is, prove that $\|f + g\| \le \|f\| + \|g\|$ and $\|\lambda f\| = |\lambda| \|f\|$ for all $f, g \in \mathcal{HK}(I)$ and $\lambda \in \mathbb{R}$.

Exercise 4.62   Let $I \subset \mathbb{R}$ and $f \in L^1(I)$. Prove that $\|f\| \le \|f\|_1$. This shows that the imbedding $L^1(I) \hookrightarrow \mathcal{HK}(I)$ is continuous.

Exercise 4.63   Suppose $f : I \to \mathbb{R}$ is Henstock-Kurzweil integrable over I. For $t \in I$, define $\widetilde{f}(t) = \chi_{[a,t]} f$ by $\widetilde{f}(t)(x) = \chi_{[a,t]}(x) f(x)$ for all $x \in I$. Show $\widetilde{f} : I \to HK(I)$ is continuous with respect to the Alexiewicz norm.

Exercise 4.64   Show $\mathcal{HK}(I)$ is separable.

Exercise 4.65   Let $I = [a,b]$ and define $\|f\|'$ by

$$\|f\|' = \sup \left\{ \left| \int_J f \right| : J \subset I \text{ is a closed subinterval} \right\}.$$

Prove that $\|f\|'$ is a semi-norm and $\|f\| \le \|f\|' \le 2 \|f\|$.

Exercise 4.66   Let $I \subset \mathbb{R}^*$ be a closed, unbounded interval. Let $\mathcal{HK}(I)$ be the vector space of Henstock-Kurzweil integrable functions on I. Prove that $\|f\|'$, defined by

$$\|f\|' = \sup \left\{ \left| \int_J f \right| : J \subset I \text{ is a closed subinterval} \right\}$$

defines a semi-norm on $\mathcal{HK}(I)$ such that

$$\|f\|' \leq \|f\|_1$$

for all $f \in L^1(I)$.

Henstock-Kurzweil integrals on $\mathbb{R}^n$

Exercise 4.67 Let $I = [0, 1] \times [0, 1]$ and $x = (x_1, x_2) \in I$. Show there is a gauge γ on I such that if $\mathcal{D}$ is a γ-fine tagged partition of I, $(z, J) \in \mathcal{D}$ and $x \in J$, then $z = x$. In other words, x must be the tag for any subinterval from $\mathcal{D}$ that contains x.

Exercise 4.68 Write the multiple integral in Example 4.127 as an iterated integral.

Chapter 5

Absolute integrability and the McShane integral

Imagine the following change in the definition of the Henstock-Kurzweil integral. Let γ be a gauge on an interval I and $\mathcal{D}$ be a γ-fine tagged partition of I. Suppose we drop the requirement that if $(t, J) \in \mathcal{D}$, then $t \in J$; in other words, suppose we allow the tag to lie outside of J. Thus, we still require that $\{J : (t, J) \in \mathcal{D}\}$ be a partition of I and that $J \subset \gamma(t)$, but now require only that $t \in I$. This is exactly what E. J. McShane (1904-1989) did (see [McS1] and [McS2]) and we next study the integral that bears his name.

Clearly, every γ-fine tagged partition of I will satisfy this new definition, but so might some other sets $\mathcal{D}$. Thus, every McShane integrable function is also Henstock-Kurzweil integrable. Further, there are Henstock-Kurzweil integrable functions which are not McShane integrable. This is a consequence of the fact that the McShane integral is an absolute integral; every McShane integrable function is absolutely integrable. This result is in sharp contrast to the Henstock-Kurzweil integral, which is a conditional integral. However, we have seen that absolutely Henstock-Kurzweil integrable functions are Lebesgue integrable and we will conclude this chapter by proving the equivalence of Lebesgue and McShane integrability.

We will use the word "free" to denote that the tag need not be an element of its associated interval. Thus, the McShane integral is based on γ-fine free tagged partitions. Not surprisingly, any Henstock-Kurzweil integral proof that does not rely on any geometric constructions will carry over to prove a corresponding McShane integral result.

5.1 Definitions

Let $I \subset \mathbb{R}^*$ be a closed interval (possibly unbounded) and let $f : I \to \mathbb{R}$. We shall always assume that f is extended to all of $\mathbb{R}^*$ by defining it to be 0 off of I and that $f(\infty) = f(-\infty) = 0$.

Definition 5.1 Let $I \subset \mathbb{R}^*$ be a closed interval. A *free tagged partition* is a finite set of ordered pairs $\mathcal{D} = \{(t_i, I_i) : i = 1, \ldots, m\}$ such that I_i is a closed subinterval of I, $\cup_{i=1}^m I_i = I$, the intervals have disjoint interiors, and $t_i \in I$. The point t_i is called the *tag* associated to the interval I_i.

The Riemann sum of a function $f : I \to \mathbb{R}$ and a free tagged partition $\mathcal{D}$ is defined to be

$$S(f, \mathcal{D}) = \sum_{i=1}^m f(t_i) \ell(I_i).$$

Definition 5.2 Let $\mathcal{D} = \{(t_i, I_i) : i = 1, \ldots, m\}$ be a free tagged partition of I and γ be a gauge on I. We say that $\mathcal{D}$ is γ-*fine* if $I_i \subset \gamma(t_i)$ for all i. We denote this by writing $\mathcal{D}$ is a γ-*fine free tagged partition* of I.

For a tagged partition, the requirement that the tag lie in the associated interval meant that a number could be a tag for at most two intervals. This is no longer the case in a free tagged partition; in fact, a single number could be a tag for *every* interval.

Example 5.3 Consider the gauge defined for the Dirichlet function $f : [0, 1] \to \mathbb{R}$ in Example 4.10 with $c = 1$. If τ is an irrational number in $[0, 1]$, then $[0, 1] \subset \gamma(\tau)$. Let $\{I_i\}_{i=1}^m$ be a partition of $[0, 1]$. Then, $\mathcal{D} = \{(\tau, I_i) : i = 1, \ldots, m\}$ is a γ-fine free tagged partition of $[0, 1]$. Note that $S(f, \mathcal{D}) = 0$ is a good approximation of the expected McShane integral of f.

We saw in Theorem 4.17 that if γ is a gauge on an interval I, then there is a γ-fine tagged partition $\mathcal{D}$ of I. Since every tagged partition is a free tagged partition, there are γ-fine free tagged partitions associated to every gauge γ and interval I.

Definition 5.4 Let $f : I \subset \mathbb{R}^* \to \mathbb{R}$. We call the function f *McShane integrable* over I if there is an $A \in \mathbb{R}$ so that for all $\epsilon > 0$ there is a gauge γ on I so that for every γ-fine free tagged partition $\mathcal{D}$ of I,

$$|S(f, \mathcal{D}) - A| < \epsilon.$$

The number A is called the *McShane integral* of f over I, and we write $A = \int_I f$.

Since we are guaranteed that γ-fine free tagged partitions exist, this definition makes sense. We will use the symbol $\int_I f$ to represent the McShane integral in this chapter.

Several observations are immediate or follow from corresponding results for the Henstock-Kurzweil integral. First, every McShane integrable function is Henstock-Kurzweil integrable and the integrals agree, since every tagged partition is a free tagged partition. Using the proof of Theorem 4.18, one sees that the McShane integral of a function is unique.

It is not hard to prove that the characteristic function of a bounded interval I is McShane integrable with $\int_{\mathbb{R}} \chi_I = \ell(I)$. In fact, if I has endpoints a and b, $a < b$, and $\epsilon > 0$, set $\gamma(t) = (a, b)$ for $t \in (a, b)$, $\gamma(a) = \left(a - \frac{\epsilon}{4}, a + \frac{\epsilon}{4}\right)$, $\gamma(b) = \left(b - \frac{\epsilon}{4}, b + \frac{\epsilon}{4}\right)$, and for $t \notin [a, b]$, let $\gamma(t)$ be an interval disjoint with $[a, b]$. Then, for every γ-fine free tagged partition, $\mathcal{D}$, $|S(f, \mathcal{D}) - (b - a)| < \epsilon$. We leave it to the reader to complete the details. See Exercise 5.2.

In the next example, we consider the analog of Example 4.43 for the McShane integral. Note that, in this case, the series $\sum_{k=1}^{\infty} a_k$ must be absolutely convergent. See the comments following the example for a discussion of the difference between the two examples.

Example 5.5 Suppose that $\sum_{k=1}^{\infty} a_k$ is an absolutely convergent series and set $f(x) = \sum_{k=1}^{\infty} a_k \chi_{[k,k+1)}(x)$ for $x \geq 1$. Then, f is McShane integrable over $[1, \infty)$ and

$$\int_1^{\infty} f = \sum_{k=1}^{\infty} a_k.$$

To prove this result, we use an argument analogous to that in Example 4.43, which we repeat here to allow the reader to more easily identify the differences.

Since the series is absolutely convergent, there is a $B > 0$ so that $|a_k| \leq B$ for all $k \in \mathbb{N}$. Let $\epsilon > 0$. Pick a natural number M so that $\sum_{k=M}^{\infty} |a_k| < \epsilon$. Define a gauge γ as in Example 4.43. For $t \in (k, k+1)$, let $\gamma(t) = (k, k+1)$; for $t = k$, let $\gamma(t) = \left(t - \min\left(\frac{\epsilon}{2^k B}, 1\right) \frac{\epsilon}{2^k B}, t + \min\left(\frac{\epsilon}{2^k B}, 1\right)\right)$; and, let $\gamma(\infty) = (M, \infty]$. Suppose that $\mathcal{D} = \{(t_i, I_i) : i = 1, \ldots, m\}$ is a γ-fine free tagged partition of $[1, \infty)$. Without loss of generality, we may assume that $t_m = \infty$ and $I_m = [b, \infty]$, so that $b > M$ and $f(t_m) \ell(I_m) = 0$. Let K be the largest integer less than or equal to b. Then, $K \geq M$.

Let $\mathcal{D}_{\mathbb{N}} = \{(t_i, I_i) \in \mathcal{D} : t_i \in \mathbb{N}\}$. If $k \in \mathbb{N}$ is a tag, then $k \leq K+1$; if $b \in \gamma(k)$, then an interval to the left of I_m could be tagged by k, and $b \in \gamma(k)$ implies $k \leq K+1$. Not all natural numbers less than or equal to b need to be tags, as was the case for the Henstock-Kurzweil integral, because an integer k between M and b is an element of $\gamma(\infty)$. For $k \in \mathbb{N}$, $\cup\{I_i : (t_i, I_i) \in \mathcal{D}_{\mathbb{N}} \text{ and } t_i = k\} \subset \gamma(k)$. Thus,

$$|S(f, \mathcal{D}_{\mathbb{N}})| = \left| \sum_{k=1}^{K+1} a_k \sum_{(t_i, I_i) \in \mathcal{D}_{\mathbb{N}}; t_i = k} \ell(I_i) \right| \leq \sum_{k=1}^{K+1} |a_k| \sum_{(t_i, I_i) \in \mathcal{D}_{\mathbb{N}}; t_i = k} \ell(I_i)$$

$$\leq \sum_{k=1}^{K+1} |a_k| \ell(\gamma(k)) < \sum_{k=1}^{K+1} |a_k| \frac{\epsilon}{2^{k-1}B} < \sum_{k=1}^{\infty} \frac{\epsilon}{2^{k-1}} = 2\epsilon.$$

Set $\mathcal{D}_k = \{(t_i, I_i) \in \mathcal{D} : t_i \in (k, k+1)\}$. For $1 \leq k < M$, $\cup_{(t_i, I_i) \in \mathcal{D}_k} I_i$ is a finite union of subintervals of $(k, k+1)$. If ℓ_k is the sum of the lengths of these subintervals, then $\ell_k \geq 1 - \frac{\epsilon}{2^k B} - \frac{\epsilon}{2^{k+1} B}$, and

$$S(f, \mathcal{D}_k) = \sum_{(t_i, I_i) \in \mathcal{D}_k} a_k \ell(I_i) = a_k \sum_{(t_i, I_i) \in \mathcal{D}_k} \ell(I_i) = a_k \ell_k.$$

Thus,

$$|S(f, \mathcal{D}_k) - a_k| = |a_k(\ell_k - 1)| \leq B\left(\frac{\epsilon}{2^k B} + \frac{\epsilon}{2^{k+1} B}\right) < \frac{\epsilon}{2^{k-1}}.$$

Note that the arguments for $\mathcal{D}_{\mathbb{N}}$ and $\mathcal{D}_k$, $1 \leq k < M$, are the same as before.

To estimate $|S(f, \mathcal{D}_k) - a_k|$ for $M \leq k \leq K$, we have

$$|S(f, \mathcal{D}_k) - a_k| = \left| \sum_{(t_i, I_i) \in \mathcal{D}_k} a_k \ell(I_i) - a_k \right|$$

$$= |a_k| \left(1 - \sum_{(t_i, I_i) \in \mathcal{D}_k} \ell(I_i)\right) \leq |a_k|,$$

the same estimate obtained for $|S(f, \mathcal{D}_K) - a_K|$ in Example 4.43. One cannot obtain a better estimate for these terms since, for $k \geq M$, $(k, k+1) \subset \gamma(\infty)$. Thus, $\cup_{(t_i, I_i) \in \mathcal{D}_k} I_i$, which is a finite union of subintervals of $(k, k+1)$, could be a set of intervals the sum of whose lengths is small. In fact, one could have $\mathcal{D}_k = \emptyset$, in which case $|S(f, \mathcal{D}_k) - a_k| = |a_k|$.

Finally, let $\mathcal{D}_\infty = \{(\infty, I_i) \in \mathcal{D}\}$. Since $f(\infty) = 0$, $S(f, \mathcal{D}_\infty) = 0$. Combining all these estimates, we have

$$
\left| S(f, \mathcal{D}) - \sum_{k=1}^{\infty} a_k \right| = \left| \sum_{k=1}^{\infty} S(f, \mathcal{D}_k) + S(f, \mathcal{D}_\mathbb{N}) + S(f, \mathcal{D}_\infty) - \sum_{k=1}^{\infty} a_k \right|
$$

$$
\leq \left| \sum_{k=1}^{M-1} \{S(f, \mathcal{D}_k) - a_k\} \right| + \left| \sum_{k=M}^{K} \{S(f, \mathcal{D}_k) - a_k\} \right|
$$

$$
+ |S(f, \mathcal{D}_\mathbb{N})| + \left| \sum_{k=K+1}^{\infty} a_k \right|
$$

$$
< \sum_{k=1}^{\infty} \frac{\epsilon}{2^{k-1}} + \sum_{k=M}^{K} |a_k| + 2\epsilon + \sum_{k=K+1}^{\infty} |a_k|
$$

$$
= \sum_{k=1}^{\infty} \frac{\epsilon}{2^{k-1}} + \sum_{k=M}^{\infty} |a_k| + 2\epsilon < 5\epsilon.
$$

It follows that f is McShane integrable over $[1, \infty)$.

As for the Henstock-Kurzweil integral, the converse of this example holds; that is, if $f(x) = \sum_{k=1}^{\infty} a_k \chi_{[k,k+1)}(x)$ is McShane integrable, then the series $\sum_{k=1}^{\infty} a_k$ is absolutely convergent. See Exercise 5.4.

Examples 4.43 and 5.5 provide an illustrative comparison between the Henstock-Kurzweil and McShane integrals. When estimating $|S(f, \mathcal{D}_k) - a_k|$ for $M \leq k \leq K$, one needs to address the fact that if $(t, I) \in \mathcal{D}$ and $I \subset (k, k+1)$, then the tag associated to I could be ∞. In that case $f(\infty)\ell(I) = 0$ and, further, this term is not a summand in $S(f, \mathcal{D}_k)$, so that $\sum_{(t_i, I_i) \in \mathcal{D}_k} \ell(I_i)$ could be much less than one. For the Henstock-Kurzweil integral, this situation could arise for at most one interval. For the McShane integral, it can happen for arbitrarily many intervals; that is, for the McShane integral, the point at ∞ may be a tag for more than one interval. This leads to the sum $\sum_{k=M}^{K} |a_k|$ in the estimate above, with arbitrarily many terms. Hence, the series must converge absolutely. This is related to the fact that the McShane integral is an absolute integral, so if f is McShane integrable then so is $|f|$. See Theorem 5.11 below.

5.2 Basic properties

In this section, we list some of the fundamental properties satisfied by the McShane integral.

Proposition 5.6   *Let $f, g : I \subset \mathbb{R}^* \to \mathbb{R}$ be McShane integrable over I.*

(1) (Linearity) If $\alpha, \beta \in \mathbb{R}$, then $\alpha f + \beta g$ is McShane integrable and

$$\int_I (\alpha f + \beta g) = \alpha \int_I f + \beta \int_I g.$$

(2) (Positivity) If $f \le g$ on I, then $\int_I f \le \int_I g$.

See Propositions 4.19 and 4.20 for proofs of these results.

Similar to the Riemann and Henstock-Kurzweil integrals, McShane integrability is characterized by a *Cauchy criterion*.

Theorem 5.7   *A function $f : I \to \mathbb{R}$ is McShane integrable over I if, and only if, for every $\epsilon > 0$ there is a gauge γ so that if $\mathcal{D}_1$ and $\mathcal{D}_2$ are two γ-fine free tagged partitions of I, then*

$$|S(f, \mathcal{D}_1) - S(f, \mathcal{D}_2)| < \epsilon.$$

See Theorem 4.29 for a proof of this result.

Using the fact that continuous functions on closed and bounded intervals are uniformly continuous there, one has

Proposition 5.8   *Let I be a closed, bounded subinterval of $\mathbb{R}$. If $f : I \to \mathbb{R}$ is continuous over I, then f is McShane integrable over I.*

See Exercise 5.5.

Using the Cauchy condition, one can prove that if f is McShane integrable over an interval I and J is a closed subinterval of I, then f is McShane integrable over J. The next result now follows.

Corollary 5.9   *Let $-\infty \le a < c < b \le \infty$. Then, f is McShane integrable over $I = [a, b]$ if, and only if, f is McShane integrable over $[a, c]$ and $[c, b]$. Further,*

$$\int_I f = \int_a^c f + \int_c^b f.$$

See Theorems 4.30 and 4.31 for details of the proof. Note that by induction, the result extends to finite partitions of $[a, b]$.

One of the key results for the Henstock-Kurzweil integral is Henstock's Lemma (Lemma 4.45). A *free tagged subpartition* of an interval $I \subset \mathbb{R}^*$ is a finite set of ordered pairs $\mathcal{S} = \{(t_i, J_i) : i = 1, \ldots, k\}$ such that $\{J_i\}_{i=1}^k$ is a subpartition of I and $t_i \in I$. We say that a free tagged subpartition is γ-fine if $I_i \subset \gamma(t_i)$ for all i.

Lemma 5.10 *(Henstock's Lemma) Let $f : I \subset \mathbb{R}^* \to \mathbb{R}$ be McShane integrable over I. For $\epsilon > 0$, let γ be a gauge such that if $\mathcal{D}$ is a γ-fine free tagged partition of I, then*

$$\left| S(f, \mathcal{D}) - \int_I f \right| < \epsilon.$$

Suppose $\mathcal{D}' = \{(x_1, J_1), \ldots, (x_k, J_k)\}$ is a γ-fine free tagged subpartition of I. Then

$$\left| \sum_{i=1}^{k} \left\{ f(x_i) \ell(J_i) - \int_{J_i} f \right\} \right| \leq \epsilon \text{ and } \sum_{i=1}^{k} \left| f(x_i) \ell(J_i) - \int_{J_i} f \right| \leq 2\epsilon.$$

The proof is the same as before.

5.3   Absolute integrability

The previous section documented the similarity between the Henstock-Kurzweil and McShane integrals. We now turn our attention to their fundamental difference. We will prove that every McShane integrable function is *absolutely integrable*

Theorem 5.11 *Let $f : I \to \mathbb{R}$ be McShane integrable over I. Then, $|f|$ is McShane integrable over I and*

$$\left| \int_I f \right| \leq \int_I |f|.$$

To prove this theorem, we will use a couple of preliminary results.

Proposition 5.12 *Let $\mathcal{D} = \{(t_i, I_i) : i = 1, \ldots, m\}$ be a free tagged partition of an interval I and let $\mathcal{J} = \{J_j : j = 1, \ldots, n\}$ be a partition of I. Then,*

$$\mathcal{D}' = \left\{ (t_i, I_i \cap J_j) : i = 1, \ldots, m, j = 1, \ldots, n, I_i^o \cap J_j^o \neq \emptyset \right\}$$

is a free tagged partition of I and $S(f, \mathcal{D}) = S(f, \mathcal{D}')$. Further, if γ is a gauge on I and $\mathcal{D}$ is γ-fine, then $\mathcal{D}'$ is γ-fine.

Proof. Let $\mathcal{F}_i = \left\{ K_{ij} = I_i \cap J_j : j = 1, \ldots, n, I_i^o \cap J_j^o \neq \emptyset \right\}$ for $i = 1, \ldots, m$ and $\mathcal{F} = \cup_{i=1}^{m} \mathcal{F}_i$. Since the intersection of two closed intervals (in $\mathbb{R}^*$) is a closed interval, each $K_{ij} \in \mathcal{F}$ is a closed interval. Consequently,

$$I = \cup_{i=1}^{m} I_i = \cup_{i=1}^{m} \cup_{K_{ij} \in \mathcal{F}_i} K_{ij}$$

decomposes I into a finite set of closed intervals. The intervals are nonoverlapping since

$$K_{ij}^o \cap K_{i'j'}^o = \left(I_i^o \cap J_j^o\right) \cap \left(I_{i'}^o \cap J_{j'}^o\right) = \left(I_i^o \cap I_{i'}^o\right) \cap \left(J_j^o \cap J_{j'}^o\right)$$

which is empty unless $i = i'$ and $j = j'$. Since $t_i \in I$ for all i, $\mathcal{D}'$ is a free tagged partition of I.

To see that $S(f, \mathcal{D}) = S(f, \mathcal{D}')$, note that $\ell(I_i) = \sum_{K_{ij} \in \mathcal{F}_i} \ell(K_{ij})$. Thus,

$$S(f, \mathcal{D}) = \sum_{i=1}^m f(t_i)\ell(I_i) = \sum_{i=1}^m f(t_i) \sum_{K_{ij} \in \mathcal{F}_i} \ell(K_{ij})$$

$$= \sum_{i=1}^m \sum_{K_{ij} \in \mathcal{F}_i} f(t_i)\ell(K_{ij}) = S(f, \mathcal{D}').$$

Finally, if $\mathcal{D}$ is γ-fine, then $(t_i, K_{ij}) \in \mathcal{D}'$ implies $K_{ij} \subset I_i \subset \gamma(t_i)$, so that $\mathcal{D}'$ is a γ-fine free tagged partition. $\qquad\square$

Notice that this result fails for tagged partitions, that is, partitions that are not free. In fact, if $c \in I = [0, 1]$, $\mathcal{D} = \{(c, [0, 1])\}$, and $\mathcal{J} = \{[0, 1/3], [1/3, 2/3], [2/3, 1]\}$, then $\mathcal{D}' = \{(c, [0, 1/3]), (c, [1/3, 2/3]), (c, [2/3, 1])\}$ is a free tagged partition (for any choice of c), but it cannot be a tagged partition because c can be an element of at most two of the intervals.

The proof of the following lemma makes crucial use of free tagged partitions. Thus, it is the first result we see that distinguishes the McShane integral from the Henstock-Kurzweil integral.

Lemma 5.13 *Let* $f : I \to \mathbb{R}$ *be McShane integrable over* I. *Let* $\epsilon > 0$ *and suppose* γ *is a gauge on* I *such that* $\left|S(f, \mathcal{D}) - \int_I f\right| < \epsilon$ *for every* γ-*fine free tagged partition* $\mathcal{D}$ *of* I. *If* $\mathcal{D} = \{(t_i, I_i) : i = 1, \ldots, m\}$ *and* $\mathcal{E} = \{(s_j, J_j) : j = 1, \ldots, n\}$ *are* γ-*fine free tagged partitions of* I, *then*

$$\sum_{i=1}^m \sum_{j=1}^n |f(t_i) - f(s_j)|\,\ell(I_i \cap J_j) < 2\epsilon.$$

Proof. Set $\mathcal{F} = \{K_{ij} = I_i \cap J_j : i = 1, \ldots, m, j = 1, \ldots, n, I_i^o \cap J_j^o \neq \emptyset\}$. Define tags t_{ij} and s_{ij} as follows. If $f(t_i) \geq f(s_j)$, set $t_{ij} = t_i$ and $s_{ij} = s_j$; if $f(t_i) < f(s_j)$, set $t_{ij} = s_j$ and $s_{ij} = t_i$. Thus, by definition, $f(t_{ij}) - f(s_{ij}) = |f(t_i) - f(s_j)|$. Let $\mathcal{D}' = \{(t_{ij}, K_{ij}) : K_{ij} \in \mathcal{F}\}$ and

$\mathcal{E}' = \{(s_{ij}, K_{ij}) : K_{ij} \in \mathcal{F}\}$. By Proposition 5.12, $\mathcal{D}'$ and $\mathcal{E}'$ are γ-fine free tagged partitions if I, so by assumption,

$$|S(f, \mathcal{D}') - S(f, \mathcal{E}')| \le \left| S(f, \mathcal{D}') - \int_I f \right| + \left| \int_I f - S(f, \mathcal{E}') \right| < 2\epsilon.$$

On the other hand,

$$\sum_{i=1}^{m} \sum_{j=1}^{n} |f(t_i) - f(s_j)| \ell(I_i \cap J_j) = \left| \sum_{K_{ij} \in \mathcal{F}} \{f(t_{ij}) - f(s_{ij})\} \ell(K_{ij}) \right|$$

$$= |S(f, \mathcal{D}') - S(f, \mathcal{E}')|,$$

which completes the proof. $\qquad\qquad\qquad\qquad\qquad\qquad\qquad\qquad\square$

In the proof above, we make use of the fact that $\mathcal{D}'$ and $\mathcal{E}'$ are free tagged partitions. The proof does not work if they cannot be free.

We can now prove Theorem 5.11.

Proof. It is enough to show that $|f|$ satisfies the Cauchy condition. Let $\epsilon > 0$ and choose a gauge γ on I such that $\left| S(f, \mathcal{D}) - \int_I f \right| < \dfrac{\epsilon}{2}$ for every γ-fine free tagged partition $\mathcal{D}$. Let $\mathcal{D} = \{(t_i, I_i) : i = 1, \ldots, m\}$ and $\mathcal{E} = \{(s_j, J_j) : j = 1, \ldots, n\}$ be γ-fine free tagged partitions of I. By Lemma 5.13,

$$|S(|f|, \mathcal{D}) - S(|f|, \mathcal{E})| = \left| \sum_{i=1}^{m} \sum_{j=1}^{n} \{|f(t_i)| - |f(s_j)|\} \ell(I_i \cap J_j) \right|$$

$$\le \sum_{i=1}^{m} \sum_{j=1}^{n} |f(t_i) - f(s_j)| \ell(I_i \cap J_j) < \epsilon.$$

The integral inequality follows from part (2) of Proposition 5.6. $\qquad\square$

Due to Theorem 5.11, it is easy to find examples of Henstock-Kurzweil integrable functions that are not McShane integrable; one merely needs a conditionally (Henstock-Kurzweil) integrable function. The function f : $[0, 1] \to \mathbb{R}$ defined by $f(0) = 0$ and $f(x) = 2x \cos \frac{\pi}{x^2} + \frac{2\pi}{x} \sin \frac{\pi}{x^2}$ for $0 < x \le 1$, which was introduced in Example 2.31, is one example of such a function. (See also Examples 4.43, 4.44 and 4.52.)

Since the McShane integral is an absolute integral, it satisfies stronger lattice properties than the Henstock-Kurzweil integral.

Proposition 5.14 *Let $f, g : I \to \mathbb{R}$ be McShane integrable over I. Then, $f \vee g$ and $f \wedge g$ are McShane integrable over I.*

Proof. Since $f \vee g = \frac{1}{2}[f + g + |f - g|]$ and $f \wedge g = \frac{1}{2}[f + g - |f - g|]$, the result follows from linearity and Theorem 5.11. $\square$

Recall that for the Henstock-Kurzweil integral, one needs to assume that both f and g are bounded above by a Henstock-Kurzweil integrable function, or bounded below by one. (See Proposition 4.67.)

5.3.1 Fundamental Theorem of Calculus

The beauty of the Henstock-Kurzweil integral is that it can integrate every derivative. Such a result cannot hold for the McShane integral. The example above, in which $f(x) = 2x \cos \frac{\pi}{x^2} + \frac{2\pi}{x} \sin \frac{\pi}{x^2}$ for $0 < x \leq 1$ and $f(0) = 0$, provides such an example. The function f is a derivative on $[0, 1]$ and hence it is Henstock-Kurzweil integrable. But it is not absolutely integrable, so it cannot be McShane integrable. In other words, not every derivative is McShane integrable. We have the following version of Part I of the Fundamental Theorem of Calculus for the McShane integral.

Theorem 5.15 *(Fundamental Theorem of Calculus: Part I) Suppose that $f : [a, b] \to \mathbb{R}$ is differentiable on $[a, b]$ and assume that f' is McShane integrable over $[a, b]$. Then,*

$$\int_a^b f' = f(b) - f(a).$$

Proof. Since f' is McShane integrable, it is Henstock-Kurzweil integrable and the two integrals are equal. By Theorem 4.16,

$$\int_a^b f' = \mathcal{HK} \int_a^b f' = f(b) - f(a).$$

$\square$

As for the Riemann and Lebesgue integrals, the assumption that f' be McShane integrable is necessary for Part I of the Fundamental Theorem of Calculus. Concerning the differentiation of indefinite integrals, the statement and proof of Theorem 4.94 yield the following result for the McShane integral.

Theorem 5.16 *Let $f : [a, b] \to \mathbb{R}$ be McShane integrable on $[a, b]$ and continuous at $x \in [a, b]$. Then, F, the indefinite integral of f, is differentiable at x and $F'(x) = f(x)$.*

In fact, the McShane integral satisfies the same version of Part II of the Fundamental Theorem of Calculus that is valid for the Henstock-Kurzweil integral, Theorem 4.95.

Theorem 5.17 *(Fundamental Theorem of Calculus: Part II) Suppose that $f : [a, b] \to \mathbb{R}$ is McShane integrable. Then, F is differentiable at almost all $x \in [a, b]$ and $F'(x) = f(x)$.*

We conclude this section by showing that every McShane integrable function can be approximated by step functions in the appropriate norm. While the result follows from previously established relationships between the McShane, Henstock-Kurzweil and Lebesgue integrals, we use a more direct proof to establish the theorem.

Theorem 5.18 *Let $f : I \to \mathbb{R}$ be McShane integrable over I and $\epsilon > 0$. There exists a step function $g : I \to \mathbb{R}$ such that $\int_I |f - g| < \epsilon$.*

Proof. Choose a gauge γ_1 of I such that $\gamma_1(t)$ is a bounded interval for all $t \in I \cap \mathbb{R}$ and $\left| S(f, \mathcal{D}) - \int_I f \right| < \epsilon/3$ for every γ_1-fine free tagged partition $\mathcal{D}$ of I. Let $\mathcal{D} = \{(t_i, I_i) : i = 1, \ldots, m\}$ be γ_1-fine. Define a step function $\varphi : I \to \mathbb{R}$ by $\varphi(t) = \sum_{i=1}^{m} f(t_i) \chi_{I_i}(t)$. Note that by construction, ∞ (or $-\infty$) must be a tag for any unbounded interval and $f(\infty) = f(-\infty) = 0$. Also, φ is McShane integrable by Exercise 5.8.

By linearity and Theorem 5.11, $|f - \varphi|$ is McShane integrable over I, so there is a gauge γ_2 on I such that $\left| S(|f - \varphi|, \mathcal{E}) - \int_I |f - \varphi| \right| < \epsilon/3$ for every γ_2-fine free tagged partition $\mathcal{E}$ of I. Set $\gamma = \gamma_1 \cap \gamma_2$.

For each subinterval I_i (from $\mathcal{D}$), let $\mathcal{E}_i$ be a γ-fine free tagged partition of I_i. Set $\mathcal{E} = \cup_{i=1}^{m} \mathcal{E}_i$, so that $\mathcal{E}$ is a γ-fine free tagged partition of I. Assume that $\mathcal{E} = \{(s_k, J_k) : k = 1, \ldots, n\}$. For each k, $1 \le k \le n$, there is a unique i_k such that $J_k \subset I_{i_k}$. Since $J_k \subset I_{i_k} \subset \gamma_1(t_{i_k})$, the set $\mathcal{F} = \{(t_{i_k}, J_k) : k = 1, \ldots, n\}$ is γ_1-fine. Since $\mathcal{E}$ is also γ_1-fine, by Lemma 5.13 we have

$$\sum_{k=1}^{n} \sum_{j=1}^{n} |f(s_j) - f(t_{i_k})| \ell(J_j \cap J_k) < \frac{2\epsilon}{3}.$$

However $\ell(J_j \cap J_k) = 0$ if $j \ne k$, so that

$$\sum_{k=1}^{n} |f(s_k) - f(t_{i_k})| \ell(J_k) < \frac{2\epsilon}{3}.$$

Since $s_k \in I_{i_k}$ implies that $\varphi(s_k) = f(t_{i_k})$,

$$S(|f - \varphi|, \mathcal{E}) = \sum_{k=1}^{n} |f(s_k) - \varphi(s_k)| \ell(J_k)$$

$$= \sum_{k=1}^{n} |f(s_k) - f(t_{i_k})| \ell(J_k) < \frac{2\epsilon}{3}.$$

Finally, $\mathcal{E}$ is also γ_2-fine, which implies

$$\int_I |f - \varphi| < S(|f - \varphi|, \mathcal{E}) + \frac{\epsilon}{3} < \epsilon,$$

as we wished to show.   □

We now establish a version of Theorem 4.62 for the McShane integral. This theorem and its proof should be contrasted to that earlier result.

Theorem 5.19   *Let $f : [a, \infty) = I \to R$ be McShane integrable over I and set $F(x) = \int_a^x f(t)\, dt$, for $a \leq x < \infty$. Then $Var(F, I) = \int_I |f(t)|\, dt$.*

Proof.   Let $a = x_0 < x_1 < \cdots < x_n = b$ be a partition of I. Then

$$\sum_{j=1}^{n} |F(x_j) - F(x_{j-1})| = \sum_{j=1}^{n} \left| \int_{x_{j-1}}^{x_j} f \right| \leq \sum_{j=1}^{n} \leq \int_I |f|$$

so taking the supremum over all such partitions shows $Var(F, I) \leq \int_I |f|$.

For the reverse inequality, assume first that f is a step function, $f = \sum_{j=1}^{n} \alpha_j \chi_{A_j}$, where $A_j = [a_j, b_j]$ and $\{A_j\}_{j=1}^{n}$ is a finite set of non-overlapping, closed subintervals with $\cup_{j=1}^{n} A_j = J \subset I$. (We may assume J is an interval by adding on terms $0\chi_{[\tilde{a}, \tilde{b}]}$ to f if necessary.) Then

$$\sum_{j=1}^{n} |F(b_j) - F(a_j)| = \sum_{j=1}^{n} \left| \int_{a_j}^{b_j} f \right|$$

$$= \sum_{j=1}^{n} |\alpha_j| \ell(A_j) = \int_J |f| = \int_I |f|.$$

Therefore, $Var(F : I) \geq Var(F : J) \geq \int_I |f|$ so that $Var(F, I) = \int_I |f|$ when f is a step function.

If f is McShane integrable over I, by Theorem 5.18 for every k there is a step function g_k such that $\int_I |f - g_k| < 1/k$. Set $G_k(t) = \int_a^t g_k$. By the first part of the proof, $Var\,(G_k, I) = \int_I |g_k|$. Since we always have the inequality $Var\,(F, I) \le \int_I |f|$, we see that

$$|Var\,(F, I) - Var\,(G_k, I)| \le Var\,(F - G_k, I) \le \int_I |f - g_k| < \frac{1}{k}.$$

Thus, $Var\,(G_k, I) \to Var\,(F, I)$ and since $\int_I |g_k| \to \int_I |f|$, we may conclude that $Var\,(F, I) = \int_I |f\,(t)|\,dt$. □

5.4 Convergence theorems

Since every McShane integrable function is Henstock-Kurzweil integrable, when considering convergence results for the McShane integral we will need to avoid the same problems that arise for the Henstock-Kurzweil integral. Thus, our conditions must eliminate the pathologies demonstrated in Examples 4.69, 4.70, and 4.71. Further, since the McShane integral is an absolute integral, it will satisfy convergence theorems stronger than the ones satisfied by the Henstock-Kurzweil integral.

We begin by introducing uniform integrability for the McShane integral and then proceed as in Section 4.7. Analogous to Definition 4.72, we have

Definition 5.20 Let $f_k : I \to R$ be McShane integrable over I for each $k \in N$. The sequence $\{f_k\}_{k=1}^{\infty}$ is called *uniformly McShane integrable* over I if for every $\epsilon > 0$ there is a gauge γ on I such that $|S\,(f_k, D) - \int_I f_k| < \epsilon$ for all k and all γ-fine free tagged partitions D of I.

We now establish the analogue of Theorem 4.73 for the McShane integral.

Theorem 5.21 Let $f_k : I \to R$. Suppose $\{f_k\}_{k=1}^{\infty}$ is uniformly McShane integrable over I and $\{f_k\}_{k=1}^{\infty}$ converges pointwise to the function f. Then f is McShane integrable over I and $\int_I |f_k - f| \to 0$.

Proof. As in the proof of Theorem 4.73, $\lim_{k \to \infty} \int_I f_k = L$ exists and f is McShane integrable with $\int_I f = L$. We do the remainder of the proof for the case when $I = [a, \infty]$; the other cases use the analogue of Theorem 5.19 for other subintervals.

Put $g_k = f_k - f$ so $g_k \to 0$ pointwise and, since f is McShane integrable, $\{g_k\}_{k=1}^{\infty}$ is uniformly McShane integrable. Let $\epsilon > 0$. There exists a gauge γ on I with $\gamma\,(t)$ bounded for every $t \in \mathbb{R}$ such that $|S\,(g_k, D) - \int_I g_k| < \epsilon$ for all k whenever D is a γ-fine free tagged partition of I. Fix such a partition $\mathcal{D} = \{(t_i, I_i) : 1 \leq i \leq m\}$ and assume that $t_1 = \infty$, $I_1 = [b, \infty]$. Let $\alpha > b$ and let $\{J_j : 1 \leq j \leq n\}$ be a partition of $[a, \alpha]$. Then $\mathcal{E} = \{(t_i, I_i \cap J_j) : 1 \leq i \leq m, \ 1 \leq j \leq n\}$ is γ-fine so Henstock's Lemma 5.10 implies

$$\left| \sum_{i=1}^{m} \sum_{j=1}^{n} \left\{ \int_{I_i \cap J_j} g_k - g_k(t_i) \ell(I_i \cap J_j) \right\} \right| \leq \epsilon.$$

Therefore,

$$\left| \sum_{j=1}^{n} \int_{J_j} g_k \right| \leq \epsilon + \left| \sum_{i=1}^{m} \sum_{j=1}^{n} g_k(t_i) \ell(I_i \cap J_j) \right|$$

$$= \epsilon + \left| \sum_{i=2}^{m} g_k(t_i) \ell(I_i \cap [a, \alpha]) \right| \quad (5.1)$$

$$\leq \epsilon + \max_{2 \leq i \leq m} |g_k(t_i)| \ell([a, b]).$$

Since $g_k \to 0$ pointwise, the last term on the right hand side of (5.1) can be made as small as desired by choosing k large enough. Hence, (5.1) implies that $Var(G_k : [a, \alpha]) \leq 2\epsilon$ for large k. Since $\alpha > b$ is arbitrary, $Var(G_k : I) \leq 2\epsilon$ for large k. By Theorem 5.19, $\int_I |g_k| \leq 2\epsilon$ for large k as desired. $\qquad\square$

Theorem 5.21 should be compared to Theorem 4.73 for the gauge integral. In particular, note that the conclusion is much stronger; indeed, $\int_I |f_k - f| \to 0$ implies $\int_J (f_k - f) \to 0$ uniformly over all subintervals J of I.

We now give an analogue of Theorem 4.75 for the McShane integral. Since the McShane integral is an absolute integral, Theorem 5.21 allows a stronger conclusion than its counterpart for the gauge integral, Theorem 4.73. Consequently, the statement and conclusion are somewhat different.

Theorem 5.22 *Let $f_k : I \to R$ be McShane integrable over I for every k and suppose that $f = \sum_{k=1}^{\infty} f_k$ pointwise on I and $\sum_{k=1}^{\infty} \int_I |f_k| < \infty$. If $s_n = \sum_{k=1}^{n} f_k$, then:*

(1) $\{s_n\}_{n=1}^{\infty}$ is uniformly McShane integrable over I;

(2) f is McShane integrable over I with $\int_I f = \sum_{k=1}^{\infty} \int_I f_k$; and

(3) $\int_I |s_n - f| = \int_I \left| \sum_{k=n+1}^{\infty} f_k \right| \to 0.$

The proof of this result is essentially the same as the proof of Theorem 4.75 so we do not repeat the argument here. However, a few remarks are in order. Since the McShane integral is an absolute integral, we do not need to assume that the $\{f_k\}$ are nonnegative as in Theorem 4.75, but we need to replace the assumption $\sum_{k=1}^{\infty} \int_I f_k < \infty$ with $\sum_{k=1}^{\infty} \int_I |f_k| < \infty$. The proof of part (1) of Theorem 5.22 then proceeds as the proof of part (1) of Theorem 4.75 except that the absolute integrability of f_k is used in the estimation of the term T_2. Conclusions (2) and (3) of Theorem 5.22 now follow from part (1) and Theorem 5.21. Since

$$\left| \sum_{k=1}^{n} \int_I f_k - \int_I f \right| = \left| \sum_{k=n+1}^{\infty} \int_I f_k \right| \leq \sum_{k=n+1}^{\infty} \int_I |f_k|,$$

the conclusion in Theorem 5.22 (3) is stronger than the conclusion in Theorem 4.75 (3).

As in Theorem 4.76 the Monotone Convergence Theorem now follows readily from Theorem 5.22.

Theorem 5.23   *(Monotone Convergence Theorem) Let $f_k : I \to R$ be McShane integrable over I and suppose that $f_k(t) \uparrow f(t) \in R$ for every $t \in I$. If $\sup_k \int_I f_k < \infty$, then:*

(1) $\{f_k\}_{k=1}^{\infty}$ is uniformly McShane integrable over I;
(2) f is McShane integrable over I; and
(3) $\lim_{k \to \infty} \int_I f_k = \int_I f = \int_I \left(\lim_{k \to \infty} f_k \right).$

Proof.   As before, set $f_0 = 0$ and for $k \geq 1$ let $g_k = f_k - f_{k-1}$. Then, $g_k \geq 0$ for all $k \geq 2$, $\sum_{k=1}^{n} g_k = f_n \to f$ pointwise, and

$$\sum_{k=1}^{\infty} \int_I g_k = \lim_n \sum_{k=1}^{n} \int_I (f_k - f_{k-1}) = \lim_n \int_I f_n = \sup_n \int_I f_n < \infty.$$

Hence, Theorem 5.22 is applicable and completes the proof. $\square$

The Dominated Convergence Theorem for the McShane integral can be derived exactly as in Chapter 4 for the gauge integral. We can, however, obtain a stronger conclusion in the Dominated Convergence Theorem for the McShane integral. We repeat the essential steps.

Definition 5.24 Let $f_k : I \to R$. The sequence $\{f_k\}_{k=1}^\infty$ is *uniformly McShane Cauchy* over I if for every $\epsilon > 0$ there exist a gauge γ on I and an N such that if $i, j \geq N$, then $|S(f_i, \mathcal{D}) - S(f_j, \mathcal{D})| < \epsilon$ whenever D is a γ-fine free tagged partition of I.

Exactly as in Proposition 4.78, we have

Proposition 5.25 *Let $f_k : I \to R$ be McShane integrable over I. Then $\{f_k\}_{k=1}^\infty$ is uniformly McShane Cauchy over I if and only if $\{f_k\}_{k=1}^\infty$ is uniformly McShane integrable over I and $\lim_{k\to\infty} \int_I f_k$ exists.*

We can now state and prove the Dominated Convergence Theorem for the McShane integral.

Theorem 5.26 *(Dominated Convergence Theorem) Let $f_k, g : I \to R$ be McShane integrable over I with $|f_k| \leq g$ on I. If $f_k \to f$ pointwise on I, then:*

(1) $\{f_k\}_{k=1}^\infty$ is uniformly McShane integrable over I;
(2) f is McShane integrable over I; and
(3) $\int_I |f_k - f| \to 0$.

The proof of part (1) of the Dominated Convergence Theorem proceeds as the proof of Theorem 4.79 (1). Parts (2) and (3) then follow immediately from Theorem 5.21.

A few remarks are in order. First, the domination assumption $|f_k| \leq g$ implies $|f_k - f_j| \leq 2g$ as in Theorem 4.79; however, the domination assumption $|f_k - f_j| \leq g$ in Theorem 4.79 allows for conditionally convergent integrals in the case of the gauge integral, whereas the assumption $|f_k| \leq g$ does not. Next, since $\left|\int_I f_k - \int_I f\right| \leq \int_I |f_k - f|$, the conclusion (3) in Theorem 5.26 implies conclusion (3) in Theorem 4.79.

We next pursue a more general a.e. version for the Monotone and Dominated Convergence Theorems as in Section 4.7 for the Henstock-Kurzweil integral.

Theorem 5.27 *Let $E \subset \mathbb{R}$. Then, E is a null set if, and only if, χ_E is McShane integrable and $\int_\mathbb{R} \chi_E = 0$.*

Proof. Suppose first that E is null and let $\epsilon > 0$. Let $\{G_j\}_{j=1}^\infty$ be a sequence of open intervals covering E and such that $\sum_{j=1}^\infty \ell(G_j) < \frac{\epsilon}{2}$.

Since the characteristic function of an interval is McShane integrable, by Proposition 5.14, $s_n = \chi_{G_1} \vee \cdots \vee \chi_{G_n}$ is McShane integrable. Since $\{s_n\}_{n=1}^{\infty}$ increases monotonically, $h = \lim_n s_n$ exists. Since s_n is a maximum of characteristic functions and $E \subset \cup_{j=1}^{\infty} G_j$, we see that $0 \leq h \leq 1$ and $\chi_E \leq h$. Note that $s_n \leq \sum_{j=1}^{n} \chi_{G_j}$, which implies that

$$\int_{\mathbb{R}} s_n \leq \sum_{j=1}^{n} \ell(G_j) \leq \sum_{j=1}^{\infty} \ell(G_j) < \frac{\epsilon}{2}.$$

By the Monotone Convergence Theorem, h is McShane integrable and $\int_{\mathbb{R}} h < \frac{\epsilon}{2}$.

Now, choose a gauge γ so that if $\mathcal{D}$ is a γ-fine free tagged partition of $\mathbb{R}^*$, then $\left| S(h, \mathcal{D}) - \int_{\mathbb{R}} h \right| < \frac{\epsilon}{2}$. Then, for any γ-fine free tagged partition, $\mathcal{D}$,

$$0 \leq S(\chi_E, \mathcal{D}) \leq S(h, \mathcal{D}) < \int_{\mathbb{R}} h + \frac{\epsilon}{2} < \epsilon.$$

Since $\epsilon > 0$ is arbitrary, χ_E is McShane integrable with $\int_{\mathbb{R}} \chi_E = 0$.

To prove the necessity, we argue as in the proof Theorem 4.42. □

Using the fact that every McShane integrable function is Henstock-Kurzweil integrable, Lemma 4.81 yields the following result.

Lemma 5.28 *Let $f_k : I \subset \mathbb{R}^* \to [0, \infty)$ be McShane integrable over I and suppose that $\{f_k(x)\}_{k=1}^{\infty}$ increases monotonically for each $x \in I$ and $\sup_k \int_I f_k < \infty$. Then, $\lim_{k \to \infty} f_k(x)$ exists and is finite for almost every $x \in I$.*

Suppose that f is McShane integrable and g is equal to f almost everywhere. Then, $E = \{x : f(x) \neq g(x)\}$ is a null set and hence $\int_{\mathbb{R}} \chi_E = 0$. Employing this fact and the Monotone Convergence Theorem allows us to prove the next lemma.

Lemma 5.29 *Let $f : I \subset \mathbb{R}^* \to \mathbb{R}$ be McShane integrable over I and suppose that $g : I \to \mathbb{R}$ is such that $g = f$ a.e. in I. Then, g is McShane integrable over I with*

$$\int_I g = \int_I f.$$

Proof. The function $h = f - g$ equals 0 a.e. in I. By linearity, it suffices to show that h is McShane integrable and $\int_I h = 0$.

Let $E = \{t \in I : h(t) \neq 0\}$. Fix $K \in \mathbb{Z}$, and for $n \in \mathbb{N}$, set

$$h_n = (|h| \wedge n) \chi_{I \cap (K, K+1]}.$$

Then, $h_n \leq n\chi_{E \cap (K, K+1]}$. By the argument in the proof of Theorem 5.27, h_n is McShane integrable over $I \cap (K, K+1]$ with $\int_{I \cap (K, K+1]} h_n = 0$. Since $\{h_n\}_{n=1}^\infty$ increases to $|h|$ pointwise, the Monotone Convergence Theorem implies that $|h|$ is McShane integrable over $I \cap (K, K+1]$ and $\int_{I \cap (K, K+1]} |h| = 0$. It now follows that h is McShane integrable over $I \cap (K, K+1]$ and $\int_{I \cap (K, K+1]} h = 0$. (See Exercise 5.3.) Since $h = \sum_{k \in \mathbb{Z}} h\chi_{I \cap (K, K+1]}$ on I, Theorem 5.22 shows that h is McShane integrable over I with $\int_I h = 0$. $\square$

We now have the necessary tools to prove a more general form of the Monotone Convergence Theorem.

Theorem 5.30 *(Monotone Convergence Theorem) Let $f_k : I \subset \mathbb{R}^* \to [0, \infty)$ and suppose that $\{f_k(x)\}_{k=1}^\infty$ increases monotonically for each $x \in I$. Suppose each f_k is McShane integrable over I and $\sup_k \int_I f_k < \infty$. Then, $\lim_{k \to \infty} f_k(x)$ is finite for almost every $x \in I$ and the function f, defined by*

$$f(x) = \begin{cases} \lim_{k \to \infty} f_k(x) & \text{if the limit is finite} \\ 0 & \text{otherwise} \end{cases}$$

is McShane integrable over I with

$$\int_I f = \lim_{k \to \infty} \int_I f_k.$$

Proof. By Lemma 5.28, $\lim_{k \to \infty} f_k(t)$ is finite a.e. in I. Let E be the null set where the limit equals ∞. Set $f(t) = \lim_{k \to \infty} f_k(t)$ if $t \notin E$ and $f(t) = 0$ if $t \in E$. Set $g_k = f_k \chi_{I \setminus E}$. Then, by Lemma 5.29, g_k is McShane integrable over I with $\int_I g_k = \int_I f_k$ and $\{g_k\}_{k=1}^\infty$ increases to f pointwise (everywhere in I). Thus, by Theorem 5.23, f is McShane integrable over I and

$$\int_I f = \lim_{k \to \infty} \int_I g_k = \lim_{k \to \infty} \int_I f_k.$$

$\square$

Recall that the proofs of Fatou's Lemma and the Dominated Convergence Theorem (Lemma 4.86 and Theorem 4.88) rely on the Monotone Convergence Theorem. Thus, those arguments imply corresponding versions for the McShane integral.

Lemma 5.31 *(Fatou's Lemma) Let $f_k : I \subset \mathbb{R}^* \to [0, \infty)$ be Mc-Shane integrable for all k, and suppose that $\liminf_{k \to \infty} \int_I f_k < \infty$. Then, $\liminf_{k \to \infty} f_k$ is finite almost everywhere in I and the function f defined by*

$$f(x) = \begin{cases} \liminf_{k \to \infty} f_k(x) & \text{if the limit is finite} \\ 0 & \text{otherwise} \end{cases}$$

is McShane integrable over I with

$$\int_I f \leq \liminf_{k \to \infty} \int_I f_k.$$

Theorem 5.32 *(Dominated Convergence Theorem) Let $f_k : I \subset \mathbb{R}^* \to \mathbb{R}$ be McShane integrable over I and suppose that $\{f_k\}_{k=1}^\infty$ converges pointwise almost everywhere on I. Define f by*

$$f(x) = \begin{cases} \lim_{k \to \infty} f_k(x) & \text{if the limit exists and is finite} \\ 0 & \text{otherwise} \end{cases}.$$

Suppose that there is a McShane integrable function $g : I \to \mathbb{R}$ such that $|f_k(x)| \leq g(x)$ for all $k \in \mathbb{N}$ and almost all $x \in I$. Then, f is McShane integrable over I and

$$\int_I f = \int_I \lim_{k \to \infty} f_k = \lim_{k \to \infty} \int_I f_k.$$

Moreover,

$$\lim \int_I |f - f_k| = 0.$$

Extensions of Fatou's Lemma analogous to Corollaries 3.98 and 3.99 hold for the McShane integral. The comparison condition for the Dominated Convergence Theorem ($|f_k(x)| \leq g(x)$) is the same as for the Lebesgue integral (Theorem 3.100), unlike the condition for the Henstock-Kurzweil integral (Theorem 4.88). This is because the Lebesgue and McShane integrals are absolute integrals, while the Henstock-Kurzweil integral is a conditional integral. The absolute integrability is also the reason why the Dominated Convergence Theorem for the McShane integral includes a stronger conclusion, that $\lim \int_I |f - f_k| = 0$, than one obtains for the Henstock-Kurzweil integral.

5.5   The McShane integral as a set function

Let $f : I \subset \mathbb{R}^* \to \mathbb{R}$ be McShane integrable and let $\mathcal{M}_I$ be the set of Lebesgue measurable subsets of I. We say that f is McShane integrable over a set $E \subset I$ if $\chi_E f$ is McShane integrable over I and define $\int_E f = \int_I \chi_E f$. If f is McShane integrable over I, we show that f is McShane integrable over every measurable set in $\mathcal{M}_I$ and that $\int f$ is countably additive. Our main result in this section is the following theorem.

Theorem 5.33   *If $f : I \to \mathbb{R}$ is McShane integrable, then the set function $\int f : \mathcal{M}_I \to \mathbb{R}$ is countably additive and absolutely continuous with respect to Lebesgue measure.*

As an immediate consequence, we see that when f is nonnegative, $\int f$ is a measure on $\mathcal{M}_I$.

Corollary 5.34   *If $f : I \to \mathbb{R}$ is nonnegative and McShane integrable, then the set function $\int f : \mathcal{M}_I \to \mathbb{R}$ is a measure on $\mathcal{M}_I$.*

The proof is a consequence of three results: f is McShane integrable over every Lebesgue measurable subset of I; the indefinite integral of f is countably additive; and, the indefinite integral of f is absolutely continuous.

Lemma 5.35   *Suppose that $f : I \to \mathbb{R}$ is McShane integrable over I. Then, f is McShane integrable over every Lebesgue measurable subset $E \subset I$.*

Proof.   Fix $\epsilon > 0$ and let γ be a gauge such that $\left| S(f, \mathcal{D}) - \int_I f \right| < \epsilon$ for every γ-fine free tagged partition $\mathcal{D}$ of I. Let E be a Lebesgue measurable subset of I. For each $k \in \mathbb{N}$, choose an open set $O_k \supset E$ and a closed set $F_k \subset E$ such that $m(O_k \setminus F_k) < \frac{\epsilon}{k2^k}$. Define a gauge γ' on I by:

$$\gamma'(x) = \begin{cases} \gamma(x) \cap O_k \text{ if } x \in E, k-1 \le |f(x)| < k \\ \gamma(x) \setminus F_k \text{ if } x \notin E, k-1 \le |f(x)| < k \end{cases}.$$

Suppose that $\mathcal{D} = \{(t_i, I_i) : i = 1, \ldots, m\}$ and $\mathcal{E} = \{(s_j, J_j) : j = 1, \ldots, n\}$ are γ'-fine free tagged partitions of E. Then,

$$\mathcal{D}' = \{(t_i, I_i \cap J_j) : i = 1, \ldots, m, j = 1, \ldots, n\}$$

and

$$\mathcal{E}' = \{(s_j, I_i \cap J_j) : i = 1, \ldots, m, j = 1, \ldots, n\}$$

are γ'-fine free tagged partitions, $S(f,\mathcal{D}) = S(f,\mathcal{D}')$ and $S(f,\mathcal{E}) = S(f,\mathcal{E}')$. Note that $\mathcal{D}'$ and $\mathcal{E}'$ use the same subintervals but have different tags. Relabelling to avoid the use of multiple subscripts, we may assume that $\mathcal{D}' = \{(t_l', K_l) : l = 1, \ldots, N\}$ and $\mathcal{E}' = \{(s_l', K_l) : l = 1, \ldots, N\}$. Then,

$$
\begin{aligned}
|S(f\chi_E, \mathcal{D}) - S(f\chi_E, \mathcal{E})| &= |S(f\chi_E, \mathcal{D}') - S(f\chi_E, \mathcal{E}')| \\
&\leq \left| \sum_{t_i' \in E} f(t_l') m(K_l) - \sum_{s_i' \in E} f(s_l') m(K_l) \right| \\
&\leq \left| \sum_{t_i' \in E, s_i' \in E} \left\{ f(t_l') m(K_l) - \int_{K_l} f \right\} \right. \\
&\quad + \left. \sum_{t_i' \in E, s_i' \in E} \left\{ \int_{K_l} f - f(s_l') m(K_l) \right\} \right| \\
&\quad + \left| \sum_{t_i' \in E, s_i' \notin E} f(t_l') m(K_l) \right| \\
&\quad + \left| \sum_{t_i' \notin E, s_i' \in E} f(s_l') m(K_l) \right| \\
&= R_1 + R_2 + R_3.
\end{aligned}
$$

By Henstock's Lemma,

$$
\begin{aligned}
R_1 &\leq \left| \sum_{t_i' \in E, s_i' \in E} \left\{ f(t_l') m(K_l) - \int_{K_l} f \right\} \right| \\
&\quad + \left| \sum_{t_i' \in E, s_i' \in E} \left\{ \int_{K_l} f - f(s_l') m(K_l) \right\} \right| \\
&\leq 2\epsilon.
\end{aligned}
$$

Next, set $\sigma_k = \{l : t_l' \in E, s_l' \notin E, k-1 \leq |f(t_l)| < k\}$. If $l \in \sigma_k$, then $K_l \subset \gamma'(t_l') \subset O_k \cap \gamma(t_l')$ and $K_l \subset \gamma'(s_l') \subset \gamma(s_k') \backslash F_k$, so that $K_l \subset O_k \backslash F_k$. Consequently, $\cup_{l \in \sigma_k} K_l \subset O_k \backslash F_k$ and $m(\cup_{l \in \sigma_k} K_l) \leq m(O_k \backslash F_k) < \frac{\epsilon}{k2^k}$.

Therefore,

$$R_2 \le \sum_{k=1}^{\infty} \sum_{l \in \sigma_k} |f(t'_l)| \, m(K_l) \le \sum_{k=1}^{\infty} \sum_{l \in \sigma_k} km(K_l)$$

$$= \sum_{k=1}^{\infty} km(\cup_{l \in \sigma_k} K_l) \le \sum_{k=1}^{\infty} k \frac{\epsilon}{k2^k} = \epsilon.$$

A similar argument shows that $R_3 \le \epsilon$, so that

$$|S(f\chi_E, \mathcal{D}) - S(f\chi_E, \mathcal{E})| \le R_1 + R_2 + R_3 \le 4\epsilon.$$

Thus, $f\chi_E$ satisfies a Cauchy condition and is McShane integrable. Since E was an arbitrary measurable subset of I, the result follows. $\square$

We show next that the indefinite integral of a McShane integrable function is countably additive.

Lemma 5.36 *If $f : I \to \mathbb{R}$ is McShane integrable, then the set function $\int f : \mathcal{M}_I \to \mathbb{R}$ is countably additive.*

Proof. Let $\{E_j\}_{j=1}^{\infty} \subset \mathcal{M}_I$ be a collection of pairwise disjoint sets and let $E = \cup_{j=1}^{\infty} E_j$. Since $E \in \mathcal{M}_I$, by Lemma 5.35, $|f| \chi_E$ is McShane integrable. Since the sets $\{E_j\}_{j=1}^{\infty}$ are pairwise disjoint, we see that $\sum_{j=1}^{k} f\chi_{E_j} \to f\chi_E$ as $k \to \infty$ and $\left|\sum_{j=1}^{k} f\chi_{E_j}\right| \le |f| \chi_E$. By the Dominated Convergence Theorem for the McShane integral,

$$\int_{\cup_{j=1}^{\infty} E_j} f = \int_E f = \lim_{k \to \infty} \sum_{j=1}^{k} \int_{E_j} f = \sum_{j=1}^{\infty} \int_{E_j} f,$$

which shows that the indefinite integral is countable additive. $\square$

Thus, the indefinite integral of a McShane integrable function f is defined on $\mathcal{M}_I$ and countably additive. When f is nonnegative, this implies that the indefinite integral defines a measure on $\mathcal{M}_I$.

We conclude by showing that the indefinite integral is absolutely continuous both as a point function and as a set function. First, we show the indefinite integral is absolutely continuous as a point function in the sense of Definition 4.113.

Lemma 5.37 *Let $I = [a, b]$, $-\infty < a < b < \infty$, and $f : I \to \mathbb{R}$ be McShane integrable. Then, $F(t) = \int_a^t f$, the indefinite integral of f, is absolutely continuous.*

Proof. Let $\epsilon > 0$. There is a gauge γ on I such that $\left| S\left(f, \mathcal{D}\right) - \int_I f \right| < \epsilon$ for every γ-fine free tagged partition $\mathcal{D}$ of I. Let $\mathcal{D}' = \{(t_i, [a_i, b_i]) : i = 1, \ldots, m\}$ be such a partition and set $M = \max\{|f(t_i)| : i = 1, \ldots, m\} + 1$ and $\delta = \epsilon/M$.

Suppose that $\{[c_j, d_j] : j = 1, \ldots, p\}$ is a collection of nonoverlapping closed subintervals of I such that $\sum_{j=1}^{p} (d_j - c_j) < \delta$. By subdividing these intervals, if necessary, we may assume that for each j, there is an $i \in \{1, \ldots, m\}$ such that $[c_j, d_j] \subset [a_i, b_i]$. For each i, set $\sigma_i = \{j : [c_j, d_j] \subset [a_i, b_i]\}$ and set $\mathcal{E} = \cup_{i=1}^{m} \{(t_i, [c_j, d_j]) : j \in \sigma_i\}$. Then, $\mathcal{E}$ is a γ-fine free partial tagged partition of I with $\sum_{i=1}^{m} \sum_{j \in \sigma_i} (d_j - c_j) < \delta$. By Henstock's Lemma,

$$\left| \sum_{j=1}^{p} \{F(d_j) - F(c_j)\} \right| \leq \left| \sum_{i=1}^{m} \sum_{j \in \sigma_i} \left\{ \int_{c_j}^{d_j} f - f(t_i)(d_j - c_j) \right\} \right|$$
$$+ \left| \sum_{i=1}^{m} \sum_{j \in \sigma_i} f(t_i)(d_j - c_j) \right|$$
$$\leq \epsilon + M\delta = 2\epsilon.$$

Thus, F is absolutely continuous. $\square$

We have shown that the point function $F : I \to \mathbb{R}$ is absolutely continuous. It is also true that the set function $\int f$ satisfies the definition of absolute continuity given in Remark 3.93. This result is an easy consequence of Theorem 5.18.

Theorem 5.38 *Let $f : I \to \mathbb{R}$ be McShane integrable over I and define F by $F(E) = \int_E f$ for $E \in \mathcal{M}_I$. Then, the set function F is absolutely continuous over I with respect to Lebesgue measure.*

Proof. Suppose that f is McShane integrable over I and fix $\epsilon > 0$. By Theorem 5.18, there is a step function s such that $\int_I |f - s| \leq \frac{\epsilon}{2}$. Let $\sum_{k=1}^{j} a_k \chi_{A_k}$ be the canonical representation of s and set $M = \max\{|a_1|, \ldots, |a_j|, 1\}$. Set $\delta = \frac{\epsilon}{2M}$ and suppose that E is a measurable subset of I with $m(E) < \delta$. Then,

$$\left| \int_E s \right| = \left| \sum_{k=1}^{j} a_k m(E \cap A_k) \right| \leq \max\{|a_1|, \ldots, |a_j|\} m(E) < M\delta \leq \frac{\epsilon}{2}.$$

Therefore,

$$\left| \int_E f \right| \leq \int_E |f - s| + \left| \int_E s \right| < \frac{\epsilon}{2} + \frac{\epsilon}{2} = \epsilon,$$

so that F is absolutely continuous with respect to Lebesgue measure. □

5.6   The space of McShane integrable functions

Let $I \subset \mathbb{R}^*$ be an interval and let $M^1(I)$ be the space of all McShane integrable functions on I. We define a semi-norm $\| \ \|_1$ on $M^1(I)$ by $\|f\|_1 = \int_I |f|$, and a corresponding semi-metric d_1 by setting $d_1(f, g) = \|f - g\|_1 = \int_I |f - g|$, for all $f, g \in M^1(I)$. It follows from (the proof of) Lemma 5.29 that $\|f\|_1 = 0$ if, and only if, $f = 0$ a.e. in I, so that $\| \ \|_1$ is not a norm and, consequently, d_1 is not a metric on $M^1(I)$. Identifying functions which are equal almost everywhere makes $\| \ \|_1$ a norm and d_1 a metric on $M^1(I)$. From Theorem 5.18, we have

Theorem 5.39   *The step functions are dense in $M^1(I)$.*

We saw in Sections 3.3.9 and 4.4.11 that the space of Lebesgue integrable functions is complete in the (semi-) metric d_1 (see the Riesz-Fischer Theorem, Theorem 3.130) while the space of Riemann integrable functions and the space of Henstock-Kurzweil integrable functions are not complete, in the appropriate (semi-) metrics. That the space of McShane integrable functions complete is a consequence of the Dominated Convergence Theorem (Theorem 5.32). We now observe that the Riesz-Fischer Theorem holds for the McShane integral.

Theorem 5.40   *(Riesz-Fischer Theorem) Let $I \subset \mathbb{R}^*$ be an interval and let $\{f_k\}_{k=1}^{\infty}$ be a Cauchy sequence in $(M^1(I), d_1)$. Then, there is an $f \in M^1(I)$ such that $\{f_k\}_{k=1}^{\infty}$ converges to f in the metric d_1.*

For a proof of this result, see Theorem 3.130.

5.7   McShane, Henstock-Kurzweil and Lebesgue integrals

Suppose that $f : I \subset \mathbb{R}^* \rightarrow \mathbb{R}$ is McShane integrable over I. Consequently, $|f|$ is McShane integrable so that both f and $|f|$ are Henstock-Kurzweil integrable over I, and f is absolutely (Henstock-Kurzweil) integrable over I. On the other hand, there are Henstock-Kurzweil integrable functions that are not McShane integrable. In Section 4.4.8, we saw that Lebesgue and absolute Henstock-Kurzweil integrability are equivalent. In this section, we prove that in the one-dimensional case McShane integrability is equivalent to absolute Henstock-Kurzweil integrability, and hence that the McShane and Lebesgue integrals are equivalent. We extend these results to higher dimensions in Section 5.5.10.

Since we will be dealing with three integrals in this section, we will identify the type of integral by letters ($\mathcal{M}$, $\mathcal{HK}$, and $\mathcal{L}$) to identify the integral being used; for example, the McShane integral of f will be denoted $\mathcal{M} \int_I f$. The crux of the matter is to prove that absolute Henstock-Kurzweil integrability implies McShane integrability.

In order to prove this result, we will employ major and minor functions, variants of the ones defined in conjunction with the Perron integral in Section 4.4.1. Let $I = [a, b]$ be a finite interval and suppose $f : I \to \mathbb{R}$.

Let γ be a gauge on I. For $a < x \le b$, we can also view γ as a gauge on $[a, x]$. Let $\pi_\gamma ([a, x])$ be the set of all γ-fine tagged partitions of $[a, x]$. Define $m_\gamma, M_\gamma : I \to \mathbb{R}^*$ by

$$m_\gamma (x) = \begin{cases} 0 & \text{if } x = a \\ \inf \{S (f, \mathcal{D}) : \mathcal{D} \in \pi_\gamma ([a, x])\} & \text{if } a < x \le b \end{cases}$$

and

$$M_\gamma (x) = \begin{cases} 0 & \text{if } x = a \\ \sup \{S (f, \mathcal{D}) : \mathcal{D} \in \pi_\gamma ([a, x])\} & \text{if } a < x \le b \end{cases} .$$

It is clear that $m_\gamma (x) \le M_\gamma (x)$ for all $x \in [a, b]$. The function M_γ is called a *major function* for f; m_γ is called a *minor function* for f.

By Exercise 4.19, we may assume that the gauge γ is defined by a positive function $\delta : I \to (0, \infty)$; that is, $\gamma (x) = (x - \delta (x), x + \delta (x))$, for all $x \in [a, b]$. We summarize our results for m_γ and M_γ in the following lemma.

Lemma 5.41 *Suppose that $f : I = [a, b] \to \mathbb{R}$ and γ is a gauge on I defined by $\delta : I \to (0, \infty)$.*

(1) *If $x - \delta (x) < u \le x \le v < x + \delta (x)$, then $M_\gamma (v) - M_\gamma (u) \ge f (x) (v - u)$.*

(2) *If $x - \delta (x) < u \le x \le v < x + \delta (x)$, then $m_\gamma (v) - m_\gamma (u) \le f (x) (v - u)$.*

(3) *$M_\gamma - m_\gamma$ is a nonnegative and increasing function on I.*

(4) If $f \geq 0$, then both M_γ and m_γ are nonnegative and increasing functions on I.

(5) Let f be Henstock-Kurzweil integrable over I and $\epsilon > 0$. Suppose that γ is a gauge on I (defined by δ) such that

$$\left| S\left(f, \mathcal{D}\right) - \mathcal{H}\mathcal{K} \int_a^b f \right| < \epsilon$$

for every $\mathcal{D} \in \pi_\gamma\left([a, b]\right)$. Then, $0 \leq M_\gamma\left(b\right) - m_\gamma\left(b\right) \leq 2\epsilon$.

Proof. To prove (1), fix u and v and let $\mathcal{D} \in \pi_\gamma\left([a, u]\right)$. Then, $\mathcal{D} \cup \left\{\left(x, [u, v]\right)\right\} \in \pi_\gamma\left([a, v]\right)$, so that

$$M_\gamma\left(v\right) \geq S\left(f, \mathcal{D} \cup \left\{\left(x, [u, v]\right)\right\}\right) = S\left(f, \mathcal{D}\right) + f\left(x\right)\left(v - u\right).$$

Taking the supremum over all $\mathcal{D} \in \pi_\gamma\left([a, u]\right)$ shows that $M_\gamma\left(v\right) \geq M_\gamma\left(u\right) + f\left(x\right)\left(v - u\right)$, which proves (1). The proof of (2) is similar. See Exercise 5.24.

For (3), fix $\epsilon > 0$ and $a \leq u < v \leq b$. By definition, we can find $\mathcal{D}, \mathcal{D}' \in \pi_\gamma\left([a, u]\right)$ such that

$$M_\gamma\left(u\right) - m_\gamma\left(u\right) \leq S\left(f, \mathcal{D}\right) - S\left(f, \mathcal{D}'\right) + \epsilon.$$

Fix $\mathcal{F} \in \pi_\gamma\left([u, v]\right)$, so that $\mathcal{E} = \mathcal{D} \cup \mathcal{F}, \mathcal{E}' = \mathcal{D}' \cup \mathcal{F} \in \pi_\gamma\left([a, v]\right)$. Thus,

$$M_\gamma\left(u\right) - m_\gamma\left(u\right) \leq S\left(f, \mathcal{D}\right) - S\left(f, \mathcal{D}'\right) + \epsilon$$
$$= S\left(f, \mathcal{E}\right) - S\left(f, \mathcal{E}'\right) + \epsilon \leq M_\gamma\left(v\right) - m_\gamma\left(v\right) + \epsilon,$$

so that $M_\gamma\left(u\right) - m_\gamma\left(u\right) \leq M_\gamma\left(v\right) - m_\gamma\left(v\right)$ and $M_\gamma - m_\gamma$ is increasing. Since it is clearly nonnegative, (3) is proved.

Part (4) follows from the fact that the nonnegativity of f implies that if $u < v$ then

$$S\left(f, \mathcal{D}\right) \leq S\left(f, \mathcal{D}\right) + f\left(x\right)\left(v - u\right) = S\left(f, \mathcal{D} \cup \left\{\left(x, [u, v]\right)\right\}\right)$$

for every $\mathcal{D} \in \pi_\gamma\left([a, u]\right)$. To prove (5), note that the hypothesis implies

$$\left| S\left(f, \mathcal{D}\right) - S\left(f, \mathcal{E}\right) \right| < 2\epsilon$$

for $\mathcal{D}, \mathcal{E} \in \pi_\gamma\left([a, b]\right)$. The result now follows from the definitions of M_γ and m_γ. $\square$

Before considering the equivalence of McShane and absolute Henstock-Kurzweil integrability, we collect a few other results.

Lemma 5.42 *Let $f : I \subset \mathbb{R}^* \to \mathbb{R}$. Suppose that, for every $\epsilon > 0$, there are McShane integrable functions g_1 and g_2 such that $g_1 \leq f \leq g_2$ on I and $\mathcal{M} \int_I g_2 \leq \mathcal{M} \int_I g_1 + \epsilon$. Then, f is McShane integrable on I.*

Proof. Let $\epsilon > 0$ and choose corresponding McShane integrable functions g_1 and g_2. There are gauges γ_1 and γ_2 on I so that if $\mathcal{D}$ is a γ_i-fine free tagged partition of I, then $|S(g_i, \mathcal{D}) - \mathcal{M} \int_I g_i| < \epsilon$ for $i = 1, 2$. Set $\gamma(z) = \gamma_1(z) \cap \gamma_2(z)$. Let $\mathcal{D}$ be a γ-fine free tagged partition of I. Then,

$$\mathcal{M} \int_I g_1 - \epsilon < S(g_1, \mathcal{D}) \leq S(f, \mathcal{D}) \leq S(g_2, \mathcal{D}) < \mathcal{M} \int_I g_2 + \epsilon < \mathcal{M} \int_I g_1 + 2\epsilon.$$

Therefore, if $\mathcal{D}_1$ and $\mathcal{D}_2$ are γ-fine free tagged partitions of I then

$$S(f, \mathcal{D}_1), S(f, \mathcal{D}_2) \in \left(\mathcal{M} \int_I g_1 - \epsilon, \mathcal{M} \int_I g_1 + 2\epsilon. \right).$$

This implies that

$$|S(f, \mathcal{D}_1) - S(f, \mathcal{D}_2)| < 3\epsilon.$$

By the Cauchy criterion, f is McShane integrable. □

This result is an analog of Lemma 4.32 on Henstock-Kurzweil integration; the proofs are the same.

As a consequence of this lemma, we show that increasing functions are McShane integrable.

Example 5.43 Let $f : I = [a, b] \to \mathbb{R}$ be increasing. Divide $[a, b]$ into j equal subintervals by setting $x_k = a + \frac{k}{j}(b - a)$, for $k = 0, 1, \ldots, j$, and $I_k = [x_{k-1}, x_k]$, for $k = 1, \ldots, j$. Set $g_1(t) = \sum_{k=1}^{j} f(x_{k-1}) \chi_{I_k}(t)$ and $g_2(t) = \sum_{k=1}^{j} f(x_k) \chi_{I_k}(t)$. Then, g_1 and g_2 are step functions and, hence, McShane integrable. Since $\ell(I_k) = \frac{b-a}{j}$,

$$0 \leq \mathcal{M} \int_I g_2 - \mathcal{M} \int_I g_1$$

$$= \sum_{k=1}^{j} f(x_k) \frac{b-a}{j} - \sum_{k=1}^{j} f(x_{k-1}) \frac{b-a}{j} = \{f(b) - f(a)\} \frac{b-a}{j}.$$

Given $\epsilon > 0$, we can make $\mathcal{M} \int_I g_2 - \mathcal{M} \int_I g_1 < \epsilon$ by choosing j sufficiently large. By Lemma 5.42, f is McShane integrable.

We are now ready to prove the equivalence of McShane and absolute Henstock-Kurzweil integrability.

Theorem 5.44 *Let $f : I = [a, b] \to \mathbb{R}$. Then, f is McShane integrable over I if, and only if, f is absolutely Henstock-Kurzweil integrable over I.*

Proof.   We have already observed that McShane integrability implies absolute Henstock-Kurzweil integrability. For the converse, by considering f^+ and f^-, it is enough to show the result when f is nonnegative and Henstock-Kurzweil integrable.

Fix $\epsilon > 0$ and choose a gauge γ on I such that $\left| S\left(f, \mathcal{D}\right) - \mathcal{HK} \int_a^b f \right| < \epsilon$ whenever $\mathcal{D}$ is a γ-fine tagged partition of $[a, b]$. Let δ correspond to γ. Extend f to $[a, b+1]$ by setting $f(t) = 0$ for $b < t \le b+1$, and extend m_γ and M_γ to $[a, b+1]$ by defining $m_\gamma(t) = m_\gamma(b)$ and $M_\gamma(t) = M_\gamma(b)$ for $b < t \le b+1$.

Define functions H_n and h_n by $H_n(t) = n\left(M_\gamma\left(t + \frac{1}{n}\right) - M_\gamma(t)\right)$ and $h_n(t) = n\left(m_\gamma\left(t + \frac{1}{n}\right) - m_\gamma(t)\right)$. By Lemma 5.41 (4), M_γ and m_γ are increasing so that Example 5.43 implies H_n and h_n are nonnegative and McShane integrable. Set $H = \liminf_{n \to \infty} H_n$ and $h = \limsup_{n \to \infty} h_n$.

By a linear change of variable (Exercise 5.6), observe that

$$
0 \le \mathcal{M} \int_a^b H_n = \mathcal{M} \int_a^b n\left(M_\gamma\left(t + \frac{1}{n}\right) - M_\gamma(t)\right) dt
$$

$$
= n\left(\mathcal{M} \int_b^{b+1/n} M_\gamma - \mathcal{M} \int_a^{a+1/n} M_\gamma\right)
$$

$$
\le n\left(\mathcal{M} \int_b^{b+1/n} M_\gamma\right) = M_\gamma(b).
$$

Thus, $\liminf_{n \to \infty} \mathcal{M} \int_a^b H_n < \infty$ so that $\liminf_{n \to \infty} H_n$ is finite almost everywhere and, by Fatou's Lemma (Lemma 5.31), there is a real-valued function $\overline{H}$ which is equal to H a.e. and such that $\mathcal{M} \int_a^b \overline{H} \le M_\gamma(b)$. If $E_1 = \{t \in [a, b] : H(t) \ne \overline{H}(t)\}$, then E_1 is null and $\overline{H} = 0$ on E_1.

Fix $t \in [a, b]$ and suppose that $n > \frac{1}{\delta(t)}$. Then, by Lemma 5.41 (1) and (2),

$$
f(t) \le n\left(M_\gamma\left(t + \frac{1}{n}\right) - M_\gamma(t)\right) = H_n(t)
$$

and

$$
f(t) \ge n\left(m_\gamma\left(t + \frac{1}{n}\right) - m_\gamma(t)\right) = h_n(t).
$$

Consequently, $h(t) \leq f(t) \leq H(t)$ for all $t \in [a, b]$. Since $H_n(t) \geq f(t) \geq h_n(t)$ for large n,

$$\liminf_{n \to \infty} \left(\overline{H} - h_n\right)^+(t) = \overline{H}(t) - \limsup_{n \to \infty} h_n(t) = \overline{H}(t) - h(t)$$

for almost every $t \in [a, b]$. Arguing as above shows that $0 \leq M \int_a^b h_n \leq m_\gamma(b)$, so that

$$M \int_a^b \left(\overline{H} - h_n\right)^+ \leq M \int_a^b \overline{H} + M \int_a^b h_n \leq M_\gamma(b) + m_\gamma(b).$$

By Fatou's Lemma applied to $\left(\overline{H} - h_n\right)^+$, there is a real-valued function $\overline{F}$ which is equal to $\overline{H} - h$ a.e. and is McShane integrable. Note that the function $\overline{h} = \overline{H} - \overline{F}$ is McShane integrable and equal to h a.e.. Let $E_2 = \{t \in [a, b] : (\overline{H} - h)(t) \neq \overline{F}(t)\}$; then E_2 is null and $\overline{F} = 0$ on E_2.

Let $E = E_1 \cup E_2$ and redefine $\overline{H}$ and $\overline{h}$ to be 0 on E. Since this only changes the functions on a set of measure 0, by Lemma 5.29, these new functions are McShane integrable with the same integral as before. Define $\overline{f}$ by

$$\overline{f}(x) = \begin{cases} f(x) & \text{if } x \notin E \\ 0 & \text{if } x \in E \end{cases},$$

so that $\overline{f} = f$ almost everywhere and $\overline{h} \leq \overline{f} \leq \overline{H}$ on $[a, b]$.

We claim that

$$M \int_a^b \left(\overline{H} - \overline{h}\right) \leq 2\epsilon.$$

In fact, by Lemma 5.41 (3), $M_\gamma - m_\gamma$ is increasing, so

$$H_n(t) - h_n(t) = n\left(\left\{M_\gamma\left(t + \frac{1}{n}\right) - m_\gamma\left(t + \frac{1}{n}\right)\right\} - \{M_\gamma(t) - m_\gamma(t)\}\right)$$
$$\geq 0.$$

Thus,

$$
\begin{aligned}
0 \le \mathcal{M} \int_a^b (H_n - h_n) \\
= \mathcal{M} \int_a^b n \left(M_\gamma \left(t + \frac{1}{n} \right) - M_\gamma (t) - m_\gamma \left(t + \frac{1}{n} \right) + m_\gamma (t) \right) dt \\
= n \left(\mathcal{M} \int_b^{b+1/n} M_\gamma - \mathcal{M} \int_a^{a+1/n} M_\gamma - \mathcal{M} \int_b^{b+1/n} m_\gamma + \mathcal{M} \int_a^{a+1/n} m_\gamma \right) \\
\le n \left(\mathcal{M} \int_b^{b+1/n} M_\gamma - \mathcal{M} \int_b^{b+1/n} m_\gamma \right) = M_\gamma (b) - m_\gamma (b) \le 2\epsilon
\end{aligned}
$$

by Lemma 5.41 (5). Now, for almost every $t \in [a, b]$,

$$
\begin{aligned}
\overline{F} (t) = \liminf_{n \to \infty} \left(\overline{H} - h_n \right) (t) = \overline{H} (t) + \liminf_{n \to \infty} \left(-h_n (t) \right) \\
= \liminf_{n \to \infty} H_n (t) + \liminf_{n \to \infty} \left(-h_n (t) \right) \le \liminf_{n \to \infty} \left(H_n - h_n \right) (t) .
\end{aligned}
$$

Define $\mathfrak{H}$ by

$$
\mathfrak{H} (x) = \begin{cases} \liminf_{n \to \infty} (H_n - h_n) (x) & \text{if } \liminf_{n \to \infty} (H_n - h_n) (x) \text{ is finite} \\ 0 & \text{otherwise} \end{cases},
$$

so that $\overline{F} \le \mathfrak{H}$ almost everywhere and $\mathfrak{H}$ is finite everywhere. By Fatou's Lemma,

$$
\mathcal{M} \int_a^b \overline{F} \le \mathcal{M} \int_a^b \mathfrak{H} \le \liminf_{n \to \infty} \mathcal{M} \int_a^b (H_n - h_n) \le 2\epsilon,
$$

as we wished to show.

It now follows from Lemma 5.42 that $\overline{f}$ is McShane integrable over I. Since $f = \overline{f}$ a.e., Lemma 5.29 shows that f is McShane integrable over I and, of course, once f is McShane integrable over I, the McShane and Henstock-Kurzweil integrals are equal. $\square$

By Theorem 5.44 and Corollary 4.92, it follows that McShane and Lebesgue integrability are equivalent. We conclude this section by giving a direct proof of this result, which uses arguments more like those found in the Lebesgue theory.

Theorem 5.45 *Let $f : [a, b] \to \mathbb{R}$. Then, f is McShane integrable if, and only if, f is Lebesgue integrable. The value of the two integrals are the same.*

Proof. Assume first that f is Lebesgue integrable over $[a, b]$. Without loss of generality, we may assume that f is nonnegative. Let $\epsilon > 0$ and by absolute continuity (see Remark 3.93) choose $\delta > 0$ so that $\mathcal{L} \int_A f < \epsilon$ whenever $A \subset [a, b]$ is measurable and $m(A) < \delta$. Set $\Delta = \min(\epsilon, \delta)$.

Let $\alpha = \min\left\{1, \frac{\epsilon}{\delta + b - a}\right\}$. Set $E_k = \{t \in [a, b] : (k-1)\alpha \le f(t) < k\alpha\}$ for $k \in \mathbb{N}$. Then, $E_k \cap E_j = \emptyset$ if $k \ne j$ and $[a, b] = \cup_{k-1}^\infty E_k$. For each k, choose an open set G_k such that $E_k \subset G_k$ and $m(G_k \setminus E_k) < \frac{\Delta}{2^k k}$. Define a gauge γ on $[a, b]$ as follows. If $t \in E_k$, then choose an open interval $\gamma(t) \subset G_k$ that contains t.

Suppose that $\mathcal{D} = \{(t_i, I_i) : i = 1, \dots, l\}$ is a γ-fine free tagged partition of $[a, b]$. We will show that

$$\left| S(f, \mathcal{D}) - \mathcal{L} \int_a^b f \right| < 3\epsilon,$$

which implies that f is McShane integrable with integral equal to $\mathcal{L} \int_a^b f$.

For $i = 1, \dots, l$, choose k_i so that $t_i \in E_{k_i}$. Then,

$$\left| S(f, \mathcal{D}) - \mathcal{L} \int_a^b f \right| \le \sum_{i=1}^l \mathcal{L} \int_{I_i} |f(t_i) - f(t)| \, dt$$

$$\le \sum_{i=1}^l \mathcal{L} \int_{I_i \cap E_{k_i}} |f(t_i) - f(t)| \, dt$$

$$+ \sum_{i=1}^l \mathcal{L} \int_{I_i \setminus E_{k_i}} f(t_i) + \sum_{i=1}^l \mathcal{L} \int_{I_i \setminus E_{k_i}} f(t) \, dt$$

$$= R_1 + R_2 + R_3.$$

If $t_i, t \in E_{k_i}$, then both $f(t_i)$ and $f(t)$ belong to the interval $[(k_i - 1)\alpha, k_i \alpha)$, so that $|f(t_i) - f(t)| < \alpha$. Thus,

$$R_1 \le \sum_{i=1}^l \alpha m(I_i \cap E_{k_i}) \le \alpha(b - a) < \epsilon.$$

To estimate R_2, since $t_i \in E_{k_i}$ and $I_i \subset \gamma(t_i) \subset G_{k_i}$, we have

$$R_2 = \sum_{k=1}^\infty \sum_{i : k_i = k} \mathcal{L} \int_{I_i \setminus E_{k_i}} f(t_i) \le \sum_{k=1}^\infty \sum_{i : k_i = k} k \alpha m(I_i \setminus E_{k_i})$$

$$\le \sum_{k=1}^\infty k \alpha m(G_k \setminus E_k) \le \sum_{k=1}^\infty k \alpha \frac{\Delta}{2^k k} < \sum_{k=1}^\infty \frac{\epsilon}{2^k} = \epsilon.$$

Finally, let $A = \cup_{i=1}^{l} (I_i \setminus E_{k_i})$. Since $I_i \subset G_{k_i}$,

$$m(A) = \sum_{i=1}^{l} m(I_i \setminus E_{k_i}) = \sum_{k=1}^{\infty} \sum_{i:k_i=k} m(I_i \setminus E_{k_i})$$

$$\leq \sum_{k=1}^{\infty} m(G_k \setminus E_k) \leq \sum_{k=1}^{\infty} \frac{\Delta}{2^k k} \leq \sum_{k=1}^{\infty} \frac{\delta}{2^k} < \delta.$$

Thus, by the choice of δ, $R_3 \leq \mathcal{L} \int_A f < \epsilon$. Combining all these estimates shows that $\left| S(f,\mathcal{D}) - \mathcal{L} \int_a^b f \right| < 3\epsilon$, proving that f is McShane integrable and $\mathcal{M} \int_a^b f = \mathcal{L} \int_a^b f$.

For the remainder of the proof, assume that f is McShane integrable over $[a,b]$ and let $F(t) = \mathcal{M} \int_a^t f$. By Theorem 4.115, it is enough to show that F is absolutely continuous on $[a,b]$ to conclude that f is Lebesgue integrable there. Fix $\epsilon > 0$ and let γ be a gauge on $[a,b]$ such that $\left| S(f,\mathcal{D}) - \mathcal{M} \int_a^b f \right| < \epsilon$ for every γ-fine free tagged partition $\mathcal{D}$ of $[a,b]$. Let $\mathcal{D}_0 = \{(t_i, I_i) : i = 1, \ldots, l\}$ be a γ-fine free tagged partition of $[a,b]$, let $M = \max\{|f(t_i)|, t = 1, \ldots, l\}$, and set $\eta = \frac{\epsilon}{M+1}$.

Suppose that $\{[y_j, z_j] : j = 1, \ldots, k\}$ is a finite collection of nonoverlapping subintervals of $[a,b]$ such that

$$\sum_{j=1}^{k} (z_j - y_j) < \eta.$$

Replacing $[y_j, z_j]$ by the nondegenerate intervals in $\{[y_j, z_j] \cap I_i\}_{i=1}^{l}$, we may assume that for each j there is an i so that $[y_j, z_j] \subset I_i$. Set $\mathcal{D}_i = \{(t_i, [y_j, z_j]) : [y_j, z_j] \subset I_i\}$, for $i = 1, \ldots, l$. Then, $\mathcal{D} = \cup_{i=1}^{l} \mathcal{D}_i$ is a γ-fine free tagged subpartition of $[a,b]$. Since

$$\left| \sum_{j=1}^{k} \int_{y_j}^{z_j} f - S(f,\mathcal{D}) \right| = \left| \sum_{j=1}^{k} \int_{y_j}^{z_j} f - \sum_{i=1}^{l} \sum_{[y_j,z_j] \subset I_i} f(t_i)(z_j - y_j) \right|$$

$$= \left| \sum_{i=1}^{l} \sum_{[y_j,z_j] \subset I_i} \left\{ \int_{y_j}^{z_j} f - f(t_i)(z_j - y_j) \right\} \right| \leq \epsilon,$$

by Henstock's Lemma,

$$\left| \sum_{j=1}^{k} (F(z_j) - F(y_j)) \right| = \left| \sum_{j=1}^{k} \int_{y_j}^{z_j} f \right|$$

$$= \left| \sum_{j=1}^{k} \int_{y_j}^{z_j} f - S(f, \mathcal{D}) \right| + |S(f, \mathcal{D})|$$

$$\leq \epsilon + M \sum_{j=1}^{k} (z_j - y_j)$$

$$< 2\epsilon.$$

Thus, F is absolutely continuous with respect to Lebesgue measure and f is Lebesgue integrable. $\qquad\square$

Remark 5.46 *Theorems 5.44 and 5.45 are valid for unbounded intervals $I \subset \mathbb{R}$. See Exercises 5.25 and 5.26.*

5.8 McShane integrals on $\mathbb{R}^n$

The McShane integral can be extended to functions defined on intervals in $(\mathbb{R}^*)^n$ in the same manner as the Henstock-Kurzweil integral. If f is defined on an interval $I \subset (\mathbb{R}^*)^n$, we assume that f vanishes at all infinite points and extend the definition of f to all of $(\mathbb{R}^*)^n$ by setting f equal to 0 off of I. (See Sections 4.4.4 and 4.4.12). In fact, the only change needed to define the McShane integral over I is to extend the definition of a free tagged partition (Definition 5.1) to the interval I in the obvious way.

Definition 5.47 Let I be a closed subinterval of $(\mathbb{R}^*)^n$ and $f : I \to \mathbb{R}$. We call the function f *McShane integrable* over I if there is an $A \in \mathbb{R}$ so that for all $\epsilon > 0$ there is a gauge γ on I so that for every γ-fine free tagged partition $\mathcal{D}$ of I,

$$|S(f, \mathcal{D}) - A| < \epsilon.$$

Since every gauge γ has at least one corresponding γ-fine tagged partition, and hence a γ-fine free tagged partition, this definition makes sense. The number A, called the *McShane integral* of f over I and denoted by $A = \int_I f$, is unique. The proof of this statement is the same as before.

Recall that every McShane integrable function is Henstock-Kurzweil integrable. Since the value of the McShane integral is unique, it must equal the Henstock-Kurzweil integral. Thus, the basic properties of the McShane integral, such as linearity, positivity and the Cauchy criterion, carry over to this setting without further proof. By Example 4.126, the characteristic function of a brick is McShane integrable; by linearity, step functions are McShane integrable. Again, the McShane integral is an absolute integral in this setting. Finally, the Monotone Convergence Theorem, Dominated Convergence Theorem and Fatou's Lemma hold for the McShane integral in $\mathbb{R}^n$.

5.9   Fubini and Tonelli Theorems

One of the main points of interest in the study of multiple integrals concerns the equality of multiple and iterated integrals. In Chapter 3, we gave conditions for the equality of these integrals for the Lebesgue integral in the Fubini and Tonelli Theorems (Theorems 3.111 and 3.112). We now establish versions of these two results for the McShane integral. These results are used later to establish the connection between the Lebesgue and McShane integrals on $\mathbb{R}^n$. In proving the Fubini Theorem for the Lebesgue integral, we used Mikusinski's characterization of the Lebesgue integral. Since we do not have such a characterization for the McShane integral, our method of proof will be quite different and more in line with the usual proofs of the Fubini and Tonelli theorems for the Lebesgue integral. (See [Ro, 12.4].)

For simplicity, we consider the case $n = 2$. We will use the notation for sections and iterated integrals that was employed in Section 3.3.8. In particular, it is enough for a function to be defined almost everywhere.

We begin with a lemma which establishes the connection between Lebesgue measure and the McShane integral.

Lemma 5.48   *Suppose that $E \subset \mathbb{R}^2$ is measurable with $m_2(E) < \infty$. Then, $m_2(E) = \int_{\mathbb{R}^2} \chi_E$.*

Proof.   First, assume that E is a brick in $\mathbb{R}^2$. The gauge defined in Example 4.126 for the Henstock-Kurzweil integral also proves that χ_E is McShane integrable and $\int_{\mathbb{R}^2} \chi_E = v(E) = m_2(E)$.

Next, assume that E is open. Then, by Lemma 3.44, E is a union of a countable collection of pairwise disjoint bricks, $\{B_i\}_{i \in \sigma}$. Since m_2 is countably additive, the Monotone Convergence Theorem implies

$$m_2(E) = \sum_{i \in \sigma} m_2(B_i) = \sum_{i \in \sigma} \int_{\mathbb{R}^2} \chi_{B_i} = \int_{\mathbb{R}^2} \sum_{i \in \sigma} \chi_{B_i} = \int_{\mathbb{R}^2} \chi_E,$$

so the result holds for open sets.

Now assume that E is a $\mathcal{G}_\delta$ set. Then, $E = \cap_{i=1}^\infty G_i$ with G_i open, $m_2(G_i) < \infty$, and $G_i \subset G_{i+1}$. By Proposition 3.34, the Monotone Convergence Theorem, and the previous result, we have

$$m_2(E) = \lim_{i \to \infty} m_2(G_i) = \lim_{i \to \infty} \int_{\mathbb{R}^2} \chi_{G_i} = \int_{\mathbb{R}^2} \lim_{i \to \infty} \chi_{G_i} = \int_{\mathbb{R}^2} \chi_E.$$

We proved in Theorem 5.27 that if $E \subset \mathbb{R}$ is a null set, then $\int_\mathbb{R} \chi_E = 0$. The same proofs works for subsets of $\mathbb{R}^n$, so the conclusion holds for null sets in $\mathbb{R}^2$.

Finally, assume that E is measurable and $m_2(E) < \infty$. Then, $E = G \setminus B$, where G is a $\mathcal{G}_\delta$ set, B is a null set, and $B \subset G$. This follows from Theorem 3.36, which is valid in higher dimensions, by setting $B = G \setminus E$, which is a null set. From the previous results, we have

$$m_2(E) = m_2(G) = \int_{\mathbb{R}^2} \chi_G = \int_{\mathbb{R}^2} \chi_G - \int_{\mathbb{R}^2} \chi_B = \int_{\mathbb{R}^2} (\chi_G - \chi_B) = \int_{\mathbb{R}^2} \chi_E.$$

This completes the proof of the lemma. $\qquad \square$

From the equivalence of the Lebesgue and McShane integrals in $\mathbb{R}$ (Theorem 5.45) and Theorem 3.114, we derive

Lemma 5.49 *Let $E \subset \mathbb{R}^2$ be measurable with $m_2(E) < \infty$. Then:*

(1) for almost every $x \in \mathbb{R}$, the sections E_x are measurable;
(2) the function $x \longmapsto m(E_x)$ is McShane integrable over $\mathbb{R}$;
(3) $m_2(E) = \int_\mathbb{R} m(E_x)\, dx$.

We now have the machinery in place to establish a Fubini Theorem for the McShane integral.

Theorem 5.50 *(Fubini's Theorem) Let $f : \mathbb{R} \times \mathbb{R} \to \mathbb{R}$ be McShane integrable. Then:*

(1) f_x is McShane integrable in $\mathbb{R}$ for almost every $x \in \mathbb{R}$;
(2) the function $x \longmapsto \int_\mathbb{R} f_x = \int_\mathbb{R} f(x, y)\, dy$ is McShane integrable over $\mathbb{R}$;

(3) the following equality holds:

$$\int_{\mathbb{R}\times\mathbb{R}} f = \int_{\mathbb{R}} \left(\int_{\mathbb{R}} f_x\right) dx = \int_{\mathbb{R}} \int_{\mathbb{R}} f(x,y)\, dy dx.$$

Proof. First, assume that f is a simple function with $f(x) = \sum_{i=1}^{k} a_i \chi_{A_i}(x)$, where the A_i's are measurable, pairwise disjoint, and $m_2(A_i) < \infty$. From Lemmas 5.48 and 5.49, (1) and (2) hold and

$$\int_{\mathbb{R}^2} f = \sum_{i=1}^{k} a_i \int_{\mathbb{R}^2} \chi_{A_i} = \sum_{i=1}^{k} a_i \int_{\mathbb{R}} m\left((A_i)_x\right) dx = \int_{\mathbb{R}} \sum_{i=1}^{k} a_i m\left((A_i)_x\right) dx$$

$$= \int_{\mathbb{R}} \sum_{i=1}^{k} a_i \int_{\mathbb{R}} \chi_{(A_i)_x}(y)\, dy dx = \int_{\mathbb{R}} \int_{\mathbb{R}} f(x,y)\, dy dx.$$

Next, assume that f is nonnegative and McShane integrable. By Theorem 3.62, there is a sequence of nonnegative, simple functions $\{f_k\}_{k=1}^{\infty}$ which increases pointwise to f. By Exercise 5.28, each f_k is McShane integrable, and from the Monotone Convergence Theorem, $\int_{\mathbb{R}^2} f = \lim_{k\to\infty} \int_{\mathbb{R}^2} f_k$. Since $\{(f_k)_x\}_{k=1}^{\infty}$ increases to f_x for every $x \in \mathbb{R}$, the Monotone Convergence Theorem implies that $\left\{\int_{\mathbb{R}} f_k(x,y)\, dy\right\}_{k=1}^{\infty}$ increases to $\int_{\mathbb{R}} f(x,y)\, dy$ for almost every x. To see that $\int_{\mathbb{R}} f(x,y)\, dy$ is finite for almost every x, note that by the Monotone Convergence Theorem,

$$\int_{\mathbb{R}} \left(\lim_{k\to\infty} \int_{\mathbb{R}} f_k(x,y)\, dy\right) dx = \lim_{k\to\infty} \int_{\mathbb{R}} \left(\int_{\mathbb{R}} f_k(x,y)\, dy\right) dx$$

$$= \lim_{k\to\infty} \int_{\mathbb{R}^2} f_k = \int_{\mathbb{R}^2} f < \infty.$$

Thus, $\int_{\mathbb{R}} f(x,y)\, dy = \lim_{k\to\infty} \int_{\mathbb{R}} f_k(x,y)\, dy$ is finite for almost every x. Consequently, from our previous work and two applications of the Monotone Convergence Theorem, we obtain

$$\int_{\mathbb{R}} \int_{\mathbb{R}} f(x,y)\, dy dx = \lim_{k\to\infty} \int_{\mathbb{R}} \int_{\mathbb{R}} f_k(x,y)\, dy dx = \lim_{k\to\infty} \int_{\mathbb{R}^2} f_k = \int_{\mathbb{R}^2} f.$$

Finally, assume that f is McShane integrable. Then, f is also Henstock-Kurzweil integrable so f is measurable by Theorem 4.128. Further, f is absolutely Henstock-Kurzweil integrable, so $f = f^+ - f^-$ with both f^+ and f^- measurable and McShane integrable. The result now follows from the case just proved. □

As was the case with the Lebesgue integral, we can use the Fubini Theorem to obtain a criterion for integrability from the existence of iterated integrals. This result is contained in the Tonelli Theorem.

Theorem 5.51 *(Tonelli's Theorem) Let $f : \mathbb{R} \times \mathbb{R} \to \mathbb{R}$ be nonnegative and measurable. If $\int_{\mathbb{R}} \int_{\mathbb{R}} f(x, y)\, dy dx$ exists and is finite, then f is McShane integrable and*

$$\int_{\mathbb{R} \times \mathbb{R}} f = \int_{\mathbb{R}} \left(\int_{\mathbb{R}} f_x \right) dx = \int_{\mathbb{R}} \int_{\mathbb{R}} f(x, y)\, dy dx.$$

The assumption in Tonelli's Theorem is that the iterated integral exists and is finite, from which one can conclude that the double integral is finite. Of course, the roles of x and y can be interchanged.

Proof. Define f_k by $f_k(x, y) = (f(x, y) \wedge k) \chi_{[-k,k] \times [-k,k]}(x, y)$. Then, each f_k is bounded, measurable and non-zero on a set of finite measure. By Exercise 5.27, each f_k is McShane integrable. From Theorem 5.50 and the Monotone Convergence Theorem, we have that $\left\{ \int_{\mathbb{R}} f_k(x, y)\, dy \right\}_{k=1}^{\infty}$ increases to $\int_{\mathbb{R}} f(x, y)\, dy$ for almost every x. By a second application of these two results, we have

$$\int_{\mathbb{R}} \int_{\mathbb{R}} f(x, y)\, dy dx = \lim_{k \to \infty} \int_{\mathbb{R}} \int_{\mathbb{R}} f_k(x, y)\, dy dx = \lim_{k \to \infty} \int_{\mathbb{R}^2} f_k = \int_{\mathbb{R}^2} f. \qquad \square$$

5.10 McShane, Henstock-Kurzweil and Lebesgue integrals in $\mathbb{R}^n$

In Section 5.5.7, we showed that in $\mathbb{R}$ the McShane and Lebesgue integrals are equivalent and that a function is Lebesgue (McShane) integrable if, and only if, it absolutely Henstock-Kurzweil integrable. In this section we extend these results to $\mathbb{R}^n$.

Theorem 5.52 *Let $f : \mathbb{R}^n \to \mathbb{R}$. Then, f is Lebesgue integrable if, and only if, f is absolutely Henstock-Kurzweil integrable.*

Proof. If f is nonnegative and measurable, the proof of Theorem 4.91 applies to $\mathbb{R}^n$ since bounded step functions which vanish outside bounded intervals in $\mathbb{R}^n$ are Henstock-Kurzweil integrable. Since any Henstock-Kurzweil integrable function is measurable by Theorem 4.128, the result follows by considering $f = f^+ - f^-$ as in the proof of Corollary 4.92. $\qquad \square$

Theorem 5.53   *Let* $f : \mathbb{R}^n \to \mathbb{R}$. *Then,* f *is Lebesgue integrable if, and only if,* f *is McShane integrable.*

Proof.   If f is McShane integrable, and hence absolutely McShane integrable, then f is absolutely Henstock-Kurzweil integrable and, therefore, Lebesgue integrable by Theorem 5.52.

Suppose that f is Lebesgue integrable. We may assume that f is nonnegative and, for convenience, that $n = 2$. Let $\mathcal{L}\int$ and $\mathcal{M}\int$ denote the Lebesgue and McShane integrals, as before. By Fubini's Theorem for the Lebesgue integral (Theorem 3.111), $\mathcal{L}\int_{\mathbb{R}^2} f = \mathcal{L}\int_{\mathbb{R}} \mathcal{L}\int_{\mathbb{R}} f(x, y)\, dy dx$. Since the Lebesgue and McShane integrals coincide in $\mathbb{R}$, $\mathcal{L}\int_{\mathbb{R}^2} f = \mathcal{M}\int_{\mathbb{R}} \mathcal{M}\int_{\mathbb{R}} f(x, y)\, dy dx$. Now, by Tonelli's Theorem for the McShane integral, f is McShane integrable and

$$\mathcal{M}\int_{\mathbb{R}^2} f = \mathcal{M}\int_{\mathbb{R}} \mathcal{M}\int_{\mathbb{R}} f(x, y)\, dy dx = \mathcal{L}\int_{\mathbb{R}^2} f.$$
□

Thus, the results of Section 5.5.7 hold in $\mathbb{R}^n$.

5.11   Exercises

Definitions

Exercise 5.1   Let γ be a gauge on $[0, 1]$ defined by $\gamma(0) = \left(-\frac{1}{4}, \frac{1}{4}\right)$, $\gamma(1) = \left(\frac{3}{4}, \frac{5}{4}\right)$, and $\gamma(t) = \left(\frac{t}{2}, \frac{1+t}{2}\right)$ for $0 < t < 1$. Give an example of a γ-fine free tagged partition tagged partition of $[0, 1]$ which is not a γ-fine tagged partition.

Exercise 5.2   Prove that the characteristic function of a bounded interval I is McShane integrable and $\int_{\mathbb{R}} \chi_I = \ell(I)$.

Exercise 5.3   Let $f, h : I \subset \mathbb{R}^* \to \mathbb{R}$. Suppose that $|f| \le h$ on I and that h is McShane integrable over I with $\int_I h = 0$. Prove that f is McShane integrable over I and $\int_I f = 0$.

Exercise 5.4   Suppose $\{a_k\}_{k=1}^{\infty} \subset \mathbb{R}$ and set $f(x) = \sum_{k=1}^{\infty} a_k \chi_{[k, k+1)}(x)$. Show that if f is McShane integrable over $[1, \infty)$, then the series $\sum_{k=1}^{\infty} a_k$ converges absolutely. For the converse, see Example 5.5.

Basic properties

Exercise 5.5   If I is a closed and bounded interval and f is continuous on I, prove that f is McShane integrable over I.

Exercise 5.6 (Translation) Let $f : [a, b] \to \mathbb{R}$ be McShane integrable over $[a, b]$ and $h \in \mathbb{R}$. Define $f_h : [a + h, b + h] \to \mathbb{R}$ by $f_h(t) = f(t - h)$. Show that f_h is McShane integrable over $[a + h, b + h]$ with $\int_{a+h}^{b+h} f_h = \int_a^b f$.

Exercise 5.7 (Dilation) Let $f : [a, b] \to \mathbb{R}$ be McShane integrable over $[a, b]$ and $h > 0$. Define $f^\tau : [\tau a, \tau b] \to \mathbb{R}$ by $f^\tau(t) = f\left(\frac{t}{\tau}\right)$. Show that f^τ is McShane integrable over $[\tau a, \tau b]$ with $\int_{\tau a}^{\tau b} f^\tau = \tau \int_a^b f$.

Absolute integrability

Exercise 5.8 Let $\varphi : I \subset \mathbb{R}^* \to \mathbb{R}$ be a step function. Prove φ is McShane integrable.

Exercise 5.9 Let $I \subset \mathbb{R}^*$ and $J \subset \mathbb{R}$ be intervals. Suppose that $g : J \to \mathbb{R}$ satisfies a Lipschitz condition (see page 35) on J and $f : I \to J$. Prove that $g \circ f$ is McShane integrable over I. [Hint: Use the proof of Theorem 5.11, the Lipschitz condition and the Cauchy criterion.]

Exercise 5.10 Let $f : \mathbb{R} \to \mathbb{R}$ be bounded and McShane integrable. For $p \in \mathbb{N}$, show that f^p is McShane integrable. [Hint: Suppose that $|f(t)| \leq M$. Use the function $g : [-M, M] \to \mathbb{R}$ defined by $g(y) = y^p$ in Exercise 5.9.]

Exercise 5.11 Let $f, g : \mathbb{R} \to \mathbb{R}$ be bounded and McShane integrable. Prove that fg is McShane integrable. [Hint: Recall that $fg = \left[(f + g)^2 - f^2 - g^2\right]/2$.]

Exercise 5.12 Let $f : [a, \infty) \to \mathbb{R}$ be McShane integrable. Prove that $\lim_{b \to \infty} \int_b^\infty |f| = 0$. [Hint: Pick γ such that $\gamma(t)$ is bounded for $t \in \mathbb{R}$ and $\left|S(|f|, \mathcal{D}) - \int_a^\infty |f|\right| < \epsilon$ whenever $\mathcal{D}$ is γ-fine free tagged partition of $[a, \infty]$. Fix such a $\mathcal{D} = \{(t_i, I_i) : i = 1, \ldots, k\}$ with $t_1 = \infty$, $I_1 = [b, \infty]$. Consider $\int_c^\infty |f|$ for $c > b$.]

Exercise 5.13 Let $f : I \to \mathbb{R}$ be McShane integrable over I. Show that $\lim_{\ell(J) \to 0} \int_J |f| = 0$. [Hint: Pick γ such that $\left|S(|f|, \mathcal{D}) - \int_I |f|\right| < \epsilon$ whenever $\mathcal{D}$ is γ-fine free tagged partition of I. Fix such a partition $\mathcal{D} = \{(t_i, I_i) : i = 1, \ldots, k\}$ and set $M = \max\{|f(t_i)| : i = 1, \ldots, k\}$. Let J be a subinterval of I. Consider $\mathcal{E} = \{(t_i, I_i \cap J) : i = 1, \ldots, k\}$ and use Henstock's Lemma to see how to choose δ so that $\ell(J) < \delta$ implies $\int_J |f| \leq 2\epsilon$.]

Exercise 5.14 Use Proposition 5.12 to prove the following variant of the Cauchy criterion. The function $f : I \to \mathbb{R}$ is McShane integrable if, and only if, for all $\epsilon > 0$ there is a gauge γ such that $|S(f, \mathcal{D}) - S(f, \mathcal{E})| < \epsilon$

for all γ-fine free tagged partitions $\mathcal{D} = \{(t_i, I_i) : i = 1, \ldots, m\}$ and $\mathcal{E} = \{(s_i, I_i) : i = 1, \ldots, m\}$, which employ the same subintervals of I.

Exercise 5.15 Use Exercise 5.14 to show that $f : I \to \mathbb{R}$ is McShane integrable if, and only if, for all $\epsilon > 0$ there is a gauge γ such that

$$\sum_{i=1}^{m} |f(t_i) - f(s_i)| \ell(I_i) < \epsilon$$

for all γ-fine free tagged partitions $\mathcal{D} = \{(t_i, I_i) : i = 1, \ldots, m\}$ and $\mathcal{E} = \{(s_i, I_i) : i = 1, \ldots, m\}$.

Exercise 5.16 Use Exercise 5.15 to show that if $f, g : I \to \mathbb{R}$ are bounded and McShane integrable, then fg is McShane integrable.

Convergence theorems

Exercise 5.17 Show that strict inequality can hold in Fatou's Lemma. [Hint: Consider $f_k = \chi_{[0,2]}$ for k odd and $f_k = \chi_{[1,3]}$ for k even.]

Exercise 5.18 Let $f : I \subset \mathbb{R}^* \to \mathbb{R}$ be McShane integrable over I. For $k \in \mathbb{N}$, define f_k, the truncation of f at k, by

$$f_k(t) = \begin{cases} -k & \text{if } f(t) < -k \\ f(t) & \text{if } |f(t)| \le k \\ k & \text{if } f(t) > k \end{cases}.$$

Show that each f_k is McShane integrable and $\int_I f_k \to \int_I f$. Further, show that such a result fails for the Henstock-Kurzweil integral.

Exercise 5.19 Suppose that f, g, and M are nonnegative and McShane integrable, and $0 \le fg \le M$. Prove that fg is McShane integrable. [Hint: Use Exercises 5.18 and 5.11.]

Exercise 5.20 Suppose that f and g are McShane integrable and g is bounded. Prove that fg is McShane integrable.

Exercise 5.21 Let $f : [0, \infty) \to \mathbb{R}$ and suppose that the function $x \longmapsto e^{-ax} f(x)$ is McShane integrable over $[0, \infty)$ for some $a \in \mathbb{R}$. Prove that $x \longmapsto e^{-bx} f(x)$ is McShane integrable over $[0, \infty)$ for every $b > a$ and the function F defined by $F(b) = \int_0^\infty e^{-bx} f(x)\, dx$ is continuous on $[a, \infty)$.

Exercise 5.22 Suppose that $f : \mathbb{R} \to \mathbb{R}$ is continuous and the function $x \longmapsto x^2 f(x)$ is bounded. Show that f is McShane integrable over $\mathbb{R}$.

Exercise 5.23   Let $f : I = [a,b] \to \mathbb{R}$ be McShane integrable and $|f| \leq c$. Suppose that $g : [-c,c] \to \mathbb{R}$ is continuous. Use Exercise 5.10 and the Weierstrass Approximation Theorem to show that $g \circ f$ is McShane integrable.

McShane, Henstock-Kurzweil and Lebesgue integrals

Exercise 5.24   Prove part (2) of Lemma 5.41.

Exercise 5.25   Extend Theorem 5.44 to unbounded intervals.

Exercise 5.26   Extend Theorem 5.45 to unbounded intervals.

Exercise 5.27   Suppose $f : \mathbb{R}^2 \to \mathbb{R}$ is measurable and bounded. If

$$m_2 \left(\{ x \in \mathbb{R}^2 : f(t) \neq 0 \} \right) < \infty,$$

prove that f is McShane integrable.

Fubini and Tonelli Theorems

Exercise 5.28   Prove that a nonnegative, simple function with support of finite measure on $\mathbb{R}^n$ is McShane integrable.

Exercise 5.29   Suppose that $f, \varphi : \mathbb{R}^2 \to [0, \infty)$, with f McShane integrable, φ simple and measurable, and $0 \leq \varphi \leq f$. Use Lemma 5.48 to show that φ is McShane integrable.

Exercise 5.30   Extend Exercise 5.6 to $\mathbb{R}^n$ using the Fubini theorem.

Bibliography

[A]   A. Alexiewicz, "Linear functionals on Denjoy integrable functions," *Colloq. Math.* **1** (1948), 289-293.

[Ban] S. Banach, "Sur le théorème de M. Vitali," *Fund. Math.* **5** (1924), 130-136.

[Bar] R. G. Bartle, *The Elements of Real Analysis*, Wiley, New York, 1976.

[BS]   R. G. Bartle and D. K. Sherbert, *Introduction to Real Analysis*, Wiley, New York, 2000.

[Be]   J. J. Benedetto, *Real Variable and Integration*, B. G. Teubner, Stuttgart, 1976.

[Br]   L. Brand, *Advanced Calculus*, Wiley, New York, 1955.

[C]   A. Cauchy, *Oeuvres Complete*, Gauthier-Villars, Paris, 1899.

[CS]   J. Cronin-Scanlon, *Advanced Calculus*, revised ed., Heath, Lexington, Ma., 1969.

[D]   G. Darboux, "Mémoire sur les fonctions discontinues," *Ann. Ecole Norm. Sup.* **4** (1875) No. 2, 57-112.

[DM]   L. Debnath and P. Mikusinski, *Introduction to Hilbert Spaces with Applications*, Academic Press, New York, 1990.

[DS]   J. D. DePree and C. W. Swartz, *Introduction to Real Analysis*, Wiley, New York, 1988.

[Fi]   E. Fischer, *Intermediate Real Analysis*, Springer-Verlag, New York, 1983.

[Fl]   T. M. Flett, *Mathematical Analysis*, McGraw-Hill, New York, 1966.

[Go]   R. Gordon, *The Integrals of Lebesgue, Denjoy, Perron and Henstock*, Amer. Math. Soc., Providence, RI, 1994.

[Gr]   I. Grattan-Guinness, *The Development of the Foundations of Mathematical Analysis from Euclid to Riemann*, MIT Press, Cambridge, 1970.

[Ha]   P. Halmos, *Measure Theory*, Van Nostrand, Princeton, 1950.

[He]   R. Henstock, "Definitions of Riemann type of variational integral," *Proc. London Math. Soc.* **11** (1961), 402-418.

[K]   J. Kurzweil, "Generalized ordinary differential equations and continuous dependence on a parameter," *Czech. Math. J.* **82** (1957), 418-449.

[Leb] H. Lebesgue, *Œuvres Scientifiques*, Volumes I and II, L'Enseignement Mathématique, Geneva, 1972.

[Lee] Lee Peng-Yee, *Lanzhou Lectures on Henstock Integration*, World Scientific, Singapore, 1989.

[LV] Lee Peng-Yee and R. Vyborny, *The Integral: An Easy Approach after Kurzweil and Henstock*, Cambridge University Press, Cambridge, 2000.

[Lew1] J. Lewin, "A truly elementary approach to the Bounded Convergence Theorem," *Amer. Math. Soc.* **93** (1986), 395-397.

[Lew2] J. Lewin, "Some applications of the Bounded Convergence Theorem for an introductory course in analysis," *Amer. Math. Soc.* **94** (1987), 988-993.

[MacN] H. M. MacNeille, "A unified theory of integration," *Natl. Acad. Sci. USA* **27** (1941), 71-76.

[Ma] J. Mahwin, *Analyse*, De Boeck and Larcier, Paris, 1997.

[McL] R. M. McLeod, *The Generalized Riemann Integral*, Mathematical Association of America, Providence, RI, 1980.

[McS1] E. J. McShane, "A Unified Theory of Integration," *Amer. Math. Monthly* **80** (1973), 349-359.

[McS2] E. J. McShane, *Unified Integration*, Acad. Press, New York, 1983.

[Mi1] J. Mikusinski, "Sur une definition de l'integrale de Lebesgue," *Bull. Acad. Polon. Sci.* **12** (1964), 203-204.

[Mi2] J. Mikusinski, *The Bochner Integral*, Academic Press, New York, 1978.

[MM] J. Mikusinski and P. Mikusinski, *An Introduction to Analysis*, Wiley, New York, 1993.

[Mu] M. Munroe, *Introduction to Measure and Integration*, Addison-Wesley, Reading, Ma., 1953.

[N] I. P. Natanson, *Theory of Functions of a Real Variable, Vol. I and II*, Frederick Ungar, New York, 1960.

[Pe] I. N. Pesin, *Classical and Modern Theories of Integration*, Academic Press, New York, 1970.

[Pf] W. F. Pfeffer, *The Riemann approach to integration*, Cambridge University Press, Cambridge, 1993.

[Ri1] G. F. B. Riemann, *Gesammelte Mathematische Werke*, H. Weber (ed.), Dover Publications, New York, 1953.

[Ri2] G. F. B. Riemann, *Partielle Differentialgleichungen und deren Anwendung auf physikalische Fragen*, K. Hattendorff (ed.), Vieweg, Braunschweig, 1869.

[Ro] H. L. Royden, *Real Analysis*, 3rd ed., Prentice Hall, Englewood Cliffs, NJ, 1988.

[Ru] W. Rudin, *Real and Complex Analysis*, 3rd ed., McGraw-Hill, New York, 1987.

[Sm] H. J. Smith, Sr., "On the integration of discontinuous functions," *Proc. Lond. Math. Soc.* **6** (1875) No. 1, 148-160.

[Sw1] C. Swartz, *Integration and Function Spaces*, World Scientific, Singapore, 1994.

[Sw2] C. Swartz, *Introduction to Gauge Integrals*, World Scientific, Singapore, 2001.

[Th] B. Thomson, "Monotone Convergence Theorem for the Riemann Integral," *Amer. Math. Monthly* **17** (2010) No. 6, 547-550.

[WZ] R. L. Wheeden and A. Zygmund, *Measure and Integral*, Marcel Dekker, New York, 1977.

[Z] A. Zygmund, *Trigonometrical Series*, Dover, USA, 1955.

Index